HANDBOOK OF EFFECTIVE TECHNICAL COMMUNICATIONS

HANDBOOK OF EFFECTIVE TECHNICAL COMMUNICATIONS

Tyler G. Hicks, P.E.
International Engineering Associates

Carl M. Valorie, Sr.
Valorie Associates

S. David Hicks   *Coordinating editor*

McGRAW-HILL BOOK COMPANY
New York St. Louis San Francisco Auckland Bogotá
Hamburg London Madrid Mexico
Milan Montreal New Delhi Panama
Paris São Paulo Singapore
Sydney Tokyo Toronto

Library of Congress Cataloging-in-Publication Data

Hicks, Tyler Gregory, date
    Handbook of effective technical communications/Tyler G. Hicks,
  Carl M. Valorie, Sr.
      p. cm.
    Includes index.
    ISBN 0-07-028781-3
    1. Communication of technical information—Handbooks, manuals,
  etc. I. Valorie, Carl M. II. Title.
  T10.5.H53 1989
  808'.066—dc 19   87-37932
                                                                           CIP

Copyright © 1989 by McGraw-Hill, Inc. All rights reserved. Printed in the United States of America. Except as permitted under the United States Copyright Act of 1976, no part of this publication may be reproduced or distributed in any form or by any means, or stored in a data base or retrieval system, without the prior written permission of the publisher.

1234567890   DOC/DOC   89321098

ISBN 0-07-028781-3

The editors for this book were Harold Crawford and Marci Nugent, and the production supervisor was Dianne Walber. It was set in Times Roman. It was composed by the McGraw-Hill Book Company Professional & Reference Division composition unit.

Printed and bound by R. R. Donnelley & Sons Company.

For more information about other McGraw-Hill materials, call
1-800-2-MCGRAW in the United States. In other countries,
call your nearest McGraw-Hill office.

Information contained in this work has been obtained by McGraw-Hill, Inc. from sources believed to be reliable. However, neither McGraw-Hill nor its authors guarantees the accuracy or completeness of any information published herein and neither McGraw-Hill nor its authors shall be responsible for any errors, omissions, or damages arising out of use of this information. This work is published with the understanding that McGraw-Hill and its authors are supplying information but are not attempting to render engineering or other professional services. If such services are required, the assistance of an appropriate professional should be sought.

To my wife Mary and my children, Carl Jr. and Mary Jo
—CARL M. VALORIE, SR.

CONTENTS

Preface xi

Section 1. Importance of Technical Writing Skills   1.1

   How Technical Writing Began   1.1
   The Field Today   1.2
   Your Future As a Technical Writer   1.2
   Engineers and Scientists As Technical Writers   1.3
   Professional Technical Writers   1.4
   Job Description   1.5
   Using This Handbook   1.7
   Steps to Better Writing   1.8
   Checklist for Clear Technical Writing   1.8
   Write Better—and Faster   1.15
   Readability Formulas   1.15

Section 2. Effective Technical Report Writing   2.1

   Why Write a Report?   2.1
   Types of Reports   2.2
   Planning Reports   2.3
   Seven Essentials of Good Reports   2.4
   Choosing the Action You Want   2.4
   Developing an Effective Outline   2.5
   Why Outlining Is Vital   2.11
   Standard Outline Guide   2.13
   Selecting an Approval-Getting Summary   2.16
   Writing for Fast and Complete Acceptance: Helpful Guidelines   2.19
   Arranging the Report for Easy Use   2.27
   Standard Report Format   2.28
   Hints on Developing a Format and Style Guide   2.35
   Applying the Format Guidelines to a Report   2.49
   Multivolume Reports   2.49
   Using Word Processors in Report Writing   2.53
   Checking and Editing the Report   2.53
   Delivering the Report   2.55
   How to Research Your Report   2.55
   Checklist for Report Writing   2.56

Section 3. How to Write Technical and Scientific Papers   3.1

   What Is a Technical Paper?   3.1
   How Technical Papers and Reports Differ   3.2
   How Papers Are Developed   3.4
   How Papers Are Processed   3.6

Writing the Paper   3.8
Submitting the Paper   3.14
Exemplary Papers   3.16
Some Typical Requirements for Submitting Papers   3.16
Which Societies Publish Papers   3.16
Checklist for Technical and Scientific Papers   3.17

Section 4. Writing Articles for Publication   4.1

Definition of an Article   4.1
Differences between Articles and Papers   4.1
Overview of Article Writing   4.2
Thirteen Types of Articles   4.3
Examples of Article Types   4.16
How to Find and Develop Article Ideas   4.17
How to Outline an Article   4.26
Typical Outlines for Technical and Scientific Articles   4.28
How to Use These Outlines   4.38
Working with Editors   4.39
How to Choose Successful Article Titles and Leads   4.47
Keeping the Middle Lean   4.52
Polishing and Submitting Your Article   4.57
In Retrospect   4.60
Checklist for Article Writing   4.60

Section 5. Creating Contract-Winning Proposals   5.1

The Need for Proposal Experts   5.1
Types of Proposals   5.2
Technical and Cost Proposals   5.10
Formats and Elements   5.15
Proposal Development Steps   5.29
Checklist for Proposal Writing   5.39

Section 6. Specification Writing   6.1

Specification Writers and Users   6.1
What Is an Engineering Specification?   6.4
Contracts and Specifications   6.4
Types of Specifications   6.5
Sources of Data   6.13
Specification Format   6.15
What Are the Specification Writing Steps?   6.24
Helpful Tips for Good Specification Writing   6.28
After the Specification Is Produced   6.29
Other Considerations   6.31
Checklist for Specification Writing   6.31

Section 7. Preparing Written Directives and Procedures   7.1

Definitions of Directives, Procedures, and Related Documents   7.2
Types of Directives and Procedures   7.9
Structure of Directives and Procedures   7.14
Language and Style Guidelines   7.27

Preparing the Document   7.30
Checklist for Directives and Procedures   7.35

Section 8.   Creating Effective Instruction Manuals and Bulletins   8.1

Features of Written Instructions   8.1
Major Types of Instruction Manuals and Bulletins   8.4
Writing to Specifications   8.8
Structure and Content of Instructions   8.10
How to Organize the Writing Job   8.30
Reminders for Common Types of Instructions   8.35
Checklist for Instruction Manuals and Bulletins   8.41

Section 9.   Writing Corporate Sales Brochures   9.1

Definition of a Sales Brochure   9.1
The Sales Brochure As a Promotional Tool   9.1
What Is a Corporate Sales Brochure?   9.2
Six Rules for Writing Effective Sales Brochures   9.4
Managing the Brochure Preparation Process   9.6
Structure and Contents   9.13
Writing the Brochure Copy   9.15
General Steps in Brochure Development   9.17
Checklist for Brochure Writing   9.17

Section 10.   How to Write Better Catalogs and Advertising   10.1

Catalogs Should Inform Users and Motivate Sales   10.1
What Is a Catalog?   10.2
Plan Your Catalog before You Write   10.2
Writing the Catalog   10.3
Catalogs Are Important   10.6
Industrial Advertising—A Quick Look   10.7
Six Rules for Writing Good Copy   10.8
Industrial Ad Headlines   10.11

Section 11.   Writing Technical and Scientific Books   11.1

Kinds of Technical Books   11.1
Who Uses Books   11.4
What Makes a Good Book   11.4
Words of Caution   11.5
How You Can Write Your Book   11.5
Qualifications for Technical Book Authorship   11.7
Finding Ideas for Technical Books   11.8
Developing Book Ideas   11.8
You Must Outline   11.9
Typical Outlines   11.10
Get Yourself under Contract   11.12
Get Ready to Write   11.15
Two Important Steps before You Write   11.15
Write, Write, Write   11.18
Handbooks   11.24
It's Up to You   11.28

x   Contents

Section 12.   How to Prepare Letters and Résumés   **12.1**

Effective Letter Writing   12.1
Ten Guidelines to Improving Business Correspondence   12.2
How to Prepare Professional Résumés   12.4
Summary Checklist   12.12

Section 13.   Grammar and Usage   **13.1**

Authorities   13.1
Spelling and Compounding   13.1
Capitals   13.15
Italics   13.29
Foreign Words   13.29
Abbreviations   13.30
Abbreviations of Frequently Used Words   13.32
Numbers and Numerals   13.37
Punctuation   13.39
Text Divisions   13.46
Good and Bad Writing   13.46
Grammatical Writing   13.51
Grammar Glossary   13.52

Section 14.   The Language of Publishers and Printers   **14.1**

Glossary   14.1
Proofreading   14.13

Index follows page **14.15**

PREFACE

This is a handbook of successful technical communications aimed at engineers, scientists, technical writers, managers, and others required to prepare technical or scientific written materials of many different types. The handbook presents clear, useful tools for writing more effectively in less time.

Today engineers, scientists, technical writers, and managers in industry, government, the military, and related organizations are required to write more than ever before. There are a number of reasons for this, including (1) greater complexity of machines and systems being designed and built, (2) the increased litigation over the performance of machines and systems, requiring a manufacturer to protect itself with carefully prepared instructions and other written materials, (3) the greater scrutiny of contracts and specifications by organizations paying for work that is done, and (4) the closer attention being paid to human factors and human safety in all new and revised designs.

As a result of these demands, the average person working in a business or industry associated with science and technology is writing more—and, in many cases, enjoying it less. To assist these people in meeting their writing loads, this handbook gives hundreds of techniques, shortcuts, and pointers. Practical throughout, the handbook can be opened to the section covering a particular writing task and quickly accessed for needed information. The many examples given are in tune with today's writing and technology.

The scope of the handbook ranges from what is probably the most common writing task of engineers and scientists—the formal technical report—to the least common—the technical book. Specific coverage of the handbook was chosen to include the most commonly met tasks faced in industry, government, the military, and similar organizations today. Other topics include major writing tasks that either the engineer or technical writer might encounter several times during a normal working life.

The handbook starts with the technical report—either formal or informal. This section shows how to research, outline, and write effective reports that get the action you seek. A number of useful examples, plus writing tips, are given in this section on report writing. The data are useful to anyone who wants to (or is required to) write a technical or scientific report that is useful and informative to its readers.

Next, the handbook considers the writing of technical and scientific papers. Since such papers often form the core literature of a field, it is important that the writer know how to prepare a paper that will be a valuable contribution to its field. This section gives many valuable tips on how to write a paper of your choice. It also shows a technical writer how to write a paper for someone else's byline.

From papers the handbook moves to technical articles. There are significant differences between articles and papers. These are pointed out. Specific writing hints and techniques show the handbook user how to write a clear article in the minimum time with almost a certainty of its acceptance by a publication of the writer's choice.

Since many industrial firms live or die according to their success in obtaining government contracts, Section 5 covers the writing of award-winning proposals. Why? Because behind every contract is a well-written proposal. This section covers both solicited and unsolicited proposals and shows the writer how to write a proposal that will win the contract sought.

Every industrial product that's manufactured or bought is covered by some sort of specification describing the items or materials that go into the product. Section 6 shows every writer how to prepare specific, clear, and easy-to-understand specifications. Since specification writing is such a widespread activity, this section will help almost every writer at some time during a busy career.

As technology and science become more complex, written directives and procedures find wider use. So Section 7 of this handbook covers these two topics. The discussion is probably the first ever to appear in print. It should be a valuable aid to all writers assigned the task of writing directives or procedures.

And the growing complexity we've mentioned several times now requires that instruction manuals be more clearly written than ever before. The same is true of product bulletins. Section 8 shows the reader—with many examples—exactly how to create useful and helpful instruction manuals and bulletins.

And with the growing complexity of technically based products, technical writers of all backgrounds are being called on more frequently to prepare corporate sales brochures. Section 9 of this handbook shows the user exactly how to prepare useful brochures for sales purposes in the least time.

Catalogs and advertising are also seeing greater input from technical writers—for the same reasons as brochures. Complete data on writing these two types of promotion are given in Section 10. Using the data given, any technical writer can produce a superior catalog or ad for technically based products or services.

The ultimate challenge in technical writing is the authorship of a technical or scientific book. And as technology moves on, more books detailing the new advances are needed by working engineers, scientists, and managers. Section 11 of the handbook shows the technical writer how to plan and write a significant book that will be a contribution to the literature of its field.

Letters and résumés are other writing tasks faced by today's technical personnel. Section 12 covers these important writing challenges. Using the information given, the technical writer will impove his or her output quickly and easily.

Since all writing involves grammar and usage, a quick review of key points is given in Section 13. And to help the technical writers deal more proficiently with printers and publishers, Section 14 details the common language used by these specialists. Knowing this language will help every technical writer get better results from contacts with printers and publishers.

This, then, is a handbook for all working engineers, scientists, technical writers, and managers who face the common—and uncommon—writing tasks of modern science and technology. It covers all the major types of writing these people will meet on their job and in their profession. The handbook was prepared because none of the currently available works in the field provide the scope and variety of coverage given here. Both of us believe that every engineer, scientist, technical writer, and manager will benefit from using the handbook. These benefits include saving time, producing better and clearer written materials, making a greater impact on superiors and customers, and building one's career more strongly and with a better-written history of what one has accomplished.

Tyler G. Hicks
Carl M. Valorie, Sr.

SECTION 1
IMPORTANCE OF TECHNICAL WRITING SKILLS

HOW TECHNICAL WRITING BEGAN

Engineers and scientists from the earliest days of recorded history have written reports, proposals, and other documents about their work. Much of the world's best-known technical writing has been done by outstanding engineers and scientists, such as Vitruvius, Agricola, Smeaton, Rankine, Parsons, Taylor, Hoover, Perry, Marks, Kent, and Rutherford. Studies show that, in general, the greater a person's engineering or scientific achievements, the larger the number of his or her published works of all kinds.

Until the start of World War II most engineers and scientists did all the technical writing related to their projects. Thus, engineers prepared instruction manuals, maintenance brochures, specifications, parts lists, and similar material. Scientists wrote reports covering their research findings, results of investigations, etc. There were few qualified technical writers in any field. The only major area in which engineers and scientists did not write extensively was industrial advertising. But even in this field engineers and scientists were often asked to check copy and verify technical facts.

With the start of World War II millions of young men and women were assigned duties covering the operation or maintenance of complex aircraft, naval vessels, tanks, and a variety of weapons. Adequate written instructions were needed. The Air Force was one of the first of the services to prepare training and instruction manuals for their flying and maintenance personnel. Other services followed. Soon equipment manufacturers began preparing comprehensive operating, maintenance, and instruction manuals as part of their contracts. Within a few years a huge volume of training and instruction literature was developed. Much of this was in the area of electronics, aircraft, submarines, and automatic weapons.

Most of the wartime technical writing was done by nonengineers and nonscientists—people with little engineering or scientific training. Some wartime writers were journalism majors who drifted from their normal work to the higher-paying, technical writing field. Others were trained technicians with writing ability who wanted to upgrade their earnings. Another group was composed of former creative writers, such as novelists and poets, who turned to technical writing for greater security and other benefits. Few engineers and scientists were employed as full-time technical writers because industry and government felt that the talents of these people were more useful when devoted to design, development, and research.

With the ending of World War II, private industry expanded its civilian output. Since many new products were complex, comprehensive instruction in their use was required. Industry turned to its wartime experience and began producing civilian technical literature that resembled, in many respects, the wartime literature. Introduction of nuclear energy, missiles, satellites, space probes, transistors, computers, and a variety of other new devices tremendously increased the need for well-prepared technical literature.

THE FIELD TODAY

Technical writing is becoming more important with every scientific and engineering advance. Today hardly any scientific projects are undertaken without lengthy reports, feasibility studies, progress analyses, and status summaries. Engineering activities often require a full range of technical reports, society papers, technical-magazine articles, and even books. In selling and marketing, the technical writer prepares catalogs, news items, equipment releases, and sales brochures.

The booming missile and space industry employs thousands of technical writers. These men and women prepare millions of written pieces for all kinds of readers—from laborers operating ditch diggers to advanced scientists studying space-flight techniques. The output of these writers varies from a single-page maintenance instruction to a volume of five hundred or more pages covering an important scientific or engineering subject. Operating and maintenance instruction manuals for some advanced missile systems run to several thousand pages, weigh 100 or more pounds, stand 5 feet high, and cost almost $1 million to prepare. And the burden on technical writers is increasing. For as our equipment becomes more complex, so do the instructions for operation, maintenance, testing, and design.

YOUR FUTURE AS A TECHNICAL WRITER

Beginners in technical writing have a bright and promising future if they develop their skills well because technical writing needs more highly qualified people than ever before. During World War II, when the need for technical writers suddenly skyrocketed out of all proportion to the supply, extensive training was bypassed. While people with little or no training were able to write and produce the needed material, some time and effort were wasted, and the written materials were not always as good as they might have been.

Today every project manager and engineering supervisor recognizes the need for *trained* technical writers. Gone is the time when *any* kind of writing background was acceptable. The beginning technical writer today must know much about the use of English, outlines, illustrations, specifications, parts lists, etc. But of all the qualifications, the most important is the ability to write clear, concise English. For this is the essence of technical writing. Unless you learn how to properly organize and clearly present written material, your chances of succeeding in technical writing are extremely small. So this book concentrates on developing your writing skills. While doing this, you will also learn much about the other phases of technical writing.

Today there are two main categories of people doing technical writing: (1)

engineers and scientists and (2) professional technical writers. It is important that you understand how each performs the task in the production of written material.

ENGINEERS AND SCIENTISTS AS TECHNICAL WRITERS

Engineers and scientists generally prepare reports, articles, papers, or books as an adjunct to their normal duties. The normal duties for which the engineer or scientist is employed *are not* writing. For example, an aeronautical engineer is hired to design a specific airplane—let's say a new jet aircraft. But during its design the engineer may write 10 reports to supervisors on various phases of the work. Some of these reports may be extremely short—so that they are in memo form. Others may be hundreds of pages in length, requiring months of preparation. Much the same is true of the scientist.

During or after the design of this jet aircraft or any other product, the engineer might decide, or might be asked, to write an article about the product for one of the technical magazines (often called *business papers*) in the field. One such paper is *Aviation Week*. Were the engineer writing an article about one feature of the product, say, the shock absorbers in the landing gear, the engineer might submit the article to *Product Engineering* or *Machine Design*. Or the engineer might, of course, have submitted such an article to *Aviation Week* instead. A scientist doing work on this or another project might submit the work to the same papers. But it is more likely that he or she would send it to the *Aeronautical Engineering Review* or some similar publication.

While working on this design project, the engineer might also decide to prepare an engineering paper for presentation before and publication by the American Society of Mechanical Engineers or the Institute of Electrical and Electronics Engineers. The engineering paper would probably differ from the article in a number of ways. The paper might be more mathematical; it might be longer; it might be far more specialized than the article. While most engineering and scientific papers are prepared on request, the engineer could decide to prepare the paper and then seek someone in an engineering or scientific society who would encourage submission of the paper.

After long or unusual experience in a field an engineer or scientist might decide to write a technical book covering some phase of his or her work. Or the employer might delegate him or her to write the book. Lastly, the engineer or scientist could be encouraged by a book publisher to write a book.

In all these examples our engineer or scientist is doing routine work first; *then* he or she writes about it. His or her primary job is seldom writing. Instead it is design, operation, maintenance, research, or some other task. But the writing is an important adjunct to the engineer's main effort. Technical writing is the means by which the engineer or scientist communicates knowledge and findings to others in the field; it is one of the most respected ways of reporting engineering and scientific developments. In a speech to scientists, Dr. Milton S. Eisenhower, president of Johns Hopkins University, stated that in a recent year reports in the physical and life sciences alone "filled 55,000 professional journals containing 1,200,000 significant articles, plus 60,000 scientific books and another 100,000 research monographs."

Since the engineer or scientist is closer to his or her work than anyone else, he or she usually is best qualified by reason of knowledge to write about his or her activities. But, unfortunately, highly developed engineering or scientific abilities are not always accompanied by well-developed writing skills. So we find that *some* engineering and scientific writing has, in the past, been poor. For this reason, and because more and more engineers and scientists are required to write as part of their job, courses in technical communication are becoming popular. Today the engineer or scientist *must* be able to write well if he or she wishes to have an outstanding career; otherwise he or she may find his or her advancement is limited. In a survey of 3,800 engineering graduates of Purdue University, about 90 percent of the engineers ranked writing and speech as "must" or "very important" subjects in their professional careers. Many industrial firms now have publication-incentive awards for technical material published by their engineers and scientists. Cash awards are common. Other firms give salary and professional experience credit for the material their people publish.

While engineers and scientists are writing more material with a greater degree of skill, they still cannot meet the tremendous needs of industry and government. Also, in recent years there has been a tendency to limit the writing of engineers and scientists to their immediate work. This leaves an enormous demand unfilled, for example, operating and maintenance instructions. To meet this demand and to help overworked engineers and scientists, the professional technical writer was recruited.

PROFESSIONAL TECHNICAL WRITERS

Engineers and scientists who write are, as we saw, usually part-time writers. Their main task is not writing—it is something else. And until recently few engineers and scientists were trained writers; they picked up their writing skills as they needed them.

Professional technical writers are, as distinguished from engineers and scientists, employed primarily to write. Any other tasks are secondary. And more and more today the professional technical writer is a trained individual. He or she is a specialist in technical communication. His or her engineering and scientific training is usually broad instead of specific. This broad background enables him or her to rapidly assimilate the engineering and scientific facts used in writing.

Today's professional technical writer handles a variety of tasks, from instructions for a home owner on how to operate his or her new washing machine to procedures for launching a rocket to the moon. In this book we avoid the smaller technical-writing tasks like operating washing machines, assembling model airplanes, or running lawn mowers. Instead we concentrate on the big jobs such as reports, instruction manuals, and books. Applying the skills learned for these big tasks, the writer can easily perform any of the smaller ones.

The modern professional technical writer is becoming more important every day. Originally assigned only operating and maintenance instruction jobs, his or her scope of activities has broadened. Now we find the professional technical writer preparing almost every instruction manual used by the Army, Air Force, Navy, Coast Guard, and Marines. Some of these manuals are outstanding contributions to the literature; some could never have been written without the help of the professional technical writer.

Projects employing the professional writer today cover such varied activities as nuclear energy, guided missiles, submarine warfare, distant early-warning systems, and space exploration. In addition, the professional technical writer has actively entered a new field—one that will ensure his or her future. He or she has become a "ghostwriter" for the engineer and scientist. This means that the professional technical writer will now engage in all the engineering and scientific writing activities we discussed earlier. So the new writer must be adept at reports, articles, society papers, books, manuals, and a raft of miscellaneous writing forms.

The emergence of the new writer can be traced to several causes. Shortages of engineers in key industries led to studies of how engineers spent their working time. Where the writing load was heavy, some of it was shifted to professional technical writers. This left more time for engineers and scientists to devote to primary tasks. And with the growing competence of technical writers, many firms found that the work done by a team of an engineer and a writer was better than either could do alone. The engineer supplied detailed technical know-how while the writer expressed this knowledge in clear, concise prose. Lastly, the writer could bring important specialized knowledge to the job. This knowledge covered items such as illustrations, tabulations, the use of color, relations with the client for whom the engineering and writing were being done, production schedules, and printing information.

JOB DESCRIPTION

The Center for Technical Publications Studies, Fordham University, held a seminar workshop to prepare a comprehensive job description for the technical writer. The job description developed by the university is reproduced below, with the permission of the center.

Read this description carefully. It covers many of the topics you will study in this book. Once you can perform all the tasks listed in this job description, you will be well on your way toward a higher level of proficiency as a technical writer.

Job Description and Performance Requirements[1]

Writes instructive or descriptive material on technical or scientific subjects, interpreting and creating an acceptable presentation of the facts or the ideas and theories of others for a given audience.

Work Performed

1. Performs research necessary to obtain complete understanding of the scope of the proposed publication and to gain a thorough technical knowledge of the subject.

    Receives a verbal or written work order for the desired publication, together with instructions on its general purpose, and any available basic reference material, such as specifications, proposals, correspondence, engineering

---

[1]From "A Report of a Study to Determine the Duties and Responsibilities Called for under the Job Entitled 'Technical Writer,'" prepared by Joseph Child and Robert Johnson, under the direction of Harold N. Schleich, The Center for Technical Publications Studies, School of General Studies, Fordham University, New York, N.Y.

reports, drawings, photos, similar publications, and supervisory or sales memoranda and notes.

Studies the supplied reference material to acquire background information on the project and to ascertain policy governing content, presentation, and quality level. May consult with engineers, other technical personnel, the publications supervisor or sales personnel to clarify technical or other details of the writing project.

Analyzes information on hand to determine whether additional research is required or whether the supplied material is sufficient and can be adapted to the publication requirement.

If additional research is required, determines the most logical sources and the best method for obtaining the required information. Performs the necessary research; may make field trips to libraries, government agencies, manufacturers, educational institutions, technical societies, etc. May confer with customer's technical staff through established lines of liaison and may observe, study, or operate the actual equipment, object, or process.

Makes suitable notes to ensure proper correlation and retention of the information obtained.

2. Organizes the proposed manuscript to provide an orderly plan for the preparation of the required text material.

Prepares a general outline; breaks the subject material into major topics, considering:

   a. The general purpose of the manuscript (catalog, magazine article, engineering report, equipment operation or maintenance manual, etc.)
   b. The specific application (formal training, guide for field operations, promotion, general information, etc.)
   c. The knowledge and skill level of the user
   d. The complexity of the subject

Arranges these major topics in logical order. Determines the logical sub-topics to be discussed or treated under each major topic and arranges these in proper sequence.

Classifies and indexes the reference material in accordance with the general outline.

Prepares detailed outline: analyzes the reference material for each topic and develops and expands ideas into further sub-topics, grouping and arranging them to achieve continuity and best subject coverage. Repeats this procedure for each topic, developing the outline for smaller and smaller portions of the manuscript, to the logical ultimate.

3. Prepares a draft of the manuscript in accordance with the detailed outline.

Writes the text, drawing upon his or her developed knowledge of the subject and desired scope, and using his or her communications skills to create an acceptable presentation of the technical data for the given audience. May conduct additional research to validate or clarify portions of the technical data. Uses a style and format for the writing set forth in applicable specifications or may select or develop a style or format best suited to the presentation. Defines new and unusual terms.

Determines the illustrations required to supplement the written material and selects the most suitable type of illustration, such as a photograph, line drawing, rendering, etc. Prepares sketches or preliminary layouts of line drawings and renderings and specifies the requirements for photographs. May supervise the photography. Assigns nomenclature to photographs by marking on overlays or other method. Requests the preparation of preliminary or final art from the art department and provides additional oral or written instructions as required.

IMPORTANCE OF TECHNICAL WRITING SKILLS 1.7

> Maintains written control and record of changes in cross references, figure references, tables, and the like during the development of text and illustrations to ensure accuracy of these details in the final manuscript.
>
> Routes the final manuscript through established channels to obtain technical editor or customer approval.

4. Revises and rewrites text to meet technical editor's and/or customer's review requirements.

> Receives the draft of the proposed publication after technical editor or customer review. Studies the corrections, comments, criticisms, or suggestions made, to determine the specific revision requirements and their effect on other portions of the text. Rewrites affected portions of the text and requests new or revised illustrations as required. Checks very closely to assure that all references and notes in other portions of the text conform to the revised portion and makes any required changes or corrections. Reviews the new or revised illustrations to ensure accuracy and conformance with the required changes. Routes the revised text and illustrations for final approval. May obtain and present factual data as a basis for not accepting changes requested by the editor or customer.

Responsibility. Responsible for the development and presentation of text and illustrations for technical publications which may cost thousands of dollars. Responsible for completing his or her work within the budgeted hours under maximum general supervision, and also responsible for meeting acceptance standards and delivery schedules for the completed manuscript. Responsible for technical accuracy of work performed by illustrators, typists, and others engaged in producing the manuscript. Responsible for determining the necessity for liaison and research in connection with the manuscript. Responsible for conducting approved liaison and research.

Job Knowledge. Must be able to interpret technical and scientific data, such as blueprints, diagrams, charts, engineering reports, and specifications for material, equipment, publications, etc. Must know research methods and techniques. Must be able to plan and organize manuscript in accordance with the requirements of specified media. Must have a comprehensive knowledge of good grammar and punctuation and be able to write clear and concise descriptive and instructional material. Must know illustration techniques and publication production methods and practices. Must have a general knowledge of the basic sciences and specialized training or experience in the technical area in which he or she is writing, i.e., aeronautics, agriculture, electronics, chemistry, mechanics, etc.

Mental Application. Must be able to discriminate between essential and nonessential data from the reader's viewpoint and thereby determine sufficiency of content. Must keep abreast of current trends and techniques for written communication and their particular application to his or her specific work. Must be able to organize major publication projects and to determine suitable work assignments for assistant writers. Must readily adapt to different writing style and format requirements and be able to carry out several projects concurrently. Must be able to arrive at decisions and judge the relative merits of these decisions with respect to their effect on the time, cost, and acceptance standards for the end product.

USING THIS HANDBOOK

This handbook is designed to meet the job requirements and training needs of the two main categories of modern technical writers—engineers and scientists doing some writing and full-time professional technical writers. Careful study of the analyses and techniques presented will give you a surer understanding of the various forms of writing used today. With this better understanding will come improved writing—clearer communication with all who read your words.

No matter what type of technical material you write—reports, articles, books, manuals, or news items—you must use words. The more skillfully you use words, the better your writing will be. In this section you find a step-by-step, clear-writing checklist designed to help you improve any kind of technical writing. This checklist will also be helpful when you review the later sections in this book.

STEPS TO BETTER WRITING

The best way for you to produce good written material is to follow a proven procedure. But until you have enough experience to develop the procedure that best suits your particular needs, you must use the experience of other writers. The checklist presented in this section summarizes the findings of a large number of technical writers, readability experts, editors, and language teachers. Use this checklist for all your major writing assignments. The list will be useful whether you are just beginning to write technical material or have written it for many years. Follow the checklist, step by step, as you prepare any major written piece.

CHECKLIST FOR CLEAR TECHNICAL WRITING

1. Before beginning to write, classify the task as:
    1.1. An engineering or scientific report
    1.2. A technical or scientific article
    1.3. An engineering or scientific paper
    1.4. An industrial manual
    1.5. A military manual
    1.6. An industrial specification
    1.7. A sales or news item
    1.8. A technical or scientific book
2. Determine the desired length of the written piece from:
    2.1. Firm or agency authorizing the work
    2.2. Editor of technical magazine or journal
    2.3. Industrial or military specification
    2.4. Advertising or public relations agency
    2.5. Prospective publisher
    2.6. Company recommendations
    2.7. The amount of data to be presented
    2.8. Requirements of a scientific or learned society
3. Define the reason for writing the piece:
    3.1. To report findings, conclusions, or recommendations
    3.2. To instruct students, employees, or military personnel
    3.3. To describe equipment, experiments, procedures
    3.4. To regulate construction or manufacture
    3.5. To persuade people to purchase or use products or services

3.6. To publicize new products or services or new information about the company, organization, or personnel
4. Classify your typical readers:
   4.1. Probable age (20 to 30 years; 30 to 40 years; 40 to 60 years)
   4.2. Education (grade school, high school, college)
   4.3. Sex (male, female, both)
   4.4. Reading habits (read widely or little)
   4.5. Occupation (student, technician, engineer, scientist)
   4.6. Purpose for reading (business, education, pleasure)
5. Collect data you will need for your writing from:
   5.1. People close to project; in the same field; specialists
   5.2. Publications—books, articles, manuals, catalogs, reports, drawings, specifications, advertisements
   5.3. Equipment—machines, buildings, devices, products
6. Get additional data by making field trips to:
   6.1. Interview technicians, engineers, scientists, or others concerned with the subject of the writing project
   6.2. Observe equipment in use
   6.3. Operate equipment
   6.4. Study recommended procedures
   6.5. Visit suppliers, contractors, vendors
   6.6. Observe tests
7. Where necessary, conduct interviews from your office by:
   7.1. Mail surveys:
      7.1.1. Use a good covering letter (Figure 1.1).
      7.1.2. Choose short, direct questions (Figure 1.2).
      7.1.3. Mail well in advance of deadline for return.
      7.1.4. Provide a postpaid or stamped return envelope.
   7.2. Telephone (local or long-distance)
      7.2.1. Prepare questions in advance.
      7.2.2. Have secretary record answers.
      7.2.3. Inform interviewee that his or her answers are being recorded.
8. Determine what specifications, if any, govern the writing:
   8.1. Industrial-writing specifications
   8.2. Military handbook or manual specifications
   8.3. Federal or state government specifications
   8.4. Trade-association specifications
9. Determine deadlines for each stage of the writing task:
   9.1. Rough outline of manuscript
   9.2. Final outline of manuscript
   9.3. Rough draft of specified portion of manuscript
   9.4. Rough draft of remainder of manuscript
   9.5. Completed final manuscript
   9.6. Completed illustrations
   9.7. Delivery of manuscript to customer, printer, or editor
   9.8. Printed manual, book, catalog, etc.
10. Estimate the writing time and cost to:
    10.1. Prepare rough outline
    10.2. Prepare final outline
    10.3. Write rough draft
    10.4. Prepare final draft
    10.5. Prepare illustrations

```
                              Adelt Technical Company
                              1309 Seventh Street
                              San Francisco CA 94567
                              May 25, 19--


        James Caldwell, Chief Engineer
        Culver Missile Associates
        Box 120
        Culver City CA 92359

        Dear Mr. Caldwell:

        My firm has requested me to prepare an illustrated
        brochure for civilian and military personnel working
        with the MX Guided Missile. You have been recommended
        to me as an outstanding engineer with a wide knowledge
        of the MX and similar missiles.

        Will you kindly complete the enclosed questionnaire,
        returning it to me by July 31st? Any additional inform-
        ation you can supply about the performance of this MX
        missile would be appreciated. This could be in the form
        of test and field reports, engineering papers, technical
        articles, etc.

        Thank you for your help.

        Very truly yours,

           Sandra Spring
        Sandra Spring,
        Mechanical Engineer

        SS:st
```

FIG. 1.1 Covering letter for mail questionnaire.

11. Establish contacts with specialists in your group for help in:
    - **11.1.** Estimating writing time and costs
    - **11.2.** Estimating illustration time and cost
    - **11.3.** Obtaining technical approval of manuscript content
    - **11.4.** Securing permission for use of material from other sources
12. Assemble your data, using modern methods:
    - **12.1.** Record interviews on a dictating machine or tape recorder.
    - **12.2.** Have a secretary take notes during interviews.
    - **12.3.** Get full cooperation from a technical library.
    - **12.4.** Use latest catalog collections—like *Sweet's, Refinery Catalog, Composite Catalog,* etc.
    - **12.5.** Secure pertinent government publications.
    - **12.6.** Use file cards or looseleaf notebooks to record data.
    - **12.7.** Obtain pertinent drawings and photos.

MISSILE DATA QUESTIONNAIRE

Please answer all questions in the space provided. If more space is needed, use a separate sheet for each question. Mark the number of the question on each sheet.

1. What is the maximum range in statute miles of the MX missile?
   _____.

2. What is the payload in pounds of the MX missile?
   _____.

3. Name the manufacturer of the following components of the MX:

   Guidance system_____.

   Power plant_____.

   Launching pad_____.

   Airframe_____.

   Nose cone_____.

4. List, in descending importance, the intended missions of the MX missile._____
   _____
   _____

5. Give the advantages of the MX missile over other missiles of similar design and mission capability.
   _____
   _____

6. State the disadvantages of the MX missile when compared with missiles of similar design and mission capability.
   _____

7. List the last 6 flights of the MX missile, stating date and location of the flight, payload, duration of flight, objective of flight, and outcome of flight._____
   _____

FIG. 1.2 Typical questionnaire for use in mail or verbal interviews.

- 12.8. Return unusable books, drawings, catalogs.
- 12.9. Keep only useful data on hand—this will save time and energy and prevent continual sorting of data.
- 13. Prepare a rough outline:
    - 13.1. Assign a temporary, working title to the subject.
        - 13.1.1. Name the device or machine or project.
        - 13.1.2. State what you will say about it or them.
        - 13.1.3. Keep the title short.   EXAMPLE:
            Performance of Microprocessors in Missiles Electronic Systems
    - 13.2. List available data in random order.   EXAMPLE:
        Types of microprocessors used
        Where used in missiles
        Need for microprocessor units
        Performance—successful, unsuccessful
        Related circuit components
        Improvements needed
        Typical circuits
        Design criteria
        Test procedures
        Predicted future performance
        Microprocessor specifications
        Materials of construction
    - 13.3. Regroup data under *Introduction, Body,* and *Conclusion.*   EXAMPLE:

        Introduction
        Need for microprocessor units
        Where used in missiles
        Types of microprocessors used

        Body
        Design criteria
        Typical circuits
        Microprocessor specifications
        Materials of construction
        Test procedures
        Performance
        Related circuit components

        Conclusion
        Improvements needed
        Predicted future performance
- 14. Prepare the final outline from the regrouped data:
    - 14.1. Decide on order of data presentation:
        - 14.1.1. Order of importance to reader
        - 14.1.2. Order of events in experiments, tests, etc.
        - 14.1.3. Order of steps in an operating or maintenance procedure
        - 14.1.4. Summary, followed by details
    - 14.2. Order-of-importance arrangement presents facts in descending order of importance (popular for articles, papers, manuals, etc.).
    - 14.3. Order-of-events arrangement presents data in chronological order (popular for some reports and articles).
    - 14.4. Order-of-steps arrangement gives step-by-step procedure for performing a specific task.

- 14.5. Summary arrangement gives major results first, followed by details (popular for technical reports and articles).
- 14.6. Any arrangement can begin with the summary, conclusions, or recommendations.
- 14.7. Use clause, sentence, or phrase entries in outline when possible—avoid single-word entries except for short writing tasks.
- 14.8. Show relative importance of entries:
  - 14.8.1. By numbers and letters: 1, 2, 3; A, B, C, etc.; (good for long outlines, numerous entries)
  - 14.8.2. By decimals, as in this checklist (good where written piece will have paragraph numbers)
  - 14.8.3. By indention (good for short outlines, few entries)
- 14.9. Insert bibliographic references you intend to use.
- 14.10. Indicate what illustrations will be used, and where.
15. Have the outline checked by competent technical personnel:
    - 15.1. Engineers, scientists, or technicians
    - 15.2. Advertising, public relations, community relations staff
    - 15.3. Where you are the technical authority, check the outline yourself
    - 15.4. Make any appropriate changes
16. Collect and evaluate illustrations for the manuscript:
    - 16.1. Investigate illustration sources:
      - 16.1.1. Journals, catalogs, books, advertisements
      - 16.1.2. Equipment drawings and photographs
      - 16.1.3. Test data charts, photographs, and drawings
      - 16.1.4. Manufacturers, civic groups, government
    - 16.2. Study the available illustrations:
      - 16.2.1. Are they helpful to the reader?
      - 16.2.2. Do they illustrate what you want to say?
      - 16.2.3. Are important parts labeled?
    - 16.3. Indicate in the outline the size of each illustration to be used (one column, two column, full page, etc.).
    - 16.4. Check the number of illustrations chosen:
      - 16.4.1. Do you have enough illustrations?
      - 16.4.2. Have you chosen too many?
      - 16.4.3. Can the illustrations be reproduced by the available printing equipment?
17. Begin your writing—take care to write convincingly:
    - 17.1. Be certain your statements are exact.
    - 17.2. Be sure you know your subject.
    - 17.3. Be careful to write for *your* readers.
    - 17.4. Be clear in everything you say.
    - 17.5. Be personal—speak with your reader in a friendly way—use crisp, lively words, etc., where these words will help build reader interest.   EXAMPLE:
      (Impersonal) The procedure to be followed in checking the clutch action consists of moving the shift lever from neutral to reverse and then to forward.
      (Personal) To check the clutch action, move the shift lever from neutral to reverse and then to forward.
    - 17.6. Try to use short sentences—readability experts recommend that the *average* sentence length not exceed 20 words.
    - 17.7. Alternate short and long sentences and paragraphs—never use a series

of short sentences or paragraphs unless you are trying to produce a special effect.

17.8. Choose words your readers comprehend—define an unknown word in the first sentence in which it occurs or in a footnote.   EXAMPLE:
> The wing planform—outline of wing when viewed from above—is triangular in the X-22 aircraft.

OR:
> The wing planform* is triangular in the X-22 aircraft.
> *Outline of wing when viewed from above.

17.9. Select active-voice verbs whenever you can.   EXAMPLE:
> To etch the specimen, *polish* one surface. Then *coat* this surface with acetic acid. *Wait* 10 minutes before removing the acid.

17.10. Active-voice verbs are useful when you write instruction and training manuals, operating instructions, and maintenance manuals.   EXAMPLE:

to be revised	*revise*
must be changed	*change*
should be checked	*check*

17.11. Work every verb you use.   EXAMPLE:
> (Poor)   Reports are to be filed here.
> (Better)   File reports here.
> (Poor)   Radar pulses, to be read accurately, must be interpreted by trained personnel.
> (Better)   Accurate radar-pulse interpretation requires trained personnel.

17.12. Be specific—avoid vaguely worded sentences, confused facts, and jumbled references:

17.12.1. Use exact names of equipment, job titles, personnel.

17.12.2. Avoid generalized references like "the unit," "given above," "as previously stated," etc.

17.12.3. Give paragraph, section, or chapter number when referring to previously presented material.

17.12.4. Refer to photos, drawings, and tables by their numbers.

17.12.5. Where numbers are not used for text or illustrations, give page on which the referenced material appears.

17.12.6. Weed out vague words—use shorter, crisper, specific words.   EXAMPLE:

for the reason that	*because*
in order to	*to*
along the lines of	*like*

17.12.7. Give step-by-step instructions.   EXAMPLE:
> (Poor)   To start the unit, turn the switch to START. When it is up to speed, check the oil level and pressure. Thereafter, check these and other items hourly.
> (Better)   To start the generator, turn the starter switch (mounted on left side of control panel) to START. When the generator reaches its rated speed of 1,800 r/min, check the bearing oil level in the sight gage and the oil pressure on the pressure gage. The sight and pressure gages are mounted on the outboard end of the main

bearing. While the generator is operating, check the oil level, oil pressure, armature current, and field current hourly.
- 17.12.8. Be simple—use clear language; do not try to impress your readers with the extent of your vocabulary.
- 17.12.9. Be brief—today's readers are busy, and they want to "get to the point" immediately.
- 17.13. Write the illustration captions.
- 17.14. Write the conclusions, summary, findings, abstract, table of contents, etc.
18. Finish the writing—then review the text and illustrations:
- 18.1. Allow some time to elapse between completion of the writing and start of the review.
- 18.2. Read the text for clarity, conciseness, and comprehensiveness.
- 18.3. Check the technical content against reliable data.
- 18.4. Check all main and secondary headings in the text and contents.
- 18.5. Evaluate the conclusions, findings, summary, contents, etc., for clarity and conciseness.
- 18.6. Check the proportion of the entire written piece—is any part too long or too short?
- 18.7. Have the manuscript read for errors in usage and grammar.
- 18.8. Check the length of the manuscript—is it too long or too short?
- 18.9. Check all references to the bibliography
- 18.10. See that all illustration and table references are correct.
- 18.11. Prepare illustrations for engraving.
- 18.12. Check the illustration captions.
- 18.13. Have the manuscript typed according to the printer's requirements.
- 18.14. Send the manuscript to the printer, publisher, or other group for handling.
- 18.15. Maintain a file of at least one copy of the manuscript and all illustrations until the piece is published.
- 18.16. File copies of the published piece with the manuscript and illustrations.

WRITE BETTER—AND FASTER

Use this checklist regularly. You will find it helps you produce better material in less time. Start now; you will benefit immediately. Most technical writers spend about 80 percent of their working time preparing to write, i.e., collecting data, organizing the outline, securing illustrations, working with the illustration staff, etc. Only about 20 percent of their time is devoted to actual writing. So if you can reduce the preparation time, you will have more hours to devote to other tasks. The checklist in this section should help you reduce preparation time by 10 percent or more.

READABILITY FORMULAS

The purpose of a readability formula is to provide a way to measure ease of reading, to tell whether a piece of writing will be easy or hard reading for a specified group of readers.

Readability formulas are designed to be used *after* a piece is written, not before the writing is begun. Many formulas have been published by various researchers. One of the best known is that of Rudolph Flesch.[2] The Flesch formulas (he developed three) use nomograms for determining *how easy to understand* a piece of writing is, and *how interesting* it is. To learn how to apply these formulas, refer to Flesch's interesting and useful book.

Other popular readability formulas are those developed by Dale-Chall, Clapp,[3] and Gunning.[4] Like the Flesch formulas, the best guides to these others are the books in which they originally appeared. The important concept to be learned in this handbook is the use of any formula—not the step-by-step calculations for a specific formula.

No readability formula yet developed will do your writing for you. Nor will any formula give you a clear, concise writing style. Writing is work—all the equations and charts you can find will not remove the irksome aspects of preparing a written piece.

Readability formulas are best used *after* you have completed the rough draft of your manuscript. Then, by applying a readability formula, you can determine how closely your writing fits the reading level of your audience. The formula gives you a reader's-eye view of your writing. For this purpose readability formulas are ideal. But never use a formula as a crutch before writing; your style and output will suffer.

If, after trying a few readability formulas, you find they are not to your liking, don't worry. Many of the most successful technical writers live formula-free lives. One, Norman Shidle, recommends: "(1) Think before you write. (2) Bring your thoughts into order. (3) Put organized thinking to work. (4) Keep moving.... Follow through. (5) Write 'sentences that march.' (6) Use words that live. (7) Use mechanical aids to clear writing. (8) Check up on yourself."[5]

If you want to develop a strong, clear writing style, apply the hints of Chester Anderson. (1) Use short paragraphs—they make for readability. (2) Avoid long and involved sentences—18 to 20 words *average* is far better than 40 to 60. (3) Watch conditional clauses and prepositional phrases—the more you have, the harder the reading. (4) Use topic sentences—they provide good road maps for paragraphs. (5) Watch transitions—be sure that one sentence follows another, and use transitional arrows to point both forward and backward. (6) Vary sentence structure—don't write all "subject-verb" sentences, or start them with such dead words as "the," "it," or "there." (7) Watch repetition of the same word or phrase. (8) Break down figures so they are understandable—and don't use too many in the narrative. (9) Make use of display, headings, enumeration, parallel construction, etc. (10) Leave plenty of white space.

---

[2]Rudolph Flesch, *The Art of Readable Writing,* Harper & Brothers, New York.
[3]John Clapp, *Accountants' Writing,* The Ronald Press, New York.
[4]Robert Gunning, *The Technique of Clear Writing,* McGraw-Hill Book Company, Inc., New York.
[5]Norman Shidle, *Clear Writing for Easy Reading,* McGraw-Hill Book Company, Inc., New York.

SECTION 2
EFFECTIVE TECHNICAL REPORT WRITING

One of the most vital writing tasks faced by engineers, scientists, and managers today is the preparation of formal and informal reports for use inside or outside their firm or organization. Technical writing consultants estimate that engineers in the United States write more than 6 million reports each year. Add to this the output of engineers in Europe, Japan, China, Canada, Latin America, and the Soviet Union and the world total probably exceeds 25 million technical reports each year.

These reports are the media for communicating research results, recommendations for specific actions, assessments of current or future conditions, and presentation of other data to key people at various levels in an organization. Technical reports thus impose two major responsibilities on professionals in engineering and science: (1) clear writing and (2) effective reading and interpretation of reports.

In this handbook our emphasis is on organizing and writing clear technical reports, which in turn aid effective reading and interpretation. In practice, both your project's merit and your expertise as an investigator will be judged by the quality of your technical report. Hence, you profit in many ways by improving your written reports. Well-written reports are easier to read and understand. The reader gets the facts faster, allowing him or her to act on the recommendations of the report. Clear reports help management reach important decisions properly. The results produced by clear reports can include increased profit, higher efficiency, faster approval of important requests, greater prestige for yourself and your organization, and more client goodwill.

WHY WRITE A REPORT?

You write a report instead of a memo or letter when you want to get action or record information. You may be seeking action of any kind—such as on a proposal for a lucrative new computer contract, on your request for more on-site safety inspections of an important field project, or on your recommendation for new robotics equipment needed for your department. Whatever the proposal is, you can write a report that gets results for you or your group. The same is true for reports which record information—such as the outcome of studies, tests, or research.

TYPES OF REPORTS

There are at least a hundred different kinds of reports popular in industry, government, and research today. Depending on their purpose, they are given various names, including: *formal, informal, research, operation, progress, construction, maintenance, market-survey, laboratory, error log, routine, preliminary, periodic, long-form, short-form, and final.* Although such terms are useful in specifying the general purpose of your report, learning how to write each kind would be a wasteful task. For regardless of the name applied to a report, the steps in writing one type of technical report are essentially the same as for any other type of report. So the writer who learns the techniques of clear report writing can apply these methods to any reporting task.

Formal and Informal Reports

A distinction made between types of reports is length. Long reports (in terms of number of pages) that require extensive research, study, investigation, or planning are often termed *long-form* or *formal reports*. Short reports made in memo, letter, or other routine form are often called *short-form* or *informal reports*.

This classification—formal and informal—is preferred by many engineers, scientists, and technical writers because it eliminates confusion that might arise from a multitude of specific terms whose meaning can vary from one technical specialty to another. Later in this section we will discuss a further classification termed the *multivolume report*.

Length

Informal reports are usually thought of as being about 6 or fewer double-spaced typewritten pages in length. Condensed versions of formal reports may also be about this long. Some writing authorities define a formal report as a detailed technical document having more than 6 double-spaced typewritten pages—others say more than 10 such pages. In this handbook we'll use the first length—6 pages—for informal reports; more than this for formal reports, recognizing that the definition is flexible.

Note that you may perform the same steps in preparing to write either a formal or an informal technical report. The difference, usually, is in the amount of time you spend preparing to write the report. Collection of data for a formal report may require several months, or more. For an informal report the needed information may be obtained in a few hours or less. But in both types of reports you generally must obtain data of some sort as part of your writing task.

When writing an informal report as a memo or letter, follow the general outline given in this section for a formal report. But delete sections and headings from the outline, as necessary. By following the general outline and examples, you will maintain a clarity and a logical structure in your report. As such, it's certain to be more effective to its readers.

To condense a formal report, extract the key findings or conclusions and present them in your condensation. Delete those topics that provide background or unnecessary qualifying information. But be sure to cast your condensed report in the format given later in this section. Then your readers will find the report clearer and easier to use.

In either type of report, you must be certain that the data you present are accurate. You must also know the purpose of the report and aim it at a specific kind of reader. These points are discussed later in this section.

Technical and Management Attitudes

Ideally, a report should be written a little differently for each reader as it passes up the various levels within (or outside of) your organization. As Figure 2.1 shows, the training, interests, and thinking of executives change substantially as

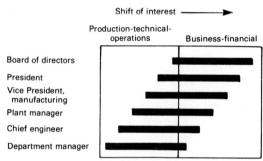

FIG. 2.1 The training, interests, and thinking of executives change substantially as a report moves upward in an organization.(W. A. Ayres and McGraw-Hill Publications Co.)

your report moves upward in an organization. At the operating levels, technicians, engineers, and production managers are more likely to give the greatest weight to technical and engineering feasibility. At the upper end of the organization, the president and board of directors will be much more interested in the business and financial implications of your report and its findings as they relate to future profits.

PLANNING REPORTS

Most reports are written to get action—to make the reader decide or to help the reader decide on a specific action. It is just as important to design your report for an answer that says "approved" as it is to build the report for easy reading and good comprehension. How do you do this?

Start by thinking through every question and problem that might arise in connection with your proposal or conclusions. Then provide acceptable answers in your report to show that you've anticipated and planned for every such possibility. Busy executives have all sorts of issues to consider—and plenty of worries of their own. If they must stop and solve a problem before they can approve your report, the easiest and quickest step for them to take is to say no.

So check your plans for every report. Do your plans answer the major questions your readers will probably have? Questions like:

- What is the subject of this report?
- Why should I be interested in the subject?
- What—briefly—is the background of the report?
- What action, if any, is recommended?
- Why is this action recommended?

Developing answers to those questions will provide useful input for your report.

SEVEN ESSENTIALS OF GOOD REPORTS

The surest way to design your report so it appeals to both the technical and financial interests of your supervisors is to follow these seven steps:

1. Choose the action you want your reader to take.
2. Develop an outline to organize your information.
3. Select an approval-getting summary.
4. Write your report in terms that will get fast acceptance.
5. Arrange your report for easy reading and use.
6. Verify facts, language, and presentation.
7. Deliver your report so it gets full attention.

Note how different this approach is from the casual, "I'll dash off a few words for the folks upstairs and get this deal approved." Certainly, some reports are written quickly—and such reports *may* produce the desired results. But to ensure consistent results from your technical reports, you must take time to plan for positive results.

CHOOSING THE ACTION YOU WANT

Use crystal-clear terms for the action or decision you want from your reader. Be certain you state exactly what you think should be done. Here are some examples of clear decisions your reader can either approve or disapprove:

The new computer should use a 32-bit microprocessor.

Two hundred worker hours should be allowed for programming.

A solid-waste collection system should be designed to recover energy from construction-phase materials.

Shale-oil research and development should be canceled because needed resources are lacking at the site.

Note that all these proposed decisions are stated in the form of a recommendation. Why? Because it is easier to state the decision you want in the form of a recommendation than in any other way. Since a recommendation is direct and concise, it is easier for your supervisor to approve. Long, involved statements of what is proposed, why it is proposed, and what it will do requires much more time for approval than a concise, clear, and direct recommendation.

But what about a report that does not have a recommendation as its conclusion? How should we indicate the action or decision the reader is to take or make? Use a similar scheme—state your thoughts in clear, strong terms. Thus, your report might say:

> Sixty hours of downtime have been saved, per quarter, through preventive maintenance.
>
> XYZ computer facility could operate more economically following installation of an improved data processing system.
>
> Hydraulic reverse-reduction gears are used in 70 percent of the engines.
>
> Improved efficiency has shortened daily programming by seven hours.

Each of those statements specifies the main thought you wish your reader to obtain from your report. There is no doubt in his or her mind as to what action to take or what decision to make. Of course, the final decision may hinge on factors not covered in your document. But you have expressed your findings clearly enough so they can easily be considered in the final decision or action.

DEVELOPING AN EFFECTIVE OUTLINE

We cannot emphasize too strongly the importance of developing an outline of your report to organize your technical data and findings. Note that this applies to every type of technical writing you do—not only to reports but to technical papers, letters, manuals, oral presentations, advertisements, and other writing tasks. The time you spend developing an outline before writing the body of your report will save much energy and wasted effort when you're writing the actual report.

Brainstorming

Some project managers and engineers find it useful to conduct a brainstorming session with others within their organization prior to drafting a final outline for a formal report. They often find that colleagues provide valuable new ideas and insights while analyzing approach, validating findings, catching errors, and offering logical organizational points for the outline and the formal report itself.

Suggestions might include: discussing particular aspects of a topic, choosing specific illustrations to support recommendations, providing technical definitions, and being alert to wording that could easily be misinterpreted or have hidden legal traps.

Many report writers work on their own—without the help of a brainstorming session. There are a number of reasons why report writers may work alone, including:

- The writer conducted research or studies with little or no outside help. Hence, he or she has sole access to the data and findings that will be presented in the report.
- The report contains confidential or classified information and findings.
- Others in the writer's group may be too busy with their own work (including report writing) to devote time to brainstorming others' reports.
- Personnel in the writer's group might not be well enough informed about the subject of the report to be able to help.

So, as a report writer, you may find it useful to work alone or work with others, or you may work on part of a report alone and part of it with others. The method you use will probably vary from one report to another and from one situation to another.

Drafting the Outline

The easiest way to begin an outline is to list, either at random or in a logical grouping, all the pertinent ideas you—either alone or with your brainstorming associates—have about the subject of your report. Once you have your ideas on paper, you can concentrate on arranging them in a suitable order for the report you plan to write. Later in this section you'll find examples of typical report outlines and formats and also guidelines for arranging your material for maximum effectiveness.

If you use help in preparing your outline, you may want to submit copies of the initial outline draft to your "think-tank" associates for review after the brainstorming is finished. Including pertinent comments and suggestions in a revised outline could improve the final report. This procedure can substantially reduce the total number of hours of work that might otherwise be required for writing certain types of reports.

If you work alone on your report, you may find it helpful to put your outline out of sight for one or more days. Then, when you have a few spare minutes, review the outline as though you had never seen it before. Major inconsistencies will usually pop out quickly, allowing you to revise your outline with a minimum of work. Taking a second look is important before you start writing and should always be part of your outline preparation. Even when you are short of time, try to allow for a few hours between the completion of your outline and your second look; doing so can be helpful in improving the outline.

Using Formats and Style Guides

While outlines for most technical writing tasks follow similar fundamental principles of organization, the form you give your outline is determined partly by the structure you wish your finished document to have. We refer to this structure as the *report format*.

Many organizations require their report writers to follow a standard format for technical reports whenever possible. Further, clients of an engineering or technical firm may occasionally require a special format for particular report projects. And if your firm considers a contract with a federal government organization such as the Department of Defense, you may be obliged to follow a report format such as that in the U.S. Government Printing Office (GPO) *Manual and Style Guide*. Further, a client might require the GPO format with modifications for a particular project.

So you *must* know the rules governing report writing in your organization *before* you begin writing. Most industrial firms that produce reports publish a format and style guide precisely for guidance of their report authors. If your firm or client does have such guidelines, be sure to follow them exactly. Supervisors will hardly pay any attention to your report if you can't follow rules.

If your firm or organization doesn't yet have formal guidelines, use the standard format given later in this section. You will also find suggestions on developing and using a style guide in your own firm.

As noted, you can save yourself a lot of trouble in report writing by deciding on the format for your finished report while developing your outline. The format determines how you present and arrange your material. Your outline provides the detailed writing plan.

Formats are useful in revising initial outline drafts to produce a comprehensive final outline. To save time, start by organizing random ideas and information about a subject roughly around the format you will use in your final version. Thus, your initial outline for an informal report might be a few lines jotted on a letter or memo you must answer. A typical outline could include headings, subheadings, and further subdivisions to provide an introduction, body, and conclusion in the finished report. For a formal technical or scientific report, your outline may run several pages and have a variety of subheadings.

Today many technical writers are using computer software to help them with their outlines. The software helps the writer sift entries as necessary to eliminate redundant or irrelevant topics. Of course, this can also be done manually from typewritten or longhand notes. The manner in which you choose to outline and write—on a typewriter, word processor, dictation machine, or in longhand—depends on your firm's facilities and how you feel most comfortable. Just remember that your primary objective—in every case—is to write a clear and concise report which will achieve the goals you set for it.

Building a Table of Contents While Outlining

Regardless of how you write, organize the headings in your outline in the order that leads your reader logically through the items your report considers. Be sure to cover any pertinent related problems or questions that might arise. This list of topics, tailored to your report format, will form the basis for your major headings in the report.

Next, add subheadings and minor topics to the draft outline, as required by the topics you'll discuss. Now add standard report elements in your main heading, i.e., title page, summary, preface, and conclusions. If you organize these according to a standard format, you'll find that you have your report's table of contents. Figure 2.2 shows an outline that achieves both of these objectives.

Once your initial outline contains all the major topics you need to present to your readers, you can draft additional brief outlines, if necessary, to tailor your material to the interests of specific individuals who will read the report. Note this is necessary only when your report will reach people at different levels in the organization. The custom-tailored document can be highly effective when you need to address separately the interests of the company president, project engineers, department managers, and others.

```
            ENGINEERING ANALYSIS OHIO POWER PLANT NO. 32

                              Contents


        Foreword

        Legal Notice

        Preface

        List of Tables and Figures



        Section                         Title
           1                             Summary
           2                             Plant Description
           3                             Alternate Fuels Capability
           4                             Primary Fuel Options
           5                             Derating
           6                             Operation and Maintenance Costs
           7                             Comparisons of Each Option
           8                             References
```

FIG. 2.2 Example of a short outline that could also be used for a table of contents in a technical report.

Writing the Final Outline

The final outline includes subtopics and details not covered in the draft version. As such, it may incorporate ideas derived from your consultations with associates and additional data gathered through follow-up study activities. Treat this outline with care—it is the explicit plan by which you will build your finished report.

Some topics lend themselves readily to division into headings, subheadings, minor subdivisions, and groupings. Others require considerable thought. Each division of the final outline should not be a mere subdivision of ideas but a practical arrangement of topics by their relative importance, or weight, in the written report.

Filling in the subtopics and lesser divisions will serve to block out the entire contents of your document. The conventional outline form follows a pattern of logical subordinations and indentions:

I. First Major Topic or Question
    A. First subtopic
        1. First subdivision of subtopic
            a. Details of subdivision
                (1) Minor items

Your final outline could follow the above pattern, but you can probably avoid extra work by patterning your outline according to the specific form you will give your final report. As noted, this could be a standard report format provided by your firm or by your client, or it could follow the typical format given in this section. When using a general format guide, be extra careful to include details required by your specific subject, such as contractual obligations, equipment, and specifications.

One more caution: Remember that every numbered and lettered division within any outline must have at least one corresponding heading elsewhere in the outline. For example, the outline fragment above would need to be followed by a second major topic or question under Roman numeral II in order to maintain logical order. Likewise, the first subtopic represented by A would have to be followed by a second subtopic under B, within the first major topic or question. The same rule applies to all the other subdivisions. Basically, there is no sense in having an I, A, 1, *a,* or (1) heading if you don't have at least a II, B, 2, *b,* or (2).

As aids to outline development, we've included some additional guidelines. Figure 2.3 is a checklist for outline preparation, and Figure 2.4 presents a final outline for a technical report along with page numbers for possible use later in a table of contents.

Note that the checklist in Figure 2.3 can be used in preparing outlines for all

---

1. *Concreteness:* Have you been concrete and specific, avoiding vague terms like *general principles, discussion,* or *study characteristics*? Are all main ideas included? Have any terms that may be liable to misinterpretation been changed or modified to eliminate ambiguity?

2. *Order:* Is the order of sections reasonable and useful? Does each idea that builds on others actually follow them? Will you be taking the reader from "where he or she is" before the report to your conclusion by logical and orderly steps? Would there be any advantage to shifting the order of any of the items?

3. *Unity:* Does each section represent a complete package of closely related ideas? Are topics that share certain traits grouped together? Are major divisions of your overall subject observed in the group arrangement?

4. *Usefulness:* Have you made a constant effort to tailor the arrangement of ideas to the interests and knowledge of your reader?

5. *Proportion:* Does the relative space devoted to each topic in the outline at least roughly indicate the weight it will carry in the finished report? Are additional breakdowns needed to reflect detailed treatment of certain points?

6. *Inclusiveness:* Have you made a place for every significant topic? Test it by reviewing your subject to see where a few obscure details would fit.

FIG. 2.3 Checklist for outline preparation. (Adapted from "Writing and Publishing Your Technical Book," F. W. Dodge Corp.)

FINAL ENGINEERING ANALYSIS
OHIO POWER PLANT NO. 32

Contents

Foreword	i
Legal Notice	ii
Preface	iii
List of Tables and Figures	iv

Section	Title	Page
1.	Summary of Comparisons	1-1
2.	Plant Description A. Equipment Descriptions B. Existing Designs C. Existing Specifications	2-1
3.	Alternate Fuels Capabilities A. Coal Technical Capabilities B. Synthetic Fuels Capabilities C. Fuel Mixture Capabilities	3-1
4.	Primary Fuel Options A. Primary Option 1: Coal Gasification B. Primary Option 2: Coal-Oil Mixture C. Primary Option 3: Coal and Oil Intermittently	4-1
5.	Derating A. Primary Option 1 B. Primary Option 2 C. Primary Option 3	5-1
6.	Operation and Maintenance Costs A. Primary Option 1 B. Primary Option 2 C. Primary Option 3	6-1
7.	Cost Comparisons of Three Options A. Selection of Options B. Additional Considerations	7-1
8.	References	8-1

FIG. 2.4 Example of a technical report outline prepared for possible use as a table of contents.

types of technical and scientific communication, including reports, papers, proposals, speeches, magazine articles, and even books.

In Figure 2.4, it is clear that the report writer has followed the checklist's six important guidelines. Perhaps most importantly, the writer has provided the foundation for a detailed investigation of each topic and subtopic while ensuring the unity of the outline's various divisions. Writers who fail to plan for this delicate balance between completeness and economy do not fare so well. Typically they succumb to the pitfall of overexpanding the outline, which leads to confusion in the organization and writing of the report itself. Further, an overly detailed outline or table of contents can make an otherwise unified set of project tasks appear fragmented.

Make certain that you include in your outline only those topics that you intend to support with substantial information in the final document. If you want your report to be successful, you will work hard to develop a comprehensive and clear publication based on the "bare-bones" outline.

Avoid the mistake of writing a long outline and merely transferring its collection of phrases and sentences to the written document. The key to effective technical writing is a strong follow-through from final outline to pages of telling text.

WHY OUTLINING IS VITAL

There are many arguments for preparing an outline for every technical or scientific piece you write. A few writers shun outlining as a useless expenditure of time and energy. But most experts agree that outlining is the best way to handle the complex and demanding tasks of technical communication.

Here are eight good reasons for using outlines:

1. To make certain you meet all the technical and contractual obligations of your report writing task
2. To help you analyze your ideas about the subject
3. To prevent omission of important points
4. To give continuity to your report
5. To help you choose the best way to organize the piece
6. To follow a standard report format or other effective report form
7. To help you tailor your report to your reader's specific interests
8. To allow you to concentrate on writing the report

Let's take a quick look at each of these reasons.

Fulfill All Requirements

Technical and scientific writing is notably complex. Moreover, it usually must conform to tight specifications. The usefulness of any technical report, then, depends heavily on the writer's ability to keep track of all requirements and to control the writing in order to meet all obligations to the company and clients. A good outline acts like a detailed road map, pointing out the best sequence of routes and turns to make, while showing how to avoid accidents and delays.

Analyze Your Ideas

If you don't plan your report before you start writing, you're certain to run into trouble. So analyze your ideas before writing. And the simplest way to do this is to list all your ideas about a subject at random, as we've seen. You don't have to struggle to get every bit of data into an initial outline; you can easily polish and reorganize the elements of the report in your revised and final versions later.

Don't Omit Important Items

Careless planning—whether it's due to laziness or lack of time—is a sure way to overlook important parts of your subject. Usually you won't remember these until the report is nearly finished. Then you will have to go back, rewrite, insert new paragraphs and try to connect them to the existing material. This seldom works. You wind up with a jerky, disconnected report that jars the reader's mind. And since your clients and supervisors are among the most sensitive readers of technical writing, you're certain to spoil their day. So, above all, outline before you write.

Give Your Report Continuity

A well-organized report reads smoothly because the writer has analyzed his or her ideas and arranged them logically. Continuity is one of the benchmarks of the experienced technical writer. If your aim is to captivate your reader and obtain his or her approval, then strive for continuity. How? Outline.

Organize Your Report

While making an outline, you must review all your existing text, illustrations, and tabular materials. When you do this properly, a pattern for reporting the project will emerge. You'll find that your project elements lend themselves to particular types of presentation—cover page, title, foreword, abstract, lists of figures and tables, text, references and the like. Once you begin organizing the material, it is easier to write the report.

Follow a Format

A thorough outline reduces the amount of work needed to go from random thoughts to a published report. Nearly all technical reports must follow some format; thus you save time and effort by basing the outline on a report format. Using an outline, you can organize your report elements, meet formatting requirements, and develop a table of contents all at the same time.

Tailor Your Report

When you write a report for a client who knows none of the details of a particular technical project, you must write differently than you would for a project engineer or business administrator who knows the details. You must tailor your document. And the easiest way to do this is to make an individual outline before writing.

Help Yourself Concentrate

Perhaps your biggest advantage when you make an outline is the freedom your mind enjoys while you write the report. Why? For several reasons. Once you outline your report and know that you've covered all the important points, you can forget every item except the one you're writing about. This allows you to focus fully on every word, sentence, and paragraph. There is no danger of starting with facts that should be in the middle of your report instead of at the beginning, so you don't have to worry.

In short, outlining greatly simplifies the writing task and increases the likelihood that your report will be approved.

STANDARD OUTLINE GUIDE

Below you'll find a standard outline guide that you can use for reports and all other major technical and scientific writing tasks. The guide reflects the findings of many professional engineering and scientific technical writers and editors. Use it to improve both the quality and the productivity of your written communications.

1. Draft the initial working outline.
    1.1. Assign a tentative title to the subject.
        1.1.1. Name the project, device, or study.
        1.1.2. Sum up, in a sentence or two, what you will say about it.
        1.1.3. Keep the title short.   EXAMPLE:
            Total Generating Cost of Power Plants.
    1.2. List available data in rough order.   EXAMPLE:
        Alternative proposals for system expansions
        Total generating costs of various options
        Contrasting cost advantages for various proposals
        Purpose of project
        Fossil-fired generating stations
        High-sulfur coal
        Low-sulfur coal
        Nuclear power generating stations
        Pressurized-water reactor
        Boiling-water reactor
        Net plant output in megawatts at operation date
        Comparative evaluation of costs
        Methods of computation
        Average annual plant capacity factors
        Annualized capital cost or fixed costs
        Levelized fuel cost
        Operating and maintenance costs
        Costs in mills per kilowatthour
    1.3. Regroup data under "Introduction," "Body," and "Conclusion."
        EXAMPLE:

> *Introduction*
> Purpose of project
> Methods of computation
> Alternative proposals for system expansions
>
> *Body*
> Annualized capital cost or fixed costs
> Levelized fuel cost
> Operating and maintenance costs
> Costs in mills per kilowatthour
> Total generating costs of various options
> Fossil-fired generating stations
> Low-sulfur coal
> High-sulfur coal
> Nuclear power generating stations
> Pressurized-water reactor
> Boiling-water reactor
> Net plant output in megawatts at operation date
>
> *Conclusion*
> Comparative evaluation of costs
> Average annual plant capacity factors
> Contrasting cost advantages for various proposals

2. Prepare the final outline from the regrouped data.
   2.1. Decide on order of presentation of facts:
      2.1.1. Order of importance to specific reader
      2.1.2. Order of events in research, experiments, etc.
      2.1.3. Order of steps in evaluation procedure
      2.1.4. Summary, followed by details
   2.2. Order-of-importance arrangement presents facts in descending order of importance (popular for articles, papers, manuals, etc.).
   2.3. Order-of-events arrangement presents data in chronological order (popular for some reports and articles).
   2.4. Order-of-steps arrangement gives step-by-step procedure for performing a specific task (good for short investigation reports, manuals, etc.).
   2.5. Summary arrangement gives major results first, followed by details (popular for many technical reports and articles).
   2.6. Any arrangement can begin with the summary, conclusions, or recommendations.
   2.7. Use clause, sentence, or phrase entries in outline when possible; avoid single-word entries except for short writing tasks.
   2.8. Show relative importance of entries:
      2.8.1. By numbers and letter: 1, 2, 3; A, B, C, etc. (good for long outlines, numerous entries).
      2.8.2. By decimals, as in this guide (good where document will have paragraph numbers or detailed sections and subsections).
      2.8.3. By indention (good for short outline, few entries).   EXAMPLE:

> *Outlining the Final Report*[1]
> Need for a written plan
>   Importance of planning

---

[1] From C. R. Anderson, A. G. Saunders, and F. W. Weeks, *Business Reports,* 3d ed., McGraw-Hill, New York, 1957.

Purposes served by outlining
Usefulness of outline
Arranging your material
  Order, unity, inclusiveness
  Logical progression of ideas
  Concreteness, proportion, usefulness
  Chronology of events
What to include in the outline
Building an outline
  Start with a list of topics
  Select main divisions
  Fit topics under main divisions
Helps in outlining
  Show rank
    By numerals and letters
    By decimals
    By indenting
  Avoid single subheads
  Avoid overlapping
  Restrict number of headings
  Subordinate reasonably
  Observe parallel construction
  Use at least two headings at each outline level
  Make each division specific
Style of the outline
  Sentence: sums up each topic in a complete sentence
  Topic: uses short phrases for quick reference
Testing the Outline
  Checking
  Brainstorming

2.9. Organize the data under a standard or prescribed format, if desired.
EXAMPLE:

A Typical Long-form Report
Title page
Summary
Table of contents
Introduction
Discussion (body of report)
Illustrations and tables
Conclusions
Recommendations
Research References
Credits
Appendixes
Index

2.10. Indicate what illustrations and tables will be used, and where.
2.11. Insert bibliographic and other references you plan to use.

3. Get the outline checked by skilled professionals.
  3.1. Engineers, scientists, technicians, managers.
  3.2. Advertising, public relations, legal experts.
  3.3. Where you are the technical authority, check outline yourself.
  3.4. Incorporate responses to make any appropriate changes.

SELECTING AN APPROVAL-GETTING SUMMARY

An accurate report summary gives your readers all they need to know to make a decision on a given subject within a few moments. Like skillful planning, this is key to an approval-getting presentation: Start your report with a summary.

The summary differs from the foreword or preface in that it aims to encourage your readers to examine the entire report, yet it presents enough of the important findings to allow them to take action and proceed to other work when necessary.

Some reports can be summarized in as short a space as one sentence. Most commonly, however, the summary consists of approximately 100 to 200 carefully chosen words and between one and three paragraphs. It sums up the main items to be covered in the body of the report, including purpose, procedure, and results. Preferably, the summary should follow the title page; but it may be placed after the preface or foreword. In long-form reports and multivolume documents, place the summary alone on a page, using individual summaries for each volume when needed.

Summary, Abstract, and Synopsis

Some technical writers prefer to use the heading "Abstract" instead of "Summary." The word *synopsis* has a similar meaning but is rarely if ever used in technical reports. All three of these serve similar functions: They don't waste the reader's time. However, the abstract differs from the summary in that it usually consists entirely of a one-paragraph statement, often in abbreviated language. Summaries are often longer, as we have noted.

It is hard to say whether an abstract or summary is better-suited to particular kinds of reports, since the choice usually depends on individual preference and on any restrictions imposed by formats that clients or managers prescribe. Some writers prefer to use a summary for formal business reports and an abstract for scientific and technical reports. The choice of headings is not as important as it once was, since readers nowadays are generally as accustomed to reading abstracts as they are to reading summaries. Few reports contain both, however. Later in this section we will look at the special characteristics of abstracts. For our purposes here, however, we'll consider abstracts as roughly interchangeable with summaries.

Do some technical writers prepare reports before writing summaries? Can a report be written without a summary? The answer to both of these is yes. Some writers find it easier to develop the summary after they have written the body of the report, even though the summary belongs at the front of the finished technical document.

We should note here, however, that some short reports carry the summary at the end, following the report body. This is done mainly for nontechnical material in informal documents. For example, a one-page business memo noting the performance of an employee during a five-day period might sum up the writer's findings in an ending summary.

In a technical report, a writer who prepares the document without a summary or abstract "up front" runs a much greater risk of getting the report rejected. Some readers will not even look at a report that does not have a summary or abstract at the beginning to catch their attention and explain what the report is about.

Why Summarize?

Ask a hundred scientists, engineers, or executives what their biggest daily chore is, and they'll almost all give the same answer—reading. Engineers, scientists, and managers in key industries say they are flooded with reports from inside and outside their firms. Few of these people have the time to read any but the most important reports reaching their desks. But almost every astute reader is glad to examine a well-prepared summary of a report. Why? Because the summary or abstract presents the salient points of the report concisely and in minimal reading time. The reader gets the facts needed for making a decision quickly. And this, as we saw earlier, is the ultimate aim of almost every report written today. Once you've selected an approval-getting summary, you are ready to write and arrange the report.

Getting Started

Start preparing your summary by considering the purpose and findings of the project you are writing about. Pinpoint the action you want your reader to take, as we discussed earlier in this section. Extract the most important points that you plan to include in your text, including methods, highlights of the study, conclusions, or recommendations. Then jot these points down in a brief (half-page) outline. Here's a typical report summary outline:

1. State the researchers' responsibilities.
   1.1. Gather information on data processing cost of system expansion alternatives.
   1.2. Compare average processing costs of options.
2. Briefly describe the examinations undertaken.
   2.1. Research methodology
   2.2. Calculation of processing costs.
3. Qualify the research approach and technical basis.
   3.1. Average usage factors
   3.2. Baseline case and reference design
4. Indicate what sections include additional definitions.
5. State conclusions or recommendations.
   5.1. Minicomputers have much lower operating costs in some regions.
   5.2. System expansion in certain regions should be based on minicomputers for greater economy.

Compose a rough draft of the summary in a few hundred words, using the facts from your outline. Don't worry about jargon or wordiness at this point, but always try to keep the summary short, accurate, pertinent, and interesting. Also, be sure to consider the specific interests and know-how of each prospective reader. Remember that you can revise the summary at any time before you submit the final version of your report for publication or review.

Polishing the Summary

After you've composed the rough draft of the summary, prepare a new version based on the first. This time, eliminate needless words or sentences and look for instances of technical or scientific jargon that prospective readers may not fully understand.

Pare the summary down to the essentials. For instance, don't write, "A study was undertaken of alternative data-processing-system expansion proposals using unit costs based on the sum of annualized capital costs, levelized power costs, and operating and maintenance costs expressed in dollars." Instead, state clearly what the project did, how it accomplished what it did, what the findings were, and any conclusions or recommendations that resulted. Use plain terms that all prospective readers will comprehend. Put your words into active, direct statements.

If you intend to submit your report to people at different levels inside or outside your firm—such as administrators, technicians and nontechnically trained clients—alter the words and approach in your polished summary to prepare additional versions for each reader. Sometimes this is all that is needed to tailor the report.

For example, in preparing a report summary for a board member or the president of your firm, you would emphasize the financial and business aspects of the data—not the technical aspects. So instead of saying, "A system expansion using nuclear power generation shows only a marginal advantage over coal-fired plants," you might write, "Actual power generating costs for nuclear power plants in the Middle Atlantic region were $____ higher than for coal-fired plants, giving nuclear plants only a marginal advantage over coal-fired plants. These higher costs were due chiefly to a 16.2 percent lower average capacity factor for nuclear stations."

Whenever possible, write the summary so that it will fit on one page or less. A length of three short paragraphs is usually sufficient to convey your message properly.

Finally, type the summary (or each polished version) double-spaced on an 8½- by 11-inch sheet. Center it on the page under the heading if you are preparing a formal report or multivolume report. In a short, informal report, you may wish to type the summary single-spaced, flush left, following the report title and author information.

The Executive Summary

Occasionally, clients or administrators may ask to see an executive summary of your project. The task of preparing an executive summary is entirely different from writing the report summaries we've discussed thus far. Since its chief function is to introduce top executives to the technical or scientific engagement, it must be written to exacting specifications.

The executive summary is delivered separately from the project report. Sometimes it may be used in oral presentations before a group of top administrators. Basically, it is a hybrid that combines the length of a short, informal report with the style and format of a formal report.

The executive summary may be several pages long and base much of its presentation on important illustrations and tables. It usually has its own title page, table of contents, introduction, body, and conclusions.

When a client or administrator requests an executive summary, plan it together with the initial outline for your technical report. Draw materials from the report as needed to summarize the salient points and conclusive recommendations in the executive summary. Be sure to include enough supporting tables and illustrations to explain thoroughly the general aspects of the project without confusing your readers.

Exhibit 2.1 is an example of an executive summary prepared for briefing a group of executives and investors whose primary interest is energy systems. Note that it summarizes important findings in tables, gives an abstract for each type of energy system described, and provides simplified schematic diagrams to increase the group's understanding of various alternatives. The exhibit has been reduced to allow reproductions for viewing with an overhead projector in an oral presentation.

Remember that this is only a sample. Your company, clients, or specific projects may specify executive summaries (if any) in different form. The data in this sample do not represent actual costs or research findings.

WRITING FOR FAST AND COMPLETE ACCEPTANCE: HELPFUL GUIDELINES

You can't drop your good intentions once the summary is written. The way you write your report can mean the difference between approval and rejection of your ideas. Since repudiation of your methods or recommendations means the time and energy spent on the report were wasted, make every effort to write the best report you can—every time.

Once you start writing, try to finish as quickly as possible. For the faster you write, the more enthusiasm you build for the project. Use the pointers in section 1 and in this section. Getting the writing done quickly allows you less time to feel sorry for yourself or to daydream about what you'd rather be doing than writing a report.

Remember: The only way to get words on paper is by writing—no amount of thinking and talking will do the writing for you. So think as you write and use your writing as a substitute for talking. You'll be amazed at how much your writing output increases.

Forget the intricate rules of grammar when you write. Get the technical facts on paper. Later, after all the data are in rough form, you can edit and improve your grammar.

You *can* start to write better reports as soon as you pick up your pen. How? Follow these guidelines:

1. Prepare report drafts from your outline.
2. Use simple language and style.
3. Aim at your probable reader.
4. Tabulate contents and main sections for easy data location.
5. Select effective illustrations and tables.
6. Dramatize your report.

Prepare Drafts from Outline

Outlines speed writing, as we've seen. Once you've developed a final version of a comprehensive outline, bringing the report to life is largely a matter of expanding upon the outline to give all the relevant data in narrative form. Remember that it is not enough merely to transfer phrases and sentences from your outline to your text. The outline is the frame around which you build your writing.

Executive Summary

for

EVALUATION OF ALTERNATE POWER SYSTEMS

Submitted to

ALLSTATE POWER AUTHORITY
53 Skyline Avenue
New York, NY 10001

Under Contract No. ASPA-1665-3

by

ARCHITECTURAL ENGINEERING, INC.
101 Montorey Drive
Spokane, Washington

August 19XX

EXHIBIT 2.1 Example of executive summary prepared for briefing a group of executives and investors whose primary interest is energy systems. The exhibit has been reduced to allow reproductions for viewing with an overhead projector. (Note that the data in this sample do not represent actual costs or research findings.)

CONTENTS	
Title	Page
Introduction and Summary	ES-2
Economic Comparison of Selected Power System	ES-3
Coal-Gasification Combined-Cycle Power Plant	ES-4
Solar Photovoltaic Power Plant	ES-5
Ocean Thermal Energy Conversion	ES-6
Liquid-Metal, Fast Breeder Reactor Power Plant	ES-7
Pressurized-Water Reactor Power Plant	ES-8
Compressed-Air-Energy Storage	ES-9
Atmospheric, Fluidized Bed Coal Power Plant	ES-10
Pressurized, Fluidized Bed Coal Power Plant	ES-11

EXHIBIT 2.1 (*Continued*).

Introduction and Summary

Architectural Engineering, Inc. (A-E), was requested to conduct evaluations of the available electric generating power systems under Contract No. SPA-1665-3. In these evaluations, A-E was required to research the available data relating to fuel costs, current labor costs, equipment costs, and other relevant data, and to present economic comparisons of several alternative power systems. In addition, A-E was required to provide comprehensive abstracts which will clearly and concisely describe all power plants selected, highlight the systems characteristics of each, and provide capital costs, in thousands of dollars, for each power plant evaluated.

Brief descriptions of the selected alternative power systems and simplified flow diagrams were prepared in order to provide an overview of each power system. Figures ES. 1 through ES. 8 provide descriptions and diagrams. These figures were reduced to accommodate the reproduction of acetate vugraphs of each figure so that they may be viewed by other technical associates at the state power authority.

Responding to the specific requirements of the above-mentioned contract, A-E conducted their technical evaluations of several viable alternative power systems for the client's consideration. The state power authority is thoroughly aware of the state's implementation plan for emissions control and the provisions for exemptions contained in the Fuel Use Act of 1978. Consequently, detailed explanations of the environmental impacts of burning high-sulfur coal prohibitions in the use of oil and gas for electric power generation require no further discussion.

The ultimate decision in selecting an additional electric power generating plant for the state rests with its power authority. However, the power authority officials requested from A-E a recommendation stating which system has the cost incentives in light of the rising fuel costs and plant costs since 1974. In reviewing all the alternatives, A-E recommends a coal-gasification, combined-cycle power plant be designed, engineered, and constructed at the proposed plant site. A-E's recommendation is substantiated by the cost incentives shown in Tables ES-1 and ES-2 when compared to the initial costs of all other plants evaluated. The coal-gasification, combined-cycle power plant can be engineered and operated at the proposed site and meet the environmental criteria, including the clean air act, established by the state.

The technical project manager and vice president of marketing in A-E shall be pleased to present any further explanations that the state power authority may require regarding the cost-comparisons economics and the plant's technical performance of the recommended alternative power system.

[*Note:* The exhibits and tabulations mentioned would follow at this point. Space limitations prevent their inclusion in this handbook.]

EXHIBIT 2.1 (*Continued*).

Use Simple Language and Style

Study the pointers in Sections 1 and 14 on effective communication and good grammar. But if you don't have time to read or can't remember all the rules well, use these two rules in all your writing: (1) Use the simple word whenever possible. (2) Prefer short sentences and paragraphs to long ones.

When choosing your phrasing and vocabulary, don't try to impress the reader with technical terms or unrelated philosophical concepts. Big words and long sentences never show off an author's skills. By the same token, steer clear of colloquialisms and slang. The narrow focus of most technical reports demands they be written in a formal style. This doesn't mean your report has to be dry or uninteresting. On the contrary, it means you must follow established rules for conveying ideas in a persuasive, engaging fashion.

Your writing should be clear and easily understood—even by the least technical reader. Writing that isn't clear makes rejection more likely. Clients and top management won't put money or time into your ideas unless they feel they understand completely everything you're talking about. Obscure language may be taken as an attempt to cover up an avenue of investigation that has been overlooked, or it may be seen as indicating poor understanding of the report writing function.

Be consistent in your use of specific terms and concepts throughout the report. Give definitions wherever necessary to make sure the reader gains a full understanding of all your material. However, try to keep repetition to a minimum. A rule of effective writing is that a writer is more likely to increase readers' knowledge of a subject when the writer assumes their ignorance of technical and scientific topics. Nonetheless, writing experts refuse to adopt a condescending tone toward readers or to load down reports with verbal fat.

Above all, unfold your topics in a logical sequence. Keep an eye toward smooth transitions between each sentence, paragraph, and section.

Aim at Your Probable Reader

Earlier we saw how you can tailor the intent and summary of a report for a given class of readers—usually engineers, scientists, or managers. Even if only 10 percent of your readers go beyond the summary, you still have to hold their attention and make them act. To do this, you must write the report so it covers their main fields of interest.

When you know the report will be read primarily by engineers, give engineering data the main emphasis. For scientists, emphasize the scientific data. For managers, emphasize the business aspects. If the same report is to be read by more than one group, break it into two or more main sections according to type of reader. For example, you might have: engineering data, scientific data, management or business data, etc. Then readers can find what interests them without wading through pages of data that may be of no value to them.

Tabulate the contents and main sections for quick use. List every major section of the report in a table of contents. Give the page number beginning each section so the reader can see at a glance how many pages are devoted to a given topic. Never slight the table of contents. Sometimes your report will be evaluated by a reading of only two items in it—the summary and the table of contents. This may seem unfair, but in the press of business activities people develop various

FIG. 2.5 Comparison of study and report sequences. *(Tom Johnson and* American Machinist.*)*

time-saving schemes for reviewing reports. The two-item reader is common in business and industry today.

Refrain from presenting data in the same order as you carried out your study of a problem. As Figure 2.5 shows, you generally conduct a study in this order: objective, detailed investigation, analysis, results. But a report should present data in the reverse order: results, analysis, detailed investigation.

Use Index Tabs

Another aid to faster use of your report is to tab each important section. Some technical writers note that index tabs increase the readability of long reports substantially. Give the tab the same title as the section it indicates; if you use a different title for the tabs, the reader will be confused. Tabs are more effective than colored pages for separating report sections because tabs are easier to locate. Every minute you save your reader in finding main topics of interest is that much more time he or she can spend actually reading the report. Of course, tabs are not needed on short reports that are only a few pages long.

Select Effective Illustrations and Tables

Solid pages of text are deadly because they are monotonous and discouraging to the readers. So open up your pages with useful illustrations and tables.

But be careful. We all know the saying about a picture and a thousand words. Like all proverbs this one is not always true—particularly for reports. A poor illustration or one not appropriate for the data in the report can waste time and space. For example, illustrations of standard test equipment like oscilloscopes, flow meters, chromatoscopes, and thermal conductivity gages waste space in a report unless they show new configurations or uses. Omit this type of illustration. In its place use illustrations showing equipment or modifications to equipment the reader is not familiar with.

There are other types of illustrations you can use in a report besides technical or scientific ones. Both semitechnical and nontechnical figures can be used if they are pertinent and convey useful information to the reader. Every illustration must be tested for its usefulness. Ask yourself, "Will my intended reader obtain any helpful information from this illustration?" If the answer is yes, include the illustration.

You can also include dramatic photos, pertinent charts and graphs, newspaper or magazine clippings (Figure 2.6), sample pages from catalogs or instruction manuals, business forms, letters from important persons or companies, memos, and excerpts from other reports. So long as the illustration or figure is relevant and helpful in conveying the report's message, use it. Be sure to obtain permission from your sources when copyrights are involved, especially if you plan to publish the report in a journal or book.

Remember not to load a report with so many illustrations that the reader has to struggle to wade through them. Exercise your judgment. There are no rules

Solar Cells: How They Work

In 1839 the French physicist Edmund Becquerel discovered that light falling on certain materials can generate a weak electric current. Intensive exploitation of this property was initiated in the 1950's by the Bell Telephone Laboratories as a possible source of power for remote telephone systems. This led to the development of the first solar cells, including those used to power early spacecraft.

The cells were formed of two layers of silicon crystal. One, the "n" layer, was "doped" with traces of a substance that enabled negatively charged electrons to escape from the crystal lattice and move freely. The other "p" layer was doped with material that allowed positive charges, or "holes," to move through the lattice.

Light falling on the cell liberated electrons in the n-layer and holes in the p-layer. If the layers were linked by a wire, the electrons flowed through it to join holes, producing an electric current.

FIG. 2.6 Press clippings can be effective in reports. (The New York Times.)

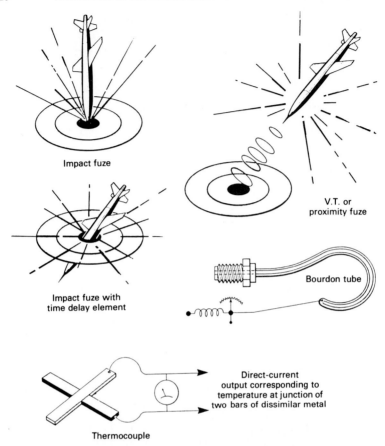

FIG. 2.7 Motion in illustrations can create greater interest while dramatizing your report. *(Department of the Air Force.)*

stating how many illustrations can be used in a report; but few technical reports should ever require more than 50. And if you can limit a report to 25 or fewer illustrations, it will be easier to read and use.

Don't overlook the opportunity to use dramatic illustrations in a report. Illustrations depicting motion, people at work, machines performing interesting tasks, etc., are more dramatic than flat, static ones as you can see in Figure 2.7. Not all your illustrations can be dramatic—use of plant layouts, circuit diagrams, and other static drawings are necessary in almost every report. But where you have a chance to choose a few dramatic illustrations, use them.

Use Photographs to Enhance the Story

Use photos wherever they will do a better, faster job of telling your story. Search manufacturer's catalogs and check with your public relations office or purchasing agent if you want to illustrate a product. Moreover, get attention with color. Use it where you want to highlight an important point—not just in photos but in section dividers, index tabs, and special drawings. Many writers find that the use of a light pastel color or two, tastefully applied, enhances the overall appeal of a report.

Dramatize Your Report

Reports can get immediate action from top management—and can make research ideas pay off. Here's an example of one that did so at a major microcomputer manufacturing firm. It got the green light from the company's top administrators. And management gave its author a budget, relieved him of other work, put him in charge—the first project of his own—and promoted him.

This was the idea: The author believed there were market possibilities for a low-priced personal computer that contained the programs and hardware commonly used by business people together in a single unit that could be carried like a briefcase and stowed under an airplane seat. Besides writing a provocative report to top management, the author added exhibit material that made it an eye-catcher. Here's a sampling of the exhibits that supported the proposal.

To emphasize the need for portable computers among corporate executives, the report used:

1. An illustrated newspaper article with the headline, "Managers Grow Dependent on Small Computers"
2. Another article with the headline, "Microcomputers Change the Face of Small Business Competition"
3. Bar graphs contrasting average paperwork output for managers using computers against those not using computers

To emphasize the salability of systems priced under $2,000, the report included:

4. A letter from Microcomputer Futures, an industry-watching group, answering an inquiry for figures on the personal computer market and sales (The letter contained a table of market projections, by price, for the next year.)

To support the engineering aspects of his proposal, he used:

5. Artist's renderings of a potential portable system, based on engineering research, including actual-size drawings of main circuit and memory boards, disk drives, keyboard, and video screen
6. System flowcharts
7. A photocopy of an article describing development and use of portable microcomputers in Japan

His final report contained 16 exhibits, all of which helped to build the story of the profit potential of the project and the means of accomplishing it.

ARRANGING THE REPORT FOR EASY USE

Good arrangement of your report is a must. In itself it demonstrates clear and orderly thinking. Most importantly, it helps your readers to get the information quickly and easily.

What's the best way to arrange a technical report? Long industrial experience shows that there are nearly as many methods as there are types of reports. However, a few techniques stand out as the most reliable for ensuring maximum effectiveness. We've already touched upon some of these—development of a comprehensive outline and use of formats. Now let's look at this organizational task in depth.

Successful arrangement of a technical report requires knowledge of the parts of the intended format. Possible parts include: cover, title page, summary or abstract, preface, table of contents, introduction, main discussion, illustrations and tables, conclusions or recommendations, references and bibliography, appendixes, index, and back cover.

Report organization also involves finding the appropriate niche among these elements for all your written materials. Further work is usually required to produce the final document according to publishing specifications for writing style, usage of words and abbreviations, type composition, indentions, etc.

In this section we'll concentrate on arranging the parts of your report. Work from your report outline or try using 3- by 5-inch file cards for organizing your facts and ideas. Write each major fact or idea on a separate card. Spread them all out on a tabletop. Then spend half an hour or so regrouping them into the best sequence for unfolding your ideas. Check the sequence against any format dictated by your client or firm or against the standard report format given below.

STANDARD REPORT FORMAT

There are many different formats in use for formal reports, depending on the requirements of specific companies, clients, and projects. However, most formats include roughly equivalent elements, and they are usually arranged along the same basic lines.

All technical and scientific reports have some form of introduction, body, and conclusion. Each of these general parts contains specific elements that appear in the printed document under individual headings. Here is a general format that has become standard in many areas of industry. Use it whenever you need a guide to arranging a formal report.

1. Transmittal letter
2. Cover and title page
3. Summary or abstract
4. Table of contents
5. Introduction
6. Main discussion
7. Conclusions or recommendations
8. References, bibliography, or credits
9. Appendixes
10. Index

Now let's take a look at each part.

Transmittal Letter

Formal reports generally are sent to clients or to individuals within the originating firm along with a letter of transmittal or cover letter. The letter notes the name of the person to whom the report is addressed and states why the report was written, who authorized it, and the date when it was first requested. It may also summarize the findings of the project and call attention to any unusual problems or recent developments related to the study.

Some firms require special letters of transmittal for all reports. If your firm has such requirements, follow them exactly. Never allow the report to leave your hands before you've checked the transmittal letter. Errors in this letter can be fatal to your document.

If your firm has no special requirements for a transmittal letter, give the matter some serious thought. Lengthy, highly complex reports containing numerous sections or multiple volumes may necessitate inclusion of a detailed letter inside the bound report. This formal letter of transmittal should precede the title page. It might explain the organization of the document and highlight areas of interest to different potential readers. Shorter, simpler reports, on the other hand, may require nothing more than a simple one- or two-paragraph cover letter clipped (not stapled) to the outside of the bound report.

Generally, keep the letter as short as possible—three paragraphs should be the most you need for the average report. Since the transmittal letter acts as a record of the delivery of the report, try to keep other details out of it, except in unusual situations like those we've noted above. If you try to summarize the full report in the transmittal letter, the person receiving the report will rarely read beyond the letter itself. So all the effort you put into your report will be wasted.

Sign the transmittal letter in ink. Then be sure the letter and report reach the person to whom they are addressed.

Cover and Title Page

The report cover may be either a title page—as in a simple report going from one person to another within the originating firm—or it may consist of a separate binding of some type, such as plastic or hard "cloth"—as in a multivolume report designed to be used for future reference. In either case, the cover carries the report title, the name of the author and originating firm, and the name of the person or firm for whom the report has been written. When the report is bound in a cover, the document must also have a title page carrying this information. Additional data often placed on a report cover include: business addresses of the author and recipient, contract numbers, and date of submittal.

Always try to pick a short, accurate title for your report. Be sure it will not mislead the reader; executives will feel you wasted their time if the title promises one thing but the report gives another. A long title can be justified if it is accurate. But, where possible, use as short a title as necessary to specify the contents of your report. Avoid extraneous words, and try to inject interest into the title by including the positive aspects of your findings. Thus, you might write "New Microcomputer System Leads to Increased Staff Productivity" instead of "Evaluation of Studies on New Microcomputer System." Certainly the first title is more interesting.

Center the title on the cover and on the title page. Use a subdued color for the binding. White is the best color for the title page, and it has been proven to make the most readable documents when combined with black type.

If your firm has its own publishing and document production experts, you may not need to worry about the material or makeup of the report cover. They will do that work for you. Nonetheless, you must make certain that all cover information is accurate, and you can help the publications staff choose an appropriate cover by informing them of the report's intended use. In Section 14 of this handbook we will further define production and publication considerations of technical writing.

Summary or Abstract

Every technical and scientific report you write should contain either a summary or an abstract, as we've already noted in our discussion of the report summary. Here we'll focus on the use of the abstract.

An abstract is sometimes preferred in complex reports intended for possible publication or storage on computerized databases, especially in fields where collections of abstracts are published for future reference. For example, many colleges, universities, major libraries, and trade associations carry abstracts of scientific and technical papers that are published periodically and permanently bound for later use. Among these publications are *Applied Mechanics Abstracts, Sociological Abstracts,* and *Chemical Abstracts.* In addition, organizations like the National Technical Information Service publish directories of scientific and technical reports, categorized by technical discipline.

Computer-based data retrieval services store many of these same publications electronically. The abstract is particularly useful for computerized information retrieval because it builds much of its statement of a report's contents on key words and phrases that are peculiar to a given topic. Using a computer terminal and retrieval service, a researcher can find reports of interest to a company or agency simply by typing the key words into the computer and giving the proper commands for seeking documents that contain the key words. For example, if the researcher wanted to find a report on quality control in the processing of aluminum ore at a particular plant, he or she would merely have to input the plant name and the phrase "aluminum quality control" with a search command. The computer would then locate any abstract having those terms or subjects.

We can expect to see increased use of the abstract in the future due to society's increasing dependency on electronic data storage. This is certainly a factor worth considering whenever you prepare an important formal report.

In the technical report, the abstract is usually a one-paragraph description of the report contents. Since it should contain key words and "search terms" from the body of the report, you should write the abstract after you have prepared your outlines and main discussion. Be as specific as possible, including only data that is essential to stating the problem under study, what was done, the most important results, and major conclusions or recommendations.

Here are some guidelines for writing abstracts:

1. Select the key points, terms, and phrases from your report by scanning your table of contents, introduction, main discussion, and conclusions for what makes the report unique.
2. The abstract or summary usually follows the title page in a technical or scien-

tific report. Remember to include a brief statement of the reason for the study and the date(s) it was conducted.
3. Where possible, use abbreviations, but only if they are in common use outside your firm.
4. Keep the abstract to one paragraph in length (about 200 words maximum), summarizing as clearly and briefly as possible your methodology, findings, and recommendations.
5. Review the abstract after you write it, cutting out any words not essential to its technical accuracy and completeness.

Here is a typical abstract of a technical report:

The total cost of generating electric power from six types of power plants is used as the basis for comparing and contrasting alternate proposals for power system expansion at State Power Authority's Smithhaven plant. A coal-gasification, combined-cycle expansion is recommended because of the relatively low total generating cost of this type of plant. Fiscal year 19XX data for costs of fuel, labor, equipment, transportation, and construction are used in calculating the total generating costs for each plant in mills per kilowatthour. Tables and simplified flow diagrams illustrate the differences between options.

Table of Contents

Your report's table of contents will say a lot about how well you have organized and presented your material. As we've noted above in the discussion on outlining, the report outline is integral to the creation of the table of contents. You can also use the 3- by 5-inch cards as guides after you've grouped them in the desired sequence.

Note that a table of contents should not necessarily contain all the headings included in the outline or cards. Use all the section headings of the report to tabulate your contents; but do not include minor subdivisions and details that were useful only to you as reminders for preparing your comprehensive presentation.

List the headings exactly as worded in the report. Give inclusive page numbers so your reader can tell how long each section of the report is. Be sure each entry is descriptive. Some busy people will "read" your report by giving the table of contents a quick glance. Misleading entries will leave them with an incorrect impression. Thus:

1. Abrasion Resistance
    a. Coating type A
    b. Coating type B
    c. Coating type C

tells the reader very little. Change this kind of entry to:

1. Evaluation of Abrasion Resistance in Three Coatings
    a. Polystyrene displayed 5 percent wear after 4 hours.
    b. Buna rubber displayed 2 percent wear after 4 hours.
    c. Polyvinyl chloride remained unworn.

The second approach to the table of contents tells the reader what to expect from the report.

Remember to include not only the sections of the main discussion in the table of contents but also the other parts of the report—abstract, introduction, conclusions, references, appendixes, and index—with page references. If you have a number of tables and illustrations in the main discussion, list these separately following the table of contents. The shorter and simpler your report is, the shorter and simpler your contents may be. Informal reports rarely contain a table of contents.

Introduction

We're taking the liberty of designating the introduction as a standard element in formal reports because it is the only opening statement most report writers ever need. Some reports do contain a preface or foreword instead of an introduction. All three serve similar functions; however, an introduction follows the table of contents and is actually considered the first part of the report body, whereas a preface or foreword can come before or after the table of contents and stand as a separate essay.

Reports intended for distribution to only one or two key executives rarely contain more than one of these types of opening statements. Longer technical and scientific documents—like textbooks, manuals, and monographs—occasionally use both an introduction and a foreword (or introduction and preface). This practice is typically reserved for the cases in which a technical authority or other well-known person or firm says a few words on why the report is needed. In such reports the foreword precedes the introduction or preface and tells how useful the report is.

Reports that will be read by a large group of people or the general public benefit from an introduction. Here the author can state briefly his or her purpose in writing, make necessary acknowledgments of help, point out difficulties and uncertainties in connection with the writing of the report, and, in general, tell readers any facts deemed pertinent to an understanding of the report. These facts might include a synopsis of the work scope, the author's background, and the firm's relevant experience. In addition, it is often helpful to present the order of discussion for the remainder of the report. You will also raise a number of questions in the reader's mind. The tone of the introduction, preface, or foreword is less formal than the report itself.

If you, your firm, or your client wish to invite comments and suggestions from readers regarding the study covered in your report, put this invitation into the introductory section. Give the name and address of the person to whom the letters should be directed. But be sure to check with your client or manager before doing so, since this may affect contractual obligations.

Main Discussion

The main discussion contains the bulk of data on the subject of your report. Typical kinds of information presented in-depth in the main discussion are: authorization, purpose and object, scope and method, limitations, definition of terms, apparatus and materials, outline of procedures, and results.

For readers this section is where the problem is defined exactly, analyzed, and solved. It is here that your readers expect to find all the analytical data to answer their questions. Support your findings through the orderly development of your argument and flawless logic. Also include in the main discussion any important charts, graphs, diagrams, tables, and quotes that will enhance your intended readers' comprehension in a quick reading. But save for the appendix the detailed illustrations, lengthy calculations, and reference materials used to develop your conclusions.

Start the main discussion following the introduction. If the report is on a project that had several phases or steps, arrange this part of your report under a series of titles that state clearly what each section is about. For example, use headings like "Equipment," "Solutions," "Procedure," "Calculation of Results," "Reliability of Findings," etc., rather than vague titles like "Discussion," "Research," and "Outcome." Better yet, develop your argument by pinpointing each key aspect of the project under telling headings. An example of this kind of heading structure is: "Basis of Report," "Study Approach," "Current Electric Bills," "Effect of 25 percent Increase in Consumption on Electric Bills," "Characteristics of Line Load," "Selection of Reactor Sizes," "Plant Location," "Pressurized Water Reactor Flowcharts," "Rate of Return and Overall Costs," and so on.

Note that the main discussion must not necessarily follow the project chronologically. It depends instead on logic, completeness, accuracy, and conclusiveness to win acceptance of your findings or recommendations.

Above all, write your main discussion in a way that leads your readers to the same conclusions you have reached. Design each report for its probable audience. To keep your writing from being condescending or too highly technical, make a list of your potential readers to guide you while preparing the discussion. Estimate the familiarity that potential readers from different disciplines are likely to have with the terms, procedures, and topics of your project. For a large audience, base your use of technical words and concepts on your estimate of the knowledge of those least familiar with the work.

Readers may lose track of the salient points in the mass of detail unless you use some amount of repetition and emphasis to increase reader retention. Too often, technical writers confine repetition and emphasis to the beginning and end of a report. That is fine—when the reader knows as much about the subject as the writer does. But this is not always the case. Therefore, be sure to:

- Define important terms and complicated concepts in the main discussion where needed
- Repeat key ideas and facts in the summary, introduction, main discussion, and conclusions
- Highlight these same ideas and facts with attention getters—subheadings, charts, and strategic positioning on each page

Conclusions or Recommendations

This part of the report may follow either the introduction or the main discussion. We prefer the latter sequence because it reinforces statements made in the summary or abstract, answers questions raised in the introduction, and draws together data from the main discussion in a concise terminal section. In our standard report format, conclusions and recommendations are cited at the beginning of the document in the summary and detailed at the end under the appropriate headings—"Conclusions" or "Recommendations."

Most likely not every report you write will require separate sections with these headings. In some reports, a synopsis of findings is all that is needed to give completeness. Thus, some reports end in a brief synopsis, some carry a terminal section with conclusions only, others end the report body with recommendations, and still others use both conclusions and recommendations separately. In any event, the report body should end with a brief summing up of the outcome of the project as cited in the main discussion, and it should give any pertinent suggestions the author has been requested to make, based on that outcome.

Generally, your presentation will be stronger if you use separate headings for conclusions and recommendations. State general conclusions or recommendations first and then proceed to specific ones. A summary of factual findings belongs under conclusions, as do any criticisms that have resulted from the study. Advisory statements—ranging from mild suggestions to concrete guidance—belong under recommendations.

Cross-reference specific conclusions, recommendations, or findings to the substantiating data in the main discussion of the report. Do this by giving the page references for specific points, or assemble your specific conclusions or recommendations under subheadings that correspond to the sections in your main discussion.

This terminal section is where repetition and emphasis pay off most noticeably. Never wait until this part of your report to introduce new material, since doing so will confuse your readers. List your conclusions and state your recommendations to remove any doubt in readers' minds as to whether or not the report answers their questions on your subject. Remember to tailor your statements to the interests of probable readers.

When possible, give your conclusions or recommendations a definite heading like "Suggested Actions," "Some Common Uses," "Probable Developments," "Work in Prospect," "Merits of the Plan," "Some Key Correlations," "Likely Cause," or "Important Precautions."

References, Bibliography, or Credits

List your research sources here, especially when you think they would help the reader. But be specific. If you referred to only one article in a magazine or to only one chapter in a book, cite that one article or chapter and give the publication information—author, title, publisher, date of publication, and page numbers.

When your research does not require reference to books, omit the bibliography and give this section a heading like "Sources," "Research Sources," "References," or "Credits." Technical and scientific reports tend to rely more heavily on actual projects and people than on books as research sources. Thus not all reports have a section of this type. However, important bibliographical references should be cited and arranged alphabetically according to a standard format.

It also pays to be generous in giving other people credit. But don't stretch this beyond the point of reasonableness. In a reference, bibliography, or credits section you have a chance to win people to your side—the people and organizations who helped you develop your report. Their approval and cooperation may be very important to you and your project in the future.

Section 14 gives guidelines for preparing these sections effectively.

Appendixes

Here is where the long tables, mathematical proofs, calculations, progress diagrams, complex flowcharts, details of comparative exhibits, and analytical spreadsheets belong. This is the reading gallery for the technical expert, not the business executive.

The appendixes leave readers free to ignore these materials if they wish. They can then concentrate on the conclusions and main discussion. Should a reader want to verify an equation, he or she can easily do so by turning to the appendix. But it is not necessary to read through every derivation, calculation, or table to understand the report. In fact, some readers will never go past such items when they encounter them in the report body. Placement of the more massive data in an appendix will make your report easier for readers to digest and thereby increase the probability that your findings and recommendations will be accepted.

Some guidelines for appendixes are:

1. Place background material that you used in your research in an appendix, arranged by individual entry and by the sequence in which you refer to it in the main discussion.
2. Give each appendix an identifying letter or number ("Appendix 1," "Appendix 2," etc.), and cite it in making related points in your text.
3. Make sure the appendixes contain only essential supporting or supplementary material.

Index

Use an index only in long, complex reports designed to be stored for future reference. Omit it from simple, routine, or short reports that can be perused in a couple of minutes. When you do use an index, prepare it carefully. It will be a big help to your readers, and to yourself at a later date when you try to find some item whose location slips your mind.

An index provides a highly detailed listing of the contents of the document in alphabetical order. It is more comprehensive than a table of contents, which lists only major topics and important subtopics in the sequence in which they appear in the document. Thus you may be able to use your report outline as the basis for building an index, adding more details after you've written and polished the report.

When feasible, check with a supervisor or publications department for guidelines before you begin indexing. Some industrial firms have staff people whose main job is to produce technical publications according to standard guidelines. Occasionally they are assigned to take care of the entire indexing task.

HINTS ON DEVELOPING A FORMAT AND STYLE GUIDE

The number of reports required in engineering and scientific fields is so large that many firms have adopted standard report formats to increase the consistency of their work and reduce confusion. As a result, a number of consulting firms and industrial organizations have developed their own format and style guides.

If your company does not already have such a guide, you can prepare one yourself. It won't take a lot of effort and should help to simplify writing and improve the overall quality of reports your firm produces. Top executives in most companies do not have enough time to write format and style guides themselves; but they usually know that the need for a guide exists. Almost invariably, an engineer or scientist who produces one will be remembered and rewarded. Even if you don't design a guide for your firm, you can prepare one to help make your own work more effective.

All you need to do is decide on the format that best meets your company's needs, then write the guide based on the information we've given you about report formats. Remember to note exceptions for special client requests, unusual projects, and other specific cases. Photocopy or print the style and format guide for distribution; then make certain everyone who needs one gets it. If you are a manager or officer in your company, you may be able to issue the guide, plus any subsequent revisions, as part or all of a management directive to your staff. Exhibit 2.2 shows portions of a typical format and style guide used for preparing engineering reports for commercial and public clients. The examples give hints on how you can design your own guide.

Many organizations use special typing paper to assist writers and typists in report preparation. This paper is sometimes referred to as a *typing mat* (Figure 2.8). It may also be known as *repro paper,* since it is often made with high-quality paper to allow direct reproduction of the typescript. Most typing mats use light blue ink to indicate margins and other typing specifications. This ink does not appear when the pages are reproduced.

Additional format examples provided here are:

- Figure 2.9: Sample of a simplified report format
- Figure 2.10: Preliminary conceptual description format
- Figure 2.11: Table of contents for a formal report
- Figure 2.12: Sample title page for a long appendix
- Figure 2.13: Separate table of contents for a long appendix
- Figure 2.14: First page of a long appendix
- Figure 2.15: Table of contents organized by code of accounts
- Figure 2.16: Code-of-accounts description from a formal engineering report
- Figure 2.17: A printed form which can help you organize a short report

PROGRAM INFORMATION REQUIREMENTS DOCUMENT

I. <u>Style Guide</u>

   A. The front matter shall consist of the following:

       1. Title page: frontispiece or reproduction of cover
       2. Contents
       3. Table list
       4. Figure list
       5. List of significant abbreviations, as appropriate

   B. The body of the report shall contain the following:

       1. Summary and conclusions, as applicable
       2. Text by section (1, 2, and 3)
       3. Decimalized paragraphs to the fourth digit
       4. Double-spaced text with three carriage spaces between paragraphs
       5. The beginning of all paragraphs typed flush left
       6. Alphabets indent five character spaces and text indented nine spaces (single-space of (a. b. c.) of text with double space between entries)
          <u>Example:</u> a. The Li target materials reprocessing line contains . . .
       7. Bulleted paragraphs single-spaced with double-spacing between paragraphs (Type bullet six spaces from margin. Type text nine spaces from margin.)

          <u>Example:</u> ● The driver fuel reprocessing line shall separate recoverable . . .
       8. Each page identified by part: i.e., on the upper-left side of the page, PART X, in uppercase

   C. The back matter shall contain as applicable:

       1. References
       2. Bibliography
       3. Appendixes (paged A-1, A-2, A-3) (See example for illustration.)

II. <u>Spacing and Headings</u>

   A. Text shall be typed double-spaced, except for alphabetized and bulleted paragraphs.

   B. Headings should stand out from the text, and the importance of each heading should be demonstrated typographically as noted below:

<u>First-Order Head</u>
(Centered on top of page)

SECTION 5 (see example)
(one line space)

<u>MAJOR ECONOMIC FACTORS</u>
(Two lines spaces and centered)

Heading styles for successive subdivisions follow:

2nd. 5.1 **GENERAL** Flush left, capitals, not underscored. Double-space after heading.

EXHIBIT 2.2 Example of a format and style guide.

3rd 5.1.1 <u>Ambient Temperatures</u> Flush left, upper and lower case, underscored. Paragraph starts new line after double space.

4th 5.1.1.1 <u>Seismology Temperatures</u> Flush left, upper and lower case, not underscored. Double-space text.

5th 5.1.1.1.1 <u>Calculation Procedure</u> Flush left, and text following on same line. Double-space text.

Type text two carriage spaces under captions except in the fifth-level headings.

All paragraphs unless otherwise directed shall begin flush left from margin line (block style).

III. Pagination

Page numbers in the report shall be Arabic. Sections should be paginated as 1-1, 2-2, etc. When printing on both sides, the right page must be odd-numbered (1, 3, 5, and so on).

IV. Equations

'quations are to be numbered consecutively on the right side of page, five spaces from right margin line. Greek letters or symbols should be traced in ink with a Leroy set.

V. Footnotes

The symbols used for table footnotes are asterisk, dagger, and double dagger. Superscript numbers should be used only to reference author(s) quoted.

VI. Abbreviations

In order to correspond to abbreviations used in other reports in this contract, the following shall be used: ft, in, Hg, psig. No degree sign (°) shall be used for F (Fahrenheit) or C (Celsius), i.e., 55 F.

VII. Illustrations and/or Exhibits

Superior-quality illustrations always enhance reports; therefore, they are to be drafted, typed, and composed, as necessary.

  A. Figures are to be typed or numbered by subsection and titled as follows:

<u>Example:</u> Figure 5-1 Effects of Time and Temperature of Bacterial Storage
Titles should appear flush left to figure.

  B. Tables are to be numbered by subsection and titled as follows:

TABLE 5-1

KEY PLANT PARAMETERS

The table number (TABLE 5-1) and title should be centered above the table, uppercase with no underscore, as shown above.

  C. Whenever the table width exceeds the margins and requires reduction, do not reduce the type table more than 23 percent (77%, mode 2).

EXHIBIT 2.2 (*Continued*).

EFFECTIVE TECHNICAL REPORT WRITING   **2.39**

VIII. Example follows:

SECTION 5
KEY PLANT PARAMETERS

5.1 GENERAL
Paragraph starts new line after double-spacing.

5.1.1 <u>Ambient Temperatures</u>
Paragraph starts new line after double-spacing.

5.1.1.1 Seismology
Paragraph starts new line after double-spacing.

5.1.1.1.1 <u>Calculation Procedure</u>   Text begins on same line as text. Avoid heading of this level if possible. Double-space text.

APPENDIX IDENTIFICATION

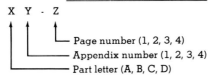

Page number (1, 2, 3, 4)
Appendix number (1, 2, 3, 4)
Part letter (A, B, C, D)

IDENTIFICATION FOR TABLES AND FIGURES

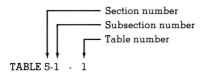

Section number
Subsection number
Table number

TABLE 5-1 - 1

Figure number

Figure 5-1   Type the word "figure" and the figure title initial cap and lowercase, i.e.,
Figure 1-1   Reactor Complex Scheme

EXHIBIT 2.2   (*Continued*).

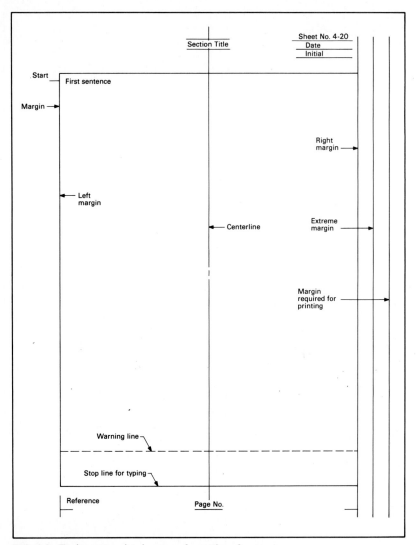

FIG. 2.8 Typing mat assists in proper formatting of a report.

EFFECTIVE TECHNICAL REPORT WRITING 2.41

| | (Heading Format) | (Between blue lines) |

(Start new section on new page) — (Top blue line on paper)
(One line space)

SAMPLE
SECTION XO
INTRODUCTION

(Centered)
(Centered)

XX BACKGROUND (Code of Accounts)
(One line space)

Flush left — As part of this effort, the Nuclear Regulatory Commission (NRC) authorized Architectural-Engineering Inc. (A-E) to investigate the . . .

(Two line spaces between paragraphs)

This report concludes that a substantial quantity of 800-MWe generating stations could operate more economically when issuing . . .

(One line space)

XX OBJECTIVES

(One line space)

XXX <u>Activities and Schedule</u> (Upper and lowercase and underscored)
(Code of Accounts)
(Two line spaces)

An activity was initiated to determine the elements of each segment of the program in order to estimate the realistic worker hours required for each task and . . .

(Two line spaces)

All other subheadings under the third-digit code-of-accounts are to be initial capital, lowercase, and underscored, i.e., <u>Gaseous Radwaste</u>.

FIG. 2.9 Sample of a simplified report format.

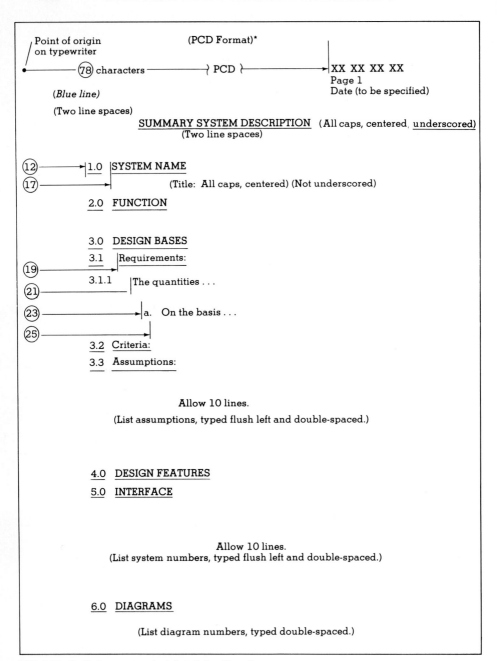

FIG. 2.10 Preliminary conceptual description format.

FORMAT EXAMPLE

1,200-MWe PRESSURIZED-WATER REACTOR

CONTENTS

Section	Title	Page
1	INTRODUCTION AND SUMMARY (Describe briefly the general plant features.)	1-1
2	SITE ENVIRONMENT AND LAND RIGHTS (Ref. Appendix E) General Discussion (State assumption, uncertainties, and any special ground rules used.)	2-1
3	STRUCTURES AND IMPROVEMENTS (Include brief description of HVAC, major structures.)	3-1
4	REACTOR PLANT EQUIPMENT (Include brief description of reactor, I&C, and safety features.)	4-1
5	TURBINE PLANT EQUIPMENT (Include description of I&C and turbine plant miscellaneous items.)	5-1
6	ELECTRIC PLANT EQUIPMENT (Include brief description of associated equipment.)	6-1
7	MISCELLANEOUS PLANT EQUIPMENT (Include a brief description.)	7-1
8	MAIN CONDENSER HEAT REJECTION SYSTEMS (Include type and brief description.)	8-1

Appendixes	Title	Page
A	Technical Description of Plant and Design	A-1
B	Equipment List	B-1
C	Drawing and Flow Diagrams	Foldouts
D	Detailed Cost Estimates	Tables
E	NUS-531 Description of Standard Hypothetical Site	E-1

FIG. 2.11 Table of contents for a formal report.

APPENDIX A

TECHNICAL DESCRIPTION OF PLANT AND DESIGN FEATURES

FIG. 2.12 Sample title page for a long appendix.

	1,135-MWe PRESSURIZED-WATER REACTOR POWER PLANT	
	TECHNICAL DESCRIPTION OF PLANT AND DESIGN FEATURES	
	APPENDIX A	
	CONTENTS	
Code of Accounts	Title	Page
	INTRODUCTION	A-1
21	STRUCTURES AND IMPROVEMENTS	A-6
22	REACTOR PLANT EQUIPMENT	A-8
23	TURBINE PLANT EQUIPMENT	A-10
24	ELECTRIC PLANT EQUIPMENT	A-12
25	MISCELLANEOUS EQUIPMENT	A-13
26	MAIN CONDENSER HEAT REJECTION SYSTEM	A-16

FIG. 2.13 Separate table of contents for a long appendix.

> # INTRODUCTION
>
> This appendix presents a detailed technical description of the Middletown Nuclear Power Generating Station utilizing a pressurized-water reactor as the nuclear steam supply system. The 1,135-MWe power plant proposed by the Public Service Company of New Hampshire at their Seabrook Station was used as the basis for the study.
>
> The appendix is divided into six major two-digit-level code of accounts as follows:
>
> | Code of Accounts | Title |
> | --- | --- |
> | 21 | STRUCTURES AND IMPROVEMENTS |
> | 22 | REACTOR PLANT EQUIPMENT |
> | 23 | TURBINE PLANT EQUIPMENT |
> | 24 | ELECTRIC PLANT EQUIPMENT |
> | 25 | MISCELLANEOUS PLANT EQUIPMENT |
> | 26 | MAIN CONDENSER HEAT REJECTION SYSTEM |
>
> Descriptions of systems and structures are limited to the three-digit level within each major code of account.
>
> These codes of account are consistent with those used in the equipment list and detailed cost estimate of this report. The format makes it convenient for the reader to quickly obtain information related to any plant system's equipment, cost, and description.
>
> A-1

FIG. 2.14 First page of a long appendix.

SAMPLE FORMAT
(For Body of Report after Summary)

1,200-MWe Pressurized-Water Reactor
Technical Description

Code of Accounts	Title	Page
10	INTRODUCTION	—
21	STRUCTURES AND IMPROVEMENTS (The listed subjects are to be described.)	—
211	Yardwork	—
212	Reactor Containment Building	—
213	Turbine Room and Heater Bay	—
214	(Not applicable)	—
215	Primary Auxiliary Building Tunnels	—
216	Waste Process Building	—
217	Fuel Storage Building	—
218A	Control Room/D-G Building	—
218B	Administration and Service Building	—
218C	(Not applicable)	—
218D	Fire Pump House Including Foundations	—
218E	Auxiliary Feedwater-Pump Building	—
218F	Manway Tunnels (RCA Tunnels)	—
218G	Electrical Tunnels	—
218H	Nonessential Switchgear Building	—
218J	Main Steam and Feedwater Pipe Enclosure	—
218K	Pipe Tunnels	—
218L	(Not applicable)	—

Note: The table of contents for the main body of the report shall contain only the two-digit code of accounts, i.e., 22, 23, 24, 25, and 26.

FIG. 2.15 Table of contents organized by code of accounts.

SAMPLE

22 REACTOR PLANT EQUIPMENT

224 RADWASTE PROCESSING

The concept of radioactive-waste processing for the plant is based on an examination of all potential pathways of radioactive release to the environment and provides processing and treatment equipment as necessary to keep the release of radioactivity to the environment as low as practicable and in compliance with 10 CFR 20 and Appendix I of CFR 50.

The transport of radioactivity from the primary coolant system to various parts of the plant during normal operation has been traced and evaluated in order to determine the performance of each process interposed between the source of radioactivity and the subsequent pathways to the environment.

There are three radwaste systems: the radioactive liquid-waste system, the radioactive gas-waste system, and the radioactive solid-waste system. All potentially radioactive liquids, gases, and solids are collected and processed according to physical and chemical properties and the radioactive concentrations. Care has been taken in design to minimize the mechanical leakage paths in these systems in order to limit unprocessed leakage.

Liquid-Waste System

The radioactive-liquid-waste system is designed to collect and process potentially radioactive liquid wastes for recycle or for release to the environment.

FIG. 2.16 Code of accounts description from a formal engineering report.

NAME: J. Q. Smith, P.E.

COMPANY OR DEPARTMENT: Consulting Engineering, Inc.

TO: A. B. Cromwell, General Manager, Aerodyne, Inc.

TITLE OF REPORT: Proposal for Improved Solid-Waste Disposal

OBJECTIVE: To improve solid industrial waste disposal system by reducing costs, eliminating present safety hazard, and removing disposal bottlenecks between source locations.

RECOMMENDATIONS

1. At sources, store waste in modular steel bins (2,000-lb capacity) instead of barrels (100-lb capacity).
2. Collect bins with fork truck instead of moving barrels by hand.
3. Dump waste into steel-lined troughs located near shipping dock.

ADVANTAGES OF RECOMMENDED APPROACH

1. Proposed system cuts labor costs by 60 percent.
2. It conforms to optimum federal safety standards.
3. It eliminates transfer bottlenecks because bins don't clutter work areas.

(TEXT OF REPORT BEGINS HERE.)

FIG. 2.17 A printed form which can help you organize a short report.

APPLYING THE FORMAT GUIDELINES TO A REPORT

As the examples show, a format and style guide should include representative examples of its own. These will help you and your staff to apply formats properly. For instance, provide a full page specifying the line number on which a new section should start and the number of carriage returns required between the word *section* and the actual title of the section; e.g., "Key Plant Parameters." Following these instructions, show exactly how many carriage returns are needed before a subsection number and title is typed; e.g., "5.1 Summary." Unless you specify these sorts of instructions clearly on a "sample page," the typist will have little or nothing to go by, and your reports may wind up looking different each time they are typed. Note that the examples we've shown are only fragments of a complete format and style guide. You may want to include many other instructions and samples, depending on the requirements of your work. Other possible samples include: page formats for special kinds of tabulated data, schematic diagrams and line drawings with figure numbers and descriptive titles, and mathematical formulas, proofs, and calculations.

If your firm uses many abbreviations and acronyms in its reports, your format and style guide can also carry a list of these with their standard meanings. Further, you can describe the characteristics of the writing style and tone you desire in your firm's reports.

For additional help in developing your own format and style guide, check the reference section of a large library; most libraries carry a number of excellent guides.

MULTIVOLUME REPORTS

So far we have discussed short-form and long-form reports, both of which generally conform to the simplified standard report format: introduction, body, and conclusion, with a summary, main discussion, recommendations, and terminal elements where appropriate. There is another type of technical report that is common in industry but that is not so straightforward. This is the multivolume report—a report that addresses a number of technical and scientific topics at length.

A multivolume report customarily contains several major parts or volumes, each consisting of a group of sections and subsections. It may or may not be printed in separate bound volumes.

Writing a multivolume report is a task that makes special organizational demands on the technical writer. The multivolume report usually takes a longer time to research and prepare. Whereas a typical formal report might be based on a few weeks of study, a multivolume report would normally stem from several months or years of extensive studies. Preparing a multivolume report typically requires periodic meetings with contributing researchers, engineers, scientists, and managers.

Figure 2.18 is a table of contents for a typical multivolume report. Note that each section of the report is organized like a formal technical report in itself. For example, each section has its own summary, body, and references. In this sam-

COST ALTERNATIVES

Legal Notice	i
Foreword	ii
Abstract	iii
Table Lists	In Each Section
Figure Lists	In Each Section

Title	Section
Summary	1

1.1   Design Criteria

1.2   Cost of NEPA Alternatives

Main Condenser Cooling Systems	2

2.1   Summary

2.2   Procedure for Development of Cooling System Cost Characteristics

2.3   Site Descriptions

2.4   Reference Plant Description and Cooling System Code of Accounts

2.5   Designs and Costs of Reference Cooling System Models

2.6   Capital Cost Estimates

2.7   Adjustments of the Costs of the Reference Models to Specific Site Conditions

2.8   Assessment of Economic Penalty Costs

2.9   Illustrative Example for Economic Penalty Assessment

2.10   Actual Utility Cost Experience

2.11   Referenced
Appendix A: Detailed Estimates of Alternate Sites

Transmission	3

3.1   Introduction

3.2   Summary

3.3   Basic Design Criteria

3.4   "Baseline Design" Description

3.5   Alternative Systems

3.6   References and Bibliography

FIG. 2.18 Table of contents for a multivolume report.

PART II

Site Access		4
4.1	Summary	
4.2	Site Characteristics and Design Approaches	
4.3	Baseline Configuration Description	
4.4	Base Configuration Cost Estimate	
4.5	Actual Utility Design and Cost Experience	
4.6	Discussion of Results	
4.7	References and Bibliography	
Flood Protection		5
5.1	Summary	
5.2	Design Criteria for Flood Protection Systems	
5.3	Riverine Site Description	
5.4	Riverine Site Cost Estimate	
5.5	North Harbor Site Description	
5.6	Cost Estimate for Flood Protection at North Harbor Site	
5.7	Regional Costs for Flood Protection Features	
5.8	Actual Utility Design and Cost Experience	
5.9	Summary of Results	
5.10	References and Bibliography	
Plant Site Earthwork		6
6.1	Summary	
6.2	Study Procedure	
6.3	Base Reference Model Description	
6.4	Site Excavation and Backfill	
6.5	Description of Alternate Reference Model	
6.6	Development of Cost Functions for Excavation and Backfill	
6.7	Dewatering for Excavation	

FIG. 2.18 *(Continued).*

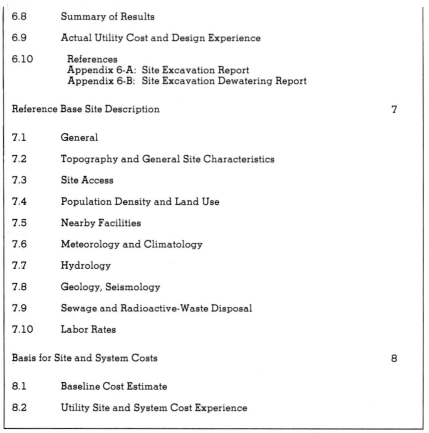

FIG. 2.18 *(Continued)*.

ple, the headings "Part I" and "Part II" are used instead of "Volume 1" and "Volume 2." Either approach would be satisfactory.

The two parts of this report were bound together in a single manuscript because it was only 300 pages long; other multivolume reports have been written that were as big as a set of encyclopedias in themselves. Designating separate volumes is particularly useful when your project involves submitting hundreds or thousands of pages of material for individual phases of a study conducted over a prolonged period of time.

If you receive an assignment to prepare a multivolume report, set aside a few hours to consider the special organizational needs and features of this extended format. *One caution:* There is a tendency among some technical writers to "overwrite," that is, to put too many details into a report. The temptation is especially strong when writing a multivolume report, since the usual parameters of the standard report format are missing.

In some ways a highly detailed approach is commendable; it shows you've done a thorough job. But you must keep the report concise. In the complexity of technical and scientific projects, researchers often believe that every fact discovered or generated is important. However, you must remind yourself that every fact that is important to you is not necessarily important to your readers; indeed, too much information can confuse them rather than increase their understanding of a topic. So try to make sure you include only the most useful and important data in your reports. Most importantly, tailor your report to meet (and not exceed) any contractual requirements applicable to the study.

Now let's look at some of the special features in our example in Figure 2-18. These should help to illustrate the unique aspects of multivolume report writing.

Each section—in addition to having its own summary, central discussion, and references and/or bibliography—has its own table and figure lists and appendix. In a single-volume formal report, there would be only one place for these items: at the back of the report. However, in the multivolume report, the contents of each section appear in tabular form at the beginning of the section. This placement improves the readability of the report by making it easier for readers to find what they need.

Despite the differences between the organization of a multivolume report and a typical formal report, certain features remain similar. The most important technical, scientific, and financial findings are summarized at the beginning of the report. Data on key topics like the engineering, feasibility, and cost of single aspects of the project are included in the relevant sections of the report in order of importance.

It should be clear from the table of contents in Figure 2.18 that engineers and scientists from several disciplines have contributed information to the multivolume report. There are sections on electrical engineering, cooling systems, flood protection, geology, waste disposal, and climatology—to name only a few disciplines involved. A report of this magnitude requires assigning individual task leaders to cover different topics and subtopics.

In short, the multivolume report requires extensive coordination of both data and people. Effective outlines of the general task and of individual responsibilities are a necessity. Brainstorming sessions like those we discussed for outline preparation are particularly helpful in preparing a multivolume report.

USING WORD PROCESSORS IN REPORT WRITING

We'll discuss how computers and word processors can assist your technical writing elsewhere in this handbook. But we'd like now to note briefly how they can aid report writing. Using a word processor can increase your productivity in writing reports; at the same time, it can help you to improve the quality of your documents and to follow a chosen format with ease. A word processor can be particularly advantageous in preparing multivolume reports. In many cases, the computer can be programmed to create documents automatically, according to a special format you choose. All you have to do is write the report and supply accurate information. You can make corrections, add paragraphs, reorder your material, and perform many other writing and editing functions without having to retype or recheck the whole document.

CHECKING AND EDITING THE REPORT

Regardless of whether you write a multivolume report, short report, or a long formal report, you'll have to verify your facts, English, and presentation before printing the final version. You or someone else will also have to edit the draft to incorporate necessary corrections and last-minute changes.

Once the report is written, review it completely before having the final draft typed. Verify every fact by comparing the data in the report with the original information. Check the English. If your grammar is less than adequate, admit this to yourself. Use the pointers in Section 13 and get someone in your office to read the report and mark errors. By making a few discreet inquiries, you can easily find one or more people near you who are well-qualified to criticize the grammatical aspects of the writing. Follow their recommendations; the report will usually be a better document if you do.

Examine the report arrangement. See that it follows the general outline given above. If necessary, you can change the sequence. But be sure there are good reasons for doing so. Of course, if your firm has a standard sequence for reports, use it; if you use a nonstandard sequence, the report has little chance of acceptance.

When reviewing the report, be as critical as possible. Every improvement you make before submitting a report is one less item that can be criticized by the person or persons on the receiving end. You cannot sit beside the reader and explain the report as it is examined. The report must stand on its own. And it should be the best you can produce. Never forget that it is far easier to criticize a piece of writing than to produce it. So we find that the world is full of critics but short of good writers—particularly of industrial reports.

The usual procedure for submission of report drafts in many industrial firms is for the person designated as study manager to submit a draft to task leaders and a representative of the client involved in the project. As study manager or technical writer, you should request comments from each task leader by a specified date.

Coordinate the comments with the preparation of a final draft. In the final stages of report preparation, many firms assign a senior technical associate experienced in the areas of the study to review the report to make certain it meets all contractual obligations. Where anything has been left out, mark the draft copy to indicate additions or insertions of material. You can do this easily by labeling the place to receive the material with a letter, and then labeling the material to be added with the same letter. Proceed through the document indicating further additions and insertions in alphabetical order.

Scrutinize all your tabular data carefully to ensure accuracy. Make certain the figures and results for similar items are consistent throughout the report, unless there is a good reason for exceptions. Note any such reasons in the text or in footnotes.

Next, review all drawings, photos, and other figures to determine that they will reproduce properly. If a particular drawing or photo is too light, try to replace it with a better copy. Coordinate the identifying numbers for each table, figure, and illustration with references in the text and table of contents. For example, be sure you haven't referred to "Table 1" when you really meant "Table 2."

In any review of your draft, keep an eye toward checking the format and style. However, it is best to conduct a separate review for this purpose. If possible, submit the document to a publications staff and an experienced technical associate to ensure that you have not missed any errors of format, style, or usage of terms and abbreviations.

Also have a senior associate review your technical conclusions for accuracy and usefulness. Doing so can help prevent the embarrassment that would result if your client were to question a particular sentence, paragraph, or finding.

DELIVERING THE REPORT

Some engineers and scientists spend months on a report and toss the finished job into interoffice mail for delivery. This is just about the worst way possible to deliver an important document. Even the best reports have trouble flying alone. If a report is worth spending many hours of your life and hundreds or thousands of dollars on, it also rates proper delivery. That's why it's smart to get yours off the ground with a personal boost.

Check to see how the report originated. If one of your supervisors requested it verbally or by memo, make an appointment to see that person. It's a good idea to wait until he or she is in a receptive mood, if you can. Be sure to avoid springing your report on the supervisors or clients when they are already distracted by problems or bad news of some kind.

Do not leave the report with the person's secretary. Bring one copy of your report to your meeting with the supervisor. Suggest that you'd be glad to sit quietly by while he or she reads the summary of your report. Perhaps the supervisor may want to ask some questions to clarify it. If, as is more probable, the person wants to read and discuss the report later, then tactfully suggest a definite date.

Be prepared to answer questions about the report with direct yes or no responses. Avoid detailed answers unless they are requested. Try to keep in mind the location of various items in the report. For example, if your supervisor or client asks, "Did you test these systems under freezing conditions?" answer, "Yes, that's covered somewhere around page 20."

Before leaving the supervisor or client, say that you'd like to have his or her opinion of the report after he or she has read it. Most people will honor such a request by sending you a note or calling you on the phone to state their feelings. If the opinion is complimentary, as it almost certainly will be if you planned and wrote the report well, you will be encouraged. Further, you can probably expect to be called into a committee or conference to answer questions about it. Here's where all your planning and research will pay off. You'll have a chance to defend your reasoning, provide additional details, and do a personal selling job.

But be sure to take the time to reread the report completely before you engage in such discussions if you're called at a later date. If a month or more passes before you're called for the talk, you'll have forgotten many of the details of the report. And nothing is so unconvincing as a scientist or engineer who can't adequately defend his or her own report.

HOW TO RESEARCH YOUR REPORT

Good reports are based on sound facts, a wide knowledge of alternate solutions to problems, creative imagination, and intelligent selection of the most promising elements. You know the expression "A man's judgment is no better than his information." If you want your information to be sound, don't sit down and write a report or proposal without doing thorough research. Besides getting all the help you can from authorities, suppliers, and your own investigation of a project, it's a good idea to tap other sources. Here's a list of some of the best sources for written information:

Technical journals: Consult the *Standard Rate and Data* at your library. If you need further information, write the editors of the magazines in your field.

Technical societies: Ask for directories of engineering and technical societies in your public or scientific library. Then write to the societies' headquarters, addressing your letters to the executive secretaries.

Trade associations: See their publications and the *Directory of National Trade Associations,* available from the U.S. Government Printing Office. Write the executive secretary of the appropriate association.

Books: The *U.S. Catalog* lists all books in print in English in 1928. Books published since 1928 are in the *Cumulative Book Index* or *Books in Print.*

Indexes of technical magazines: See the *Industrial Arts Index,* issued monthly and yearly. Also see the *Engineering Index, Chemical Abstracts, Science Abstracts, Electronics Abstracts, Agricultural Index, Biological Abstracts, Quarterly Cumulative Index Medicus,* etc. Many of these contain summaries from which you can obtain enough information to decide if reading the original is necessary.

Other indexes: Three popular ones are: *Readers' Guide to Periodical Literature,* issued monthly and yearly; *Vertical File Index,* issued in pamphlet form; *New York Times Index,* which covers newspaper-reported items.

Indexes of patents: Large libraries have bound volumes of the patents issued each year, arranged by numbers, called the *Annual Index of Inventors.* The Search Room in the U.S. Patent Office in Washington, D.C., is the only place where patents are available and organized by subject. Anyone can use this room.

Public libraries: Ask the librarians—they can be a big help. Also see the general catalog of the library.

Special Libraries Association: Ask your librarian about this association.

Authors of articles or books: Write them if you think they can help you. Be sure to address them in care of their publishers, not their job or university connection. (People change jobs faster than they do publishers.)

Famous people: If you have need for their help, write them with questions. Their cooperation and answers will be a surprise and help to you. For addresses, ask at local newspapers and libraries. Also, see *Who's Who.*

Directories: Check these: *Moody's Industrials, Thomas' Register of American Manufacturers,* and city directories.

Government publications: Write the Superintendent of Documents, Government Printing Office, North Capitol Street, NW, Washington, D.C. 20000. Ask for titles and prices of publications in your field. Many are nominally priced; some are free.

CHECKLIST FOR REPORT WRITING

Here is a tried and proven four-step scheme to follow when you write any kind of formal report. Use it and you won't overlook important parts of your report.

Step 1: Prepare Your Material

A. Collect information, facts, illustrations.
B. Check to see you have all detailed information needed.
C. Decide on the purpose of your report.
   Who will read it?
   Why do they want it?
   What do they require?
   How will they use it?
D. Draft a key sentence, stating your main objectives for the report.

Step 2: Plan Your Report

A. Classify your material using a rough outline and/or topic cards as described earlier.
B. Make a complete list of major and minor subject headings.
C. Prepare a comprehensive final outline.

Step 3: Write Your Report

A. *Summary or abstract:* State the subject, purpose, general findings, and conclusions.
B. *Introduction:* Note the reason for the report, why it is different than other studies in the field, how it is organized, and any special problems or uncertainties.
C. *Main discussion:* Describe the project, equipment, procedures, and results.
D. Write your conclusions or recommendations.
E. Prepare a table of contents.
F. Assemble and organize the appendixes and references.
G. For very long reports, prepare an index.
H. Write your transmittal letter.

Step 4: Criticize Your Report

A. Examine the report as a whole; the format; the proportioning of parts.
B. Check the agreement of the title, table of contents, introduction, and abstract or summary.
C. Check the agreement of the conclusions and introduction.
D. Check the agreement of the headings with the table of contents, and of the figures and tables with lists.
E. Check the proportion of paragraphs and sentences.
F. Examine details of text for transitions from topic to topic or part to part. Also examine sentence structure and wording.

SECTION 3
HOW TO WRITE TECHNICAL AND SCIENTIFIC PAPERS

As an engineer, scientist, or technical writer, sooner or later you'll probably be asked to write a technical paper. Or perhaps you'll develop your own idea for a paper and want to submit it to an organization or journal. Regardless of the source of motivation, you'll profit in many ways by learning to write papers effectively. For technical papers are different from reports and other writing tasks, and they require a special approach.

WHAT IS A TECHNICAL PAPER?

A technical or scientific paper is a comprehensive presentation of the results of study, research, experimentation, testing, or experience in a given area of technology or science. The subject of a paper can vary from the most advanced theories of a field to the practical everyday aspects of a job.

Technical and scientific papers comprise the permanent literature of almost all learned fields. This form of technical writing is sometimes termed the "learned paper" or "scholarly paper" to distinguish it from the more practical approach used in many reports and articles.

Engineering, scientific, and technical societies publish papers for the benefit of their membership and the field of their activity. When a paper appears in print or is presented orally, it provides a forum for discussion, debate, and continued testing of specific ideas by the author's peers. Most professional organizations thus regard their publications as major contributions to technology and science.

Our technology is advancing so rapidly that only by studying the ideas of others can we move beyond the limits of our own work. And more and more engineers and scientists are writing papers today than ever before. The number of papers published by one well-known engineering society rose more than 60 percent in a four-year period. And with more professionals writing papers today, you must strive harder to make your material excel. So if you want to publish a paper—and you should if you have suitable material—you must develop the necessary skills. In this section we'll see how you can prepare your papers better and with less waste of time and energy.

An outstanding paper helps you understand your work better, stands as a permanent record of your activities, contributes to your professional standing, and widens your acquaintance with the other workers in your chosen field. Take a look at the complete index of papers published by any professional society, and the name of almost every important person in the field will appear at least once. Scientists and engineers actively seek the prestige, acclaim, and publicity that a good paper brings to them. In many firms and universities, publication of scholarly papers ranks high when a person is considered for promotion. So don't overlook opportunities to make an important contribution to your field by writing papers.

HOW TECHNICAL PAPERS AND REPORTS DIFFER

Your technical paper may differ from a report in six ways: (1) content, (2) approach to subject, (3) method of presentation, (4) length, (5) depth of coverage, and (6) the message for its readers.

Content

In general, little limitation is placed on the content of a paper by the sponsoring organization. When a supervisor or client asks you to write a report, you must address only the issue that the supervisor or client has specified. However, technical papers often cover an extremely limited phase of a subject. For example, you could write a paper on the effect of threads on bolt strength. Or you might write a paper on the operating experience with dual-circulation boilers using 72 percent make-up. Other subjects for scientific and technical papers might be recent investigations of the potential for fuel use of oils derived from the Jojoba plant, and the case for a new ocean-going ground-effect prototype defense craft.

Note the main characteristics of these subjects—detailed treatment of a specific idea. Most of these subjects are too narrow or theoretical for the usual technical report. Many technical papers take a broad look at the theory behind a single problem. Most reports examine the practical applications of a system, device, or technique used to solve a problem or several problems.

In choosing an idea for a technical paper, you need not be concerned with the number of readers who want to examine separately the engineering, business, or administrative aspects of a problem. As long as there is one reader having the same or a similar problem, writing and publication of your paper can be justified.

Of course technical papers can cover a wide area too. But even here you'll find a difference between papers and reports. In writing a technical paper to cover a broad area, you'd make every effort to include all the references you can find. In a report on the same topic you'd probably concentrate on the work done by your own researchers and limit yourself to the most important references.

Approach to the Subject

Many technical papers try to examine every phase of a narrowly defined idea. Reports shun optional theories, confining their attention to solutions that have been proposed and are readily available. The reader of your paper is often most interested in a complete study of an idea, its background, and all possible options. Report readers usually look for a direct, immediate solution to a problem. So your approach to a subject in a paper must be different from your approach to a subject in a report.

Method of Presentation

Papers are often presented formally at a national or local meeting of a sponsoring professional society, such as the American Society of Civil Engineers or the American Chemical Society. Reports, on the other hand, are submitted directly to a supervisor or client following a review by experts within the originating firm.

Papers are usually published in preprint form by the society. A *preprint* is a preliminary copy of a paper published before the meeting at which it will be presented. Preprints serve as the basis for written or oral discussion of the author's ideas. Authors are often allowed to make corrections to those preprints that are later to be published in the society's journal. In some cases, a report may be converted to an article or paper for publication if your client or supervisor has scrutinized the report and approved it; but once a report has been submitted, there is rarely any opportunity to correct errors.

The author of a technical or scientific paper usually "reads" the paper to society members at an official group meeting. Oral and written discussion of the paper is actively encouraged among peers. The author may respond to comments in oral or written form. This portion of the discussion is often termed the *author's closure*.

Many societies publish the paper, discussion, and closure in their journals. Thus the author has the benefit of public presentation by himself, public discussion, and journal publication. Most scholarly and professional journals are regarded as key information sources by researchers because of the high quality of the data presented. By contrast, many reports are written exclusively for use within one firm or one department.

Length

Most societies place a wordage limitation on their papers. The society may recommend that no paper exceed 4,000 words, for example. A limit may also be placed on the number of illustrations that may be used in the paper. But a report can often use as many words and illustrations as are needed to describe a problem and recommend a solution. We'll be examining some typical requirements for papers submitted to professional societies shortly.

Depth of Coverage

Although many technical and scientific papers focus on the latest research and development concerning a given subject, papers may also trace a topic from its beginnings centuries ago to future projections. But in a report you must generally confine your treatment to those items that are of immediate practical use to your readers.

Your Message

Many reports urge the reader to adopt a certain technique, install a particular system, or change some kind of existing procedure. There is an immediacy about a report—you deal with active problems. Many papers do not urge any action by their readers. Instead, the paper reports past and present thinking. The conclusions you draw might in no way influence your readers to change their thinking about a problem. Of course, not every report tries to alter current viewpoints—some merely examine present events and technology. But reports more often than papers try to help the reader solve problems in new or different ways.

HOW PAPERS ARE DEVELOPED

There are two ways in which technical and scientific papers are developed: (1) A division or committee of a society or other organization solicits an author to prepare a paper on a certain subject. (2) The author writes a paper and contributes it, unsolicited, to an engineering or scientific society. With either origin, the steps in planning and writing a paper are the same. Some organizations may pay the author a small honorarium upon submission or publication of the paper, but this is an exception rather than the rule.

Solicited Papers

Here the society, through one of its divisions or committees, chooses you to write a paper, usually covering a phase of your work. The more specialized your work, the greater the chance of your being asked to prepare a paper.

For a solicited paper, the sponsoring organization usually states the general subject matter it would like the author to cover. The author, however, is allowed to make the final choice of what he or she discusses, provided the topic is in the general area requested. Thus a request for a paper on crankcase oils for high-load diesel bearings might result in the paper "A New Synthetic Crankcase Lubricant for High-Load Diesel Bearings."

Authors of solicited papers will find that the society is almost always willing to accept some alteration in the suggested topic. The reason for this is that all societies realize that an author has a far better knowledge of his or her subject than any outsider. So don't hesitate to suggest minor alterations in the topic. Such changes prevent errors in the society's programs and may allow another author to present a paper allied with yours.

It is well not to turn down requests from societies for papers on your work. You'll probably find that most organizations will give up after you've turned down three or more invitations to write a paper. Since a solicited paper has a better chance of acceptance, you're losing a good opportunity by turning down the request.

But take some time to think over any request for a paper on a specific subject. Don't agree to write the paper unless you know the subject well. If you're more familiar with another phase of the subject, tell the society. And if the suggested paper is totally out of your field of interest, gracefully decline the invitation. But to stay on the list of prospective authors, give the society a list of the topics you can write about. Do this in the same letter or conversation in which you decline the initial request. This gives the society a chance to consider papers on the topics with which you are most familiar.

Remember that your paper will be read by many of your coworkers in the field. So be sure that you choose the appropriate phase of your subject. Otherwise you may find that your paper is not as well received as it should be.

Contributed Papers

Here you decide, on your own, to write a paper on a subject with which you are most familiar. Although you write the paper without the aid of the society, you should choose a topic the society would be likely to cover. If in doubt, query the publications committee, editor, or other official charged with paper procurement and presentation. Describe the contemplated paper, its length, number of illustrations, etc. From such a description it is easy for the society to decide if the paper will be publishable. Your inquiry will also help you to avoid wasting time and energy writing a paper that could conflict with one the society has already requested from another author. Naturally, the society will give preference to a paper it solicits.

Speak to almost any society member and you will find that good ideas for papers run in cycles. Thus, for example, with the upsurge in computerized information processing in all areas of industry, professional organizations are now receiving numerous offers of papers in this area. So you might find that a paper on any phase of computerization in your field could conflict with papers that have already been solicited. This is why it is best to contact members of the society before choosing the final subject for your paper. Many members have strong ideas on good subjects. If you choose one of these subjects, or a variation of one, your contributed paper has a better chance of acceptance.

It is a good idea to write an unsolicited paper every few years. Even if it isn't accepted, the society becomes aware of you and your desire to contribute to your field by writing papers. You may find that the society will solicit a paper from you a year or so after your contributed paper was turned down. But the exact procedure varies considerably from one society to another.

Use the hints given at the beginning of this section when you choose a topic for a paper. Probably the best source of ideas for technical papers is your everyday work. If this is at all unusual or highly specialized, a paper on some phase of the work is probably justified.

You need not be a member of a society to write a paper for it. But most societies are more sympathetic to papers submitted by members than by nonmembers. You should, of course, be a member of the society representing your profession. But you need not be a member of related societies unless such membership offers advantages.

Technical Ghostwriting

Up to this point we have discussed paper writing with the assumption that you, as an engineer, scientist, or technical writer, are the prospective author—the person who will do the writing. However, some engineers and scientists prefer to concentrate on doing the research for a paper themselves while delegating the actual writing task to a professional writer. In such cases, the technical writer is sometimes referred to as a *ghostwriter;* yet the engineer or scientist is still considered the author. He or she is still the one whose name will appear on the finished document—the one who volunteers to contribute a paper or is solicited by the society to write a paper and supply most or all of the scientific and technical information.

Is technical ghostwriting ethical? Most experts agree that it is not only ethical but also preferable in cases where the author does not have the time or expertise to write the paper alone. The technical writer can enable engineers or scientists to focus energies on research and development in those areas for which they are best-qualified. In addition, the professional writer can turn out an acceptable paper faster than most engineers and scientists can write an unacceptable one.

This division of authorship is roughly analogous to the situation where an inventor hires an expert machinist to build a model of an invention. When the patent is granted, it is issued to the inventor, not the machinist, regardless of how important the machinist's role in building the model has been. No ethics have been breached as long as the machinist is paid adequately for his or her work. Specialization is basic to our system. There is no rule requiring the inventor to do his or her own machine work.

Nowadays more technical papers are being written by professional writers under the close supervision of engineers and scientists than ever before. Almost invariably, the sponsoring society gives full credit for the preparation of the paper to the engineer or scientist; the technical writer's name may not even be known to the society. Hence the term *ghostwriting*. Regardless of this trend toward increased use of ghostwriters, nearly every engineer and scientist will benefit by knowing how to write papers unaided. Therefore, we shall continue to devote most of our discussion to the techniques involved in authoring the technical or scientific paper. Certainly every technical writer should learn the skills involved in cooperating with engineers and scientists to prepare a learned paper.

HOW PAPERS ARE PROCESSED

All large technical societies follow a specific routine in handling papers. Figure 3.1 shows the procedure used by the American Society of Mechanical Engineers. While other societies may use slightly different procedures, there is a general similarity. Understanding what ASME does will help you to understand your own society better. Its procedure is as follows:

Origin

The paper is either solicited or contributed, as shown at the top of Figure 3.1. In either case you inform the editorial department of the society of your intention to submit the paper. With a solicited paper, write the editorial department, informing them of the tentative title of your paper, its approximate length, and the date you expect to submit it. If you wish to present the paper at a certain meeting of the society, ask the editorial department if your proposed delivery date allows enough time for review and preprinting.

An unsolicited paper can be submitted directly to the editorial department with a cover letter. Explain why you wrote the paper, when and at what meeting you'd like to present it, and what segment of the society membership it will interest.

Review

No matter what the origin of the paper, it will be reviewed by technical specialists in your field. They will examine it for technical content and writing clarity and

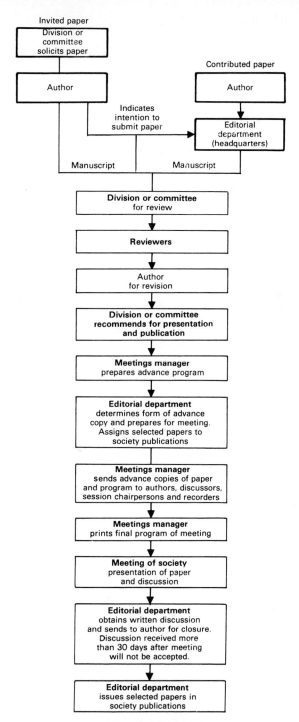

Fig. 3-1 Steps in submission to and publication of a technical paper by the American Society of Mechanical Engineers.

check for commercial bias. Your paper will be returned for revision if the reviewers find this necessary.

If this happens, the wisest move is to follow the recommendations of the society. You can object, of course, if the changes recommended would make the paper incorrect or alter its character completely. But minor changes should be made quickly and the paper returned as soon as possible. Otherwise the paper may not be accepted for publication.

Publication

Once your paper is revised and has been accepted, the production department begins its job. In most societies today your paper is preprinted before the meeting at which you present it. Preprints may be reproduced on a copying machine, typeset, or duplicated on a word processor, depending on the practice of the society. Then the paper may be selected for publication in one of the society's journals.

Meeting Arrangements

While your paper is being published, arrangements for the meeting are being made. Figure 3.1 shows the steps. Be sure to cooperate fully with the society during these arrangements. You can make the job of the conference chairperson much easier if you submit all your copy on time, fill out the required forms correctly, and follow the society recommendations on the preparation of text, illustrations, and tables.

Discussion

Your paper will probably evoke both written and oral discussion from your audience. Be prepared to answer your discussors. If they ask for data that you don't have when you present your paper, prepare an answer immediately after you leave the meeting room. For the sooner you do this important part of presenting a paper, the easier it will be.

WRITING THE PAPER

The steps in writing a technical or scientific paper are essentially the same as in writing a report or article. Briefly, you should:

1. Learn the requirements of the sponsoring society.
2. Prepare an outline of the paper's contents.
3. Write the paper, starting with a rough draft.
4. Review and check all facts.
5. Submit the paper to the society.

Let's examine each of the steps.

Many of the following suggestions are based on data published by the leading engineering and scientific societies—the American Society of Mechanical Engineers, the American Association for the Advancement of Science, the American Society of Civil Engineers, and the National Academy of Sciences–National Research Council.

Learn Your Requirements

Check with the sponsoring society to learn its requirements for length, illustrations, number of copies of the manuscript, etc. Some authors are so concerned with the content of their paper that they overlook these important details.

There are times when an author will be excused from the normal requirements of the society. Usually, if an author has made an especially timely and important discovery or other original contribution, the society may suspend its rules somewhat. And if the contribution involves a critical issue of the safety or welfare of the public, all rules may be relaxed. But both cases are rare. So it is better to plan on observing all the requirements of the society.

Some general characteristics of most technical and scientific papers are as follows:

Style. Papers tend to be rather formal, like formal reports, although they may be converted to a more conversational form for oral presentation. In any event, a good paper follows the same general rules of conciseness and clarity that apply to other kinds of writing.

Of course, a formal style is no reason for writing in a verbose or pedantic way. Neither are the goals of conciseness and clarity reasons for lowering the intellectual level of your presentation. As the author of a technical or scientific paper, you are concerned mainly with conveying information to other professionals in your field, not with popularizing data for the general public. That is more commonly a function of magazine articles and books.

It's okay to use complex technical terms or concepts in a paper where necessary, provided that your audience understands them. If you must introduce new words, define them immediately after you use them the first time. Write for the average engineer or scientist—not for the specialist.

Generally, try to state your ideas in simple terms and expressions. Keeping sentences, thoughts, and definitions short helps to maintain the attention of your audience—readers and listeners. Avoid personal bias when using the first person or referring to individuals by name. And use product names only in your acknowledgments, if possible.

Spelling and Usage. Use a standard lexicon like *Webster's New International Dictionary* as a guide to spelling and preferred usage. Check the spelling of all technical terms and names before submitting your paper.

Length. Some societies specify the minimum and maximum lengths allowed for their papers. For example, the ASME suggests that the text not exceed 4,000 words (about 14 pages of double-spaced typescript). The paper can have references and quotes to support your points, but avoid long quotations. Instead, refer to the source of the data.

Use illustrations and tables where they help clarify your text and where they demonstrate results. Omit detailed drawings, lengthy test data and calculations, and photos which, though interesting, are not necessarily important to the understanding of the subject. Manuscripts that do not meet the requirements of the society are usually returned for revision and condensation. When you don't think you can treat your subject adequately within the limits set by the society, check with the sponsoring division or committee. You may be able to obtain permission to present a longer paper.

Approvals and Clearance. As with reports and articles, it is your responsibility to secure all company approvals necessary. And never overlook government or commercial clearance on classified material. When you have obtained any needed approvals or clearances, in writing, make a note on your manuscript that these have been secured. Keep the approvals or clearances in your file.

Outline the Contents

Once you know the limits on length, illustrations, etc., prepare an outline of the paper. Do this by using the suggestions given in Section 2 on reports. First, draft a rough outline of your main ideas and salient points. As noted in Section 2, you can do this by making a random list of topics on a sheet of paper or on 3- by 5-inch cards. Then rearrange the notes to put everything into a logical sequence. Try to place the items that your audience will be most familiar with at the beginning. Second, add subtopics and supportive materials, such as illustrative tables, to the list.

The usual order of the main sections in a technical paper is:

A. Title
B. Author's name, job title, company connection, address
C. Abstract
D. Body of paper
E. Appendixes
F. Acknowledgments
G. Bibliography
H. Tables
I. Captions for illustrations
J. Illustrations, that is, photos, line drawings, charts, etc.

Most of your effort will be spent on the paper abstract, body, appendixes, tabulations, and captions. The acknowledgments, bibliography, and illustrations require time and energy. But the procedures for these are simple compared with writing the bulk of the paper.

The body of the paper requires more time to outline than any other part. Start the body outline with a short statement of the reasons for the writing of the paper. Then insert entries in a familiar-to-unfamiliar pattern. This is usually the best sequence for a technical or scientific paper. The body normally ends with the author's conclusions, though these are sometimes replaced by recommendations.

A good guide to remember when you do any kind of technical writing is to reveal clearly the direction in which your development of the topic is heading; never keep this direction secret from your readers.

Try to have your final outline reviewed by an astute technical or scientific associate and skilled writer if you feel the review will be kept confidential. The timeliness and complexity of important scientific issues requires care in the handling of new information. Without your care and vigilance, your material could end up being publicized in some way other than you intended and could possibly become distorted or misused in the process.

Write the Paper

With your outline in its final form, you are ready to begin the actual writing.

Title. Make the title of your paper explicit, descriptive, and as short as possible. Use an explanatory subtitle to clarify the meaning of the main title if this seems necessary.

Author Information. The name of the author (or authors) should appear immediately below the title of the paper, at the top of the first page. Insert the author's title and business connection as a footnote in the lower left-hand corner of the first page, or as otherwise directed by the sponsoring society.

Abstract. Begin the actual writing by preparing an abstract. Limit it to about 100 words, or as the society suggests. Do not try to condense the entire substance of your paper into the abstract. Instead, give a clear statement of the object, scope, and findings presented by the paper. Since many readers use the abstract to determine if the paper will be useful to them, be careful to see that it is concise and accurate. Libraries, computerized data retrieval services, and technical magazines may use the abstract exactly as you prepared it if it is well done.

Here are two typical abstracts for a technical paper:

> In the past few years, corrosion of high-temperature superheaters and reheaters has become an increasingly serious problem on coal-fired boilers. The nature and occurrence of corrosion observed on several boilers in the Chicago area which burn high-alkali coal are described. Some of the results obtained from an experimental study of the factors involved and the control measures applied to prevent further serious loss of tube metal are presented.[1]

> Electrodynamic apparatus is described for producing and identifying torsional resonances in a small cast-iron sample. Shear modulus is computed from the resonant frequencies. A specimen is taken from the mold of each crankshaft. Accuracy of the procedure is discussed. More than 1,000 tests have been made, and the relation of shear modulus to other measurable physical properties is demonstrated.[2]

Preparing the abstract before writing the body of the paper enables you to organize your thoughts for the writing task ahead. In many papers it may be necessary to alter the abstract slightly after the paper is finished. This small disadvantage is far outweighed by the time and energy saved in organizing your thoughts before writing the entire paper.

---

[1]From P. Sedor, E. K. Diehl, and D. H. Barnhart, *External Corrosion of Superheaters in Boilers Firing High-Alkali Coals,* ASME paper.

[2]From J. D. Swannack and R. J. Maddock, *A Dynamic Shear Modulus Apparatus and Production Test Results from a Cast Crankshaft Alloy,* ASME paper.

Body of the Paper. Begin writing the body of the paper as soon as the abstract is finished. Start with familiar facts, and lead the reader to those phases of the subject that are new or unfamiliar to the average reader.

Try to keep a typical reader in mind while writing the paper body. Remember that almost every member of a professional society is a college or university graduate. So your writing need not be pitched at the high school level, as it is in many training manuals. You can help to define your typical reader by studying back issues of the society's publications. Read both the editorial and advertising content of three or more issues so that you acquire a good knowledge of the interests and activities of a typical member.

Follow the outline while writing the paper body. You'll probably have to make some minor changes in the outline arrangement as you write. This is normal; most writers do make slight changes in their outline as they write. But make sure your paper states its purpose at the start. Follow the purpose or aim with a description of the problem or theory, the means used to develop a solution, and other related information important to the results and conclusions.

Use subheads to indicate the major divisions of the report body. Concentrate on presenting the data as concisely as possible. There is a trend today toward shorter, less complex technical and scientific papers. So if you can give all the needed information in the body of the paper without using an appendix, do so.

With the body of the paper written, put it aside while you work on the other parts. Use the following instructions.

Appendixes. Avoid the use of an appendix if possible. Where an appendix is necessary, use it for derivations of equations, long tabulations, detailed descriptions of apparatus, etc. As a general guide, use the appendix for all useful information that is not essential to the body presentation of the subject. Use two or more appendixes where there is a sharp difference between the data, such as between derivations of equations and descriptions of equipment. Use a number or letter and title for each appendix, as "Appendix 1: Derivation of Laminar-Flow Equation" or "Appendix A: Description of Missile Test Cell."

Acknowledgments. Give credit to individuals, firms, agencies, etc., who helped in supplying data, background material, illustrations, or other information. Check carefully with the author of the paper to see that you haven't overlooked any important sources of material. As a general rule, be generous with acknowledgments. You stand to lose far more by not giving credit for help than by making the slight effort needed to acknowledge all sources.

Bibliography and References. Follow the recommended style of the society when preparing references. If there are no specific requirements, refer to literature cited as follows: (1) If you have only four or five references, use footnotes in the text. (2) If you have more than five references, use a bibliography at the end of your paper. Number references serially. In the text the number for these references should appear in parentheses. With the first reference, use the following footnote: Numbers in parentheses refer to similarly numbered references in the bibliography.

Use the following format in your references to books: author (with his or her initials), name of periodical (not abbreviated), volume, number, year, page numbers.

Use extreme care when preparing your references. A single error in volume number, year, or page number can waste much time. Keep in mind the reader who may order copies of an article or book you've referred to in your paper; picture his or her annoyance if the reference were wrong.

Tables. Type tables of six to eight lines as part of the text unless the society instructs otherwise. But locate the table so that it does not run over onto a second page. Type large tables on separate sheets and put them at the end of the text. This allows easier and faster handling by the printer. Further, with the paper prepared in this order, the editorial department of the society can, if necessary, rearrange the illustrations. (Since some societies have switched to a two-column format for papers, some rearranging of tables and figures may be necessary.) Illustrations are located as near their first text reference as possible. This positioning reduces the amount of page flipping the reader must do.

Give each table a suitable descriptive heading. Number the tables consecutively and refer to them in the text by number. Avoid references like "the following table" or "the table on page 3," because the document may be printed in a slightly different form than your manuscript. Generally, give the source of each table—immediately below short tables or in footnotes or bibliographic entries for long tables.

Captions. Some societies place the captions for illustrations on a separate page by themselves. Each illustration then has an identifying number under it. Other societies insert the caption under the illustration, as in normal technical journal format. Follow the practice recommended by the sponsoring society. See Section 13 for a comprehensive discussion of illustrations.

Photographs. Submit only clear, sharp glossy prints—color prints are often preferred nowadays. Do not mount or paste photos to the manuscript. Instead, gather all photos and drawings and place them between stiff cardboard at the end of the text. Lightly mark the figure number and your name on the back of each photo. Use a felt-tip pen so the face of the print is not marred. Do not use paper clips on photos or film.

Other Illustrations. Most societies employ editorial experts and artists to produce manuscripts according to the exact specifications for publication in their journals. But you should prepare all your illustrations yourself, neatly and precisely. Use black ink and heavy white paper or tracing cloth for graphs, charts, line drawings, and sketches. Blueprints, photostats, and black-line prints are rarely suitable for a quality manuscript. But color slides and templates for use with an overhead projector can be highly effective in an oral presentation. Mark the figure number and your name on the back of each drawing. Never paste your illustrations to the text pages. And do not draw on the text pages. Follow your society's recommendations.

Review and Check All Facts

Once you have written your preliminary draft and included the supporting materials, turn to other activities for a day or more. Try to forget the paper and its content. The longer you can ignore the paper, the better are your chances for making a realistic appraisal of the paper when you return to it.

Read the paper quickly, at one sitting if possible. Many writers find it helpful to read their material aloud to themselves, especially for papers that will be presented in public. Doing so should give you a "feel" for the coherence and flow of ideas in the paper. Note on a separate pad those words, sentences, and paragraphs that do not appeal to you. As you read, verify the references to illustrations and tables. Try to have at least one reference to each illustration, in the sequence of its presentation.

Rewrite those portions of the text needing improvement. See Sections 1 and 14 for suggestions. Check all illustrative problems to see that the solutions are correct. Errors in illustrative examples are pounced on by readers. Comments by discussors of the paper are likely to be caustic if errors are detected.

Have the paper carefully reviewed by any authors with whom you've collaborated on the project, or by a close associate or supervisor, if feasible. Remember: any statement, opinion, or error in the paper is the responsibility of the person whose name appears as the author of document. As an engineer or scientist, you will be the individual responsible for the correctness of the technical or scientific data presented. As a writer, you will be responsible for the English, grammar, style, and other nonfactual aspects of the paper.

Where a paper has more than one author, prepare a copy for each author. Be certain that they discuss with you every question that arises. Many study leaders request that the author or authors of a paper return the draft they read, marked by them as follows: "Read and approved; [author's initials]; [date]. The study leader then retains this copy on file for future reference.

As we've noted, whenever there is any possibility that a manuscript may contain classified or security data, it is wise to have the paper checked for official clearance. Clearance should be the author's responsibility and should include written permission for release of the information.

Once the manuscript is approved by all necessary personnel, have it typed to the society's standards. See Exhibit 3.1 and Section 14 for additional hints on manuscript preparation. Read the final version carefully, comparing it with the approved manuscript.

SUBMIT THE PAPER

Many societies require that you submit the manuscript of your paper at least four months in advance of the meeting at which it will be presented. This much time is needed for further review, revision, recommendation for presentation and publication, editing, typesetting, printing, and distributing to prospective discussors. Don't flirt with the deadline. Otherwise, you may find your paper overlooked. *Remember:* Editorial deadlines creep up faster than most people realize.

Write a cover letter, using the suggestions given in Section 2 on reports. Mail the paper flat, using stiff cardboard to protect the document and illustrations. Register the package at the post office so it can be traced more easily if lost.

When submitting your manuscript, include a list of persons you'd like to invite to discuss your paper. The society may send advance copies of your paper to these people if you want it to do so.

Publication by Others

Many societies urge their authors to seek the widest possible audience for their papers. They occasionally encourage publication of a paper by others once it has

Scope of the Journal

This journal is devoted to the advancement of computer engineering through the dissemination of original papers disclosing new technical knowledge and basic applications of such knowledge. The technology of computer engineering is understood here to embrace selected aspects of microelectronics, circuitry, data processing, mathematics, machine language, programming, systems analysis, materials and applications, operations, and related areas of research and development. The selection of papers to be printed will be governed by the pertinence of the topic to the field of computer engineering, by the current or probable future significance of the research, and by the importance of distributing the information to the members of the society and to the profession at large.

Information for Authors

Manuscripts must be as brief as the proper presentation of the ideas will allow. Exclusion of dispensable material and conciseness of expression will influence the editors' acceptance of a manuscript. In terms of standard-size, double-spaced, typed pages, a typical maximum length is 22 pages of text (including equations), 1 page of references, 1 page of abstract, and 12 illustrations. Fewer illustrations permit more text, and vice versa. Greater length will be acceptable only in exceptional cases.

Short manuscripts, not more than one-quarter of the maximum length stated for full articles, may qualify for publication as Technical Notes or Technical Comments. They may be devoted to new developments requiring prompt disclosure or to comments on previously published papers. Such manuscripts are published within a few months of the date of receipt.

Sponsored manuscripts are published occasionally as a society service to the industry. A manuscript that does not qualify for publication, according to the above-stated requirements as to subject, scope, or length, but which nevertheless deserves widespread distribution among computer engineers, may be printed as an extra part of the journal or as a special supplement, if the author or the author's sponsor will reimburse the society for actual publication costs. Estimates are available on request. Acknowledgment of such financial sponsorship appears as a footnote on the first page of the article. Publication is prompt since such papers are not in the ordinary backlog.

All manuscripts must be double-spaced on only one side of the paper with wide margins to allow for instructions to the printer. Include a 100- to 200-word abstract. State the authors' positions and affiliations in a footnote on the first page. Equations and symbols may be handwritten or typewritten: clarity for the printer is essential. Greek letters and unusual symbols should be identified in the margin. If handwritten, distinguish between capital and lowercase letters, and indicate subscripts and superscripts. References are to be grouped at the end of the manuscript and are to be given as follows: for journal articles: authors first, then title, journal, volume, year, page numbers; for books: authors first, then title, publisher, city, edition, and page or chapter numbers. Line drawings must be clear and sharp to make clear reproductions. Use black ink on white paper or tracing cloth. Lettering should be large enough to be legible after reduction. Photographs should be glossy prints, not matte or semimatte. Each illustration must have a legend; legends should be listed in order on a separate sheet.

Manuscripts must be accompanied by written assurance as to security clearance in the event the subject matter lies in a classified area or if the paper originates under government sponsorship. Full responsibility rests with the author. Preprints of papers presented at society meetings are automatically considered for publication.

Submit manuscripts in duplicate (original plus first copy, with two sets of illustrations) to the Managing Editor, *Computer Engineering Journal*, 100 Fifth Avenue, Santa Barbara, California 93100.

EXHIBIT 3.1 Typical requirements for publication of scholarly papers in a society journal.

been presented at a society meeting or published by the society. So you may find that the paper you visualized as suitable only for society use can be published and lauded outside the society. Check with the society to learn the exact procedure to follow.

Authors' Expenses

The dues paid to a society by its members do not provide sufficient funds to pay the expenses you incur in preparing a paper and attending a meeting to present the paper. You, or your company, must pay all these expenses. As noted earlier, technical societies hardly ever pay an honorarium for papers they publish.

EXEMPLARY PAPERS

Many societies present annual awards for the best technical and scientific papers. Study of these papers can provide you with much useful information. Thus, papers that receive awards usually cover important topics and are well-written.

Study such papers and note the length and content of the abstract, body, and appendixes. See how many illustrations are used in each paper. Observe how the author used the illustrations to support the findings or conclusions of the paper. Then do the same in your paper.

SOME TYPICAL REQUIREMENTS FOR SUBMITTING PAPERS

As we've noted throughout this section, most societies that publish scholarly papers have rules for preparing and submitting the documents. Whenever you arrange to write a paper for a professional society or association, ask for a copy of the latest publication groundrules. Exhibit 3.1 shows the requirements for papers submitted to a typical scientific journal—this one happens to be on computer engineering. Both the scope of the journal and "Information for Authors" are given so you can see how some societies regard their publications and papers. The example has been adapted from the actual requirements of professional engineering and scientific societies.

WHICH SOCIETIES PUBLISH PAPERS?

If you are a member of one or more technical or scientific societies, you will know which publish papers and what types of papers they prefer. For listings of professional societies in many different fields, see the *Learned Societies Directory,* published by the American Council of Learned Societies, 345 East 45th St, New York, New York, and *Scientific and Technical Societies of the United States and Canada,* published by the National Academy of Sciences, National

Research Council, Washington, D.C. Further, *The Encyclopedia of Associations*, published by Gale Research, Inc., Chicago, Illinois, also lists professional organizations by subject.

CHECKLIST FOR TECHNICAL AND SCIENTIFIC PAPERS

1. Develop potential topics.
    a. From your own work.
    b. By collaborating with your associates.
    c. By determining typical subjects treated in journals in your field.
2. Determine the origin of the paper.
    a. Solicited.
    b. Contributed.
3. Decide who will write the paper.
    a. Yourself.
    b. You and other engineers or scientists.
    c. You and a professional writer.
4. Arrange to write the paper.
    a. Review the prospective paper with your collaborators, if any.
    b. Determine a sponsoring society.
    c. Query the society on your topic.
    d. Obtain manuscript requirements from the society.
    e. Gather data, illustrations, and supporting references.
5. Prepare an outline for the paper.
    a. Draft the outline by randomly listing main points and reference data.
    b. Study books, journals, and magazines in the field.
    c. Refine outline using suggestions given in Section 2.
6. Have collaborators, supervisor, or society check and approve the outline, if this is practical.
    a. Try to be with them when they review the outline.
    b. Revise the outline according to the reviewers' suggestions.
7. Write an abstract of the paper.
    a. Limit the abstract to about 100 words.
    b. State the paper's object, scope, and findings.
8. Write the body of the paper.
    a. Arrange your facts in a familiar-to-unfamiliar pattern.
    b. Keep a typical reader in mind.
    c. Be objective while writing.
    d. Use short sentences and paragraphs.
    e. Follow your outline.
    f. Use subheads to show division of information.
9. Prepare supporting sections.
    a. Appendixes.
    b. Acknowledgments.
    c. Bibliography.
    d. Tables.
    e. Illustrations.
    f. Illustration captions.
10. Have the paper reviewed.

a. After a time lapse from writing, review the paper yourself.
  b. Make copies and have collaborators read and approve the paper.
  c. Have a qualified person, such as your supervisor, review the paper for clearance of classified information and get a written release.
  d. Verify technical correctness.
11. Submit the paper.
  a. Allow time after submission for a review by the society.
  b. Make the changes suggested if doing so is possible without changing the paper's intent or giving incorrect data.

SECTION 4
WRITING ARTICLES FOR PUBLICATION

DEFINITION OF AN ARTICLE

For the purposes of this book we'll define a *technical article* as an organized presentation of facts and data to inform, educate, and assist the reader in the performance of his or her job. Every well-planned technical or scientific article has a specific purpose and audience. Once you know your purpose in writing and the audience you wish to reach, your writing task is simpler because then you can choose the best form in which to present your material.

The usual scientific article differs somewhat from the technical article, but the steps in writing for publication are basically the same for both types. Unlike many technical articles, most scientific articles are directed at readers who seek to increase their general knowledge about a subject. This interest contrasts with that of the typical reader of a technical magazine who is seeking information to use in solving on-the-job problems. Of course, some scientific articles also offer practical applications for the experienced specialist.

While a scientific article may have a more general slant than most technical articles, the potential benefits to the author are similar. Writing technical or scientific articles enables you to contribute knowledge to your field as you learn, gain prestige, help your firm, and earn extra income.

DIFFERENCES BETWEEN ARTICLES AND PAPERS

Articles may differ from papers published in technical and scientific fields in several ways: in writing style, depth of coverage, mathematics usage, and more. But the chief distinction between articles and papers is in the type of publication for which they are written.

Papers—with their scholarly style, in-depth coverage, and frequent use of advanced mathematics—are typically tailored to the journals of professional societies and voluntary learned organizations. Articles commonly appear in magazines, newspapers, and business publications aimed at the members of one or more trades or professions. In addition, many large corporations, including IBM, GE, and 3M, publish articles in "in-house" magazines for employees and customers. Articles frequently have a less rigid style than learned papers, offer quick solutions, and use applied mathematics to solve the problems discussed.

A few publications defy this classification, however. For example, some professional societies publish scholarly papers and informal articles together in a periodical they may call either a "journal" or a "magazine." This is not unusual in engineering and scientific fields, which demand a high level of both academic learning and skill in everyday applications. Thus, you must (1) choose the correct publication for each article you write and (2) have a working knowledge of writing articles as a task that is distinct from writing learned papers.

OVERVIEW OF ARTICLE WRITING

Here is a quick overview of the steps in writing articles for publication. Use it as a guide whenever you need help putting your ideas into article form.

I. Choose the type of article suitable for your target audience.
  A. Study the 13 types listed in this section.
  B. Decide which type is most similar to your idea.
  C. Determine what magazines might publish your article.
    1. Study six issues of each magazine in your field.
    2. Note the number of articles in each issue.
    3. Note the kinds of articles featured.
    4. Note the length of each article.
    5. Note the kinds of illustrations used.
  D. Choose appropriate characteristics for writing your article.
    1. Length.
    2. Form.
    3. Style.
    4. Number of illustrations.
II. Prepare an outline of your article.
  A. List ideas randomly.
  B. Study the sample outlines in this section.
  C. Reorder the list to put the most important points first.
  D. Follow the sample outline for your type of article.
III. Collect illustrations, tables, and charts.
  A. Note how many illustrations you need.
  B. Obtain all your illustrations, tables, and charts.
  C. Get any permission needed for use of the materials.
  D. Get general approval for the article from your firm.
IV. Contact the magazine editor about using your article.
  A. Use a query letter, telephone call, or contact-person.
    1. Introduce yourself and your article idea.
    2. Emphasize why people will want to read the article.
  B. If the editor requests, send sample materials, including one of the following:
    1. Final outline of the article.
    2. Some of the illustrations.
    3. An article excerpt.
V. Write your first draft, concentrating on technical facts.
  A. Develop a catchy title and beginning (the lead).
  B. Keep the middle lean and accurate (the body).

C. Reemphasize your most important points (the end).
VI. Polish and submit.
   A. Read the article quickly after putting it aside for a week.
   B. Check for faults, mark errors, and correct.
      1. Verify technical accuracy.
      2. Eliminate unnecessary technical terms.
      3. Use action verbs and descriptive words.
      4. Link up with your readers' experience.
      5. Make sure the final manuscript and illustrations are neat.
   C. Send your manuscript to the magazine editor.

Now let's examine article writing step by step.

THIRTEEN TYPES OF ARTICLES

A number of different methods are used to classify technical and scientific articles. Probably one of the simplest classifies articles according to the kind of information they present. You will find that in most technical magazines the typical contributed article belongs to one of these 13 types. The exact number of kinds used varies with the magazine's editorial viewpoint, method of handling, and publishing schedule.

The 13 kinds of articles used most commonly in engineering, scientific, and business magazines today are:

1. System and plant descriptions
2. Process descriptions
3. Program descriptions
4. Design procedures
5. Product functions
6. Calculation methods
7. Graphical solutions
8. Operating procedures
9. Maintenance procedures
10. Questions and answers
11. Management techniques
12. Departmental features
13. News and equipment releases

In using this or any other classification of modern technical and scientific articles, you will find some overlap. Thus an article on design procedures might also be classified as a graphical-solution article if charts are used as part of the procedure. But not all design-procedure articles use charts. And not all graphical-solution articles solve design problems. So it is necessary, in any classification system, that we have enough specific categories to cover the variety of articles being published.

Now let's look at the characteristics of each kind of article. We'll also give a few examples of each type.

Systems and Plant Descriptions

These articles describe a particular installation of some kind. In the past, most articles of this type covered entire physical plants and were called "plant stories." But the rapid growth in the importance of computer technology and systems has caused articles on computer-based systems to rival or exceed plant stories in popularity. Many technical and scientific magazines publish both kinds of articles. The system or plant description usually emphasizes:

A. Name and location of the system or plant
B. Details of the equipment configuration
C. What the system or plant will do
D. What advantages will result
E. Effect of the system or plant on business

When the system or plant is an unusual design, a different procedure is sometimes used. Instead of emphasizing the equipment, the story concentrates on design analyses and related considerations. This focus gives the reader a better understanding of the problems encountered and how they were solved.

Some examples are: descriptions of the installation of a new laboratory-device control system, power plant, computerized hospital radiology department, chemical processing plant, maintenance facility, aircraft factory, and air-conditioning system.

Note that the system or plant description can cover an entire plant or a portion of a plant, a new or an established facility, or any other kind of industrial or commercial installation.

The usual illustrations in a system or plant story are:

A. Overall photo of site and structures (for plant stories)
B. System flowchart
C. Photos of the hardware or other equipment
D. Equipment details

Most of the illustrations used in the system or plant description today are photos. Drawings are used to show the flow cycle and special design details in various paths and branches within the installation. With good illustrations, the amount of text that is needed is relatively small. Today some articles consist merely of photos, drawings, and captions.

Choose photos for system or plant descriptions with special care. In general, equipment should be clean and neatly arranged before it is photographed. There is usually little problem preparing computer hardware for photos. But when the pictures are to show details of the internal parts of a plant—such as pipes and wiring—special painting may be necessary. Where actual construction processes are to be shown, however, special preparation may be unnecessary. The workaday appearance of the equipment under these conditions sets a mood for readers, drawing them into the story.

With more and more attention being given to readability, simplicity in page layout, and visual appeal in magazines, the function of photos has changed somewhat. Besides showing the equipment being described, the photos should, if possible, be pleasing to the eye. They should have what editors call "atmosphere." So it is often worthwhile to have photos for system or plant descriptions taken by a professional magazine or industrial photographer. Effective photos have many uses—in articles, ads, exhibits, records, catalogs, and reports.

Process Descriptions

This kind of article tells how a certain process works. Here too computerization has affected technical and scientific writing. The typical process-description article can discuss anything from a mechanical process used in chemical, petroleum, and manufacturing industries, to electronic data processing, storage, and retrieval. The process-description article features:

A. Process name and use
B. Materials and equipment
C. Steps in the process
D. Operating conditions
E. Finished product
F. Rate of output
G. Costs or savings

Many process descriptions deal with a continuous, production-line type of operation. Others cover noncontinuous operations. The requirements for illustrations are similar to those for the system or plant description. This type of article is used for a wide variety of products; it applies wherever something is changed from one form to another.

Some examples are: how a particular body of information is processed by a small printing company, improved processes for manufacturing chemical products, canning and bottling of foodstuffs, dissemination of customized stock market reports by a commercial database publisher, and improved techniques for the deburring of machine parts within an industrial plant.

Program Descriptions

In this type of article, readers learn the details of a specific planning or control technique. Most program-description articles focus on computer programs that define the operations necessary for a computer to achieve particular results, but other program descriptions discuss noncomputerized programs used for organizational or scheduling purposes. The program description typically includes:

A. Program name and applications
B. Equipment requirements, if any
C. Details of program functions
D. Steps in running the program
E. Cost and benefits of use
F. Review of documentation

Since the program-description article generally discusses a technique for handling certain tasks, illustrations are often limited to flowcharts showing the critical paths and decisions involved in the program. Some articles also include figures that illustrate, in simplified form, the steps required for a user of the program to reach a desired result.

Some examples are: software packages for planning and managing engineering projects on a microcomputer, procedures and forms that streamline the task of process cost accounting in a manufacturing plant, planning and control techniques for aerospace research experiments, and new programs for calculating company payroll.

Design Procedures

This kind of article tells readers how to design a product, plant, program, device, or system. It features:

A. Requirements to be met
B. Methods available to meet them
C. Reasons for choosing method described
D. Steps to be followed
E. Special precautions

The exact order in which these items are presented will vary, depending on the subject matter and beliefs of the article's author.

Design procedures described in many technical and scientific magazines are mathematical. But other types of procedures are often described too. Where the design procedure described is extensive, more than one article may be needed to give all the steps.

Some examples are: design of a heat exchanger, direct digital control system, textile weave, circuit board, railroad tunnel, walkway, or building.

Illustrations used in the design-procedure article include:

A. Photo of item designed
B. Calculation charts
C. Detail drawings

Product Functions

Here you tell the reader how a given product or material does its job. There is almost an unlimited range of types of products you can cover. To be of maximum help to the reader, the article should discuss both the advantages and disadvantages of the product, giving the materials of construction, application ranges, capacities, and typical uses. Articles of this kind also:

A. Establish the need for the product
B. Show how the product satisfies the need
C. Describes how the reader benefits from the product
D. Gives precautions in using the product

When writing this type of article, try to discuss a product in terms of the designs available from several manufacturers. Using data from only one firm is likely to produce a severely limited article or one that appears biased. Depending on the type of publication, editors are sometimes inclined to be critical of a story limited to one manufacturer's product.

Of course where the product discussed is the only one built, the article must be limited to it. If any substitutes at all are available, include them briefly, even if they are slightly outdated.

It is extremely important in product-function articles that you keep the reader's interest foremost at all times. Some writers tend to overlook the reader's needs and concentrate on how much space they can devote to glowing descriptions of the product. This is a fatal mistake. No experienced magazine editor will accept material of this nature until it has undergone major revision to put it into terms that will interest readers. Besides, readers attach greater credibility to articles that discuss both the pros and cons of particular products.

Some examples are: how safety valves work, how high-speed diagnostic tools can be used in the small electrical shop, how a computer software package aids in database management, and how fire-resistant fabrics are used in children's clothing.

Illustrations used in product-function articles include:

A. Photos of product in use
B. Cross-section or assembly drawings
C. Installation views (photos)
D. Installation details (drawings or photos)
E. Product details (photos or drawings)

Calculation Methods

In this kind of article you show the reader how to compute some value. These articles are very popular in technical magazines, where they typically show readers how to solve a common computation problem met on their jobs. It is the type of article that appeals vividly to engineers, technicians, and scientists. It features:

A. Value to be computed
B. Factors in the computation
C. Steps to be followed
D. At least one illustrative example
E. Limitations of the method
F. Special precautions

Remember: Show your readers an easier way to compute something, and they'll clip and save your article for years. Calculation methods you describe can vary from the simplest to the most complex, from computations done with pencil and paper to those that require complex computer hardware and software. Regardless of the application, a well-prepared calculation article is as useful to an engineer as a hammer is to a carpenter.

To give your article maximum usefulness, tell the reader exactly what problems can be solved using the computing technique. Use a step-by-step procedure from raw data to solution. Present at least one, preferably two, sample problems.

When using two problems, be sure to vary the data widely from one to the other. Then the reader can get the most experience from each.

Where the problem or solution calls for a sketch, use the simplest one possible. Readers of calculation articles are usually eager to learn. The easier you make it for them, the more they will get from your article. Some calculation articles resemble program-description, design-procedure, or graphical-solution articles because the calculation utilizes a specific program to develop a design and because charts are often used in all or part of the solution.

Some examples are: calculation of hazardous-waste emissions from multiple sources, stresses in special beams, heat loss through random air currents, vehicle operating costs, labor hours for particular computer programming tasks, and project results of potential energy-conservation measures.

Illustrations used in calculation-method articles include:

A. Charts or graphs
B. Photo of item discussed

Graphical Solutions

The purpose of the graphical-solution article is to present and explain a chart, graph, or other device designed to solve a particular problem. For a long time, this has been a neglected type of article. It is unfortunate that this is so, because graphical solutions save much valuable time for the user.

A graphical solution can be presented in a number of different forms, including nomograms, intersection charts, tabulations, and two-dimensional plots of data. The objective is to save time by helping the user to make a routine calculation with the least effort. Features are:

A. Statement of the problem
B. Explanation of how the chart saves time
C. Brief description of chart
D. Equations used in chart
E. Solution of typical problem
F. Precautions and limitations

Be extremely careful to choose suitable data ranges for the chart. The chart will be almost useless if it is plotted for values between 1 and 10 when, in fact, the usual range of actual values is about 10 to 100, for example.

Always give the equation on which the chart is based. The user is lost if he or she does not know the equation and the units in which it is expressed. Many careful engineers and scientists refuse to use a graphical solution unless they know the equation on which it is based. The only illustrations usually used in this type of article are the charts and sketches needed to solve the problem presented.

Some examples are: nomograms for determining the amount of boiler blowdown in a typical heating plant and intersection charts for determining press-fit forces. Figure 4.1 shows a typical nomogram.

Operating Procedures

In this kind of article you tell the reader how to run or operate a specific machine, program, system, or device. To give the reader the widest amount of information,

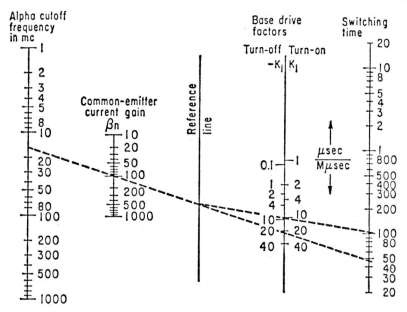

FIG. 4.1 Typical nomogram suitable for use in a magazine article. (*Product Engineering*.)

your operating instructions should cover products available from more than one manufacturer. Be careful, however, to note differences from one product to the next. Features are:

A. Practical approach
B. Step-by-step operating procedures
C. Safety considerations
D. Routine checks
E. Operating duties

Operating-procedure articles can be of major assistance to personnel in industries using equipment designed along the lines of that described in the article. In many cases, you may be able to point out useful shortcuts and other operational improvements over the steps described in a manufacturer's original operating manuals or product documentation. A variation of this type of article describes in detail the operating experiences of a new or unique system or plant. From the experiences presented, the readers derive information useful to their jobs or to their general background knowledge.

Some examples are: articles on the operation of diagnostic programs for troubleshooting minicomputers, the use of laser anemometers, tips on increasing the efficiency of environmental control systems, and the start-up of a diesel-powered generator in subzero temperatures. Variations on this type of article might deal with operating experiences in a waste processing facility or with solutions to problems in an overburdened switchboard.

Illustrations used in articles on operating procedures include:

A. Photos of machine, system, or device
B. Photos of operating procedures
C. Detail drawings of unit
D. Charts of characteristics or limits

Maintenance Procedures

In this type of article you tell your readers how to maintain, overhaul, and repair a particular class of equipment. By examining the magazines in your field, you will probably find a few different approaches to the topic of maintenance. The different approaches usually aim at two main classes of personnel and thus comprise two types of maintenance-procedure articles.

The first approach discusses maintenance scheduling, costs, staffing, management, and related issues. This approach aims its discussion at the concerns of readers in management positions.

The second approach is directed at the specific steps to be followed in maintaining a given machine, system, or other equipment. Such an article aims its discussion more at the technician doing the actual maintenance work than at the engineer or manager directing the work. Of course, people from both positions could read either type of article and benefit from the information presented.

Many industry-specific trade and technical magazines today are paying greater attention to the first type of maintenance article due to increased research and scientific analysis of maintenance procedures in various industries.

The maintenance-procedure article typically features:

A. Strong how-to slant
B. Step-by-step procedures
C. Use of illustrations wherever possible
D. Clear, specific instructions

The secret of success for any kind of maintenance article is the same as for the product-function piece. The article should be general enough to cover the products of more than one manufacturer. Only then will you attract a large number of readers.

Where a single product is so widely used that many people are interested in its care, the article can logically deal with it alone. But this is relatively rare in industry today.

Before writing any maintenance-procedure article, be sure to determine what products you will discuss, what type of readers you will address, and what kind of maintenance you will concentrate on. Many industries today divide maintenance procedures into: (1) preventive maintenance and (2) corrective maintenance. Preventive procedures comprise a plan or service intended to prevent malfunctions in equipment through cleaning, testing, and periodic replacement of components. Corrective procedures seek to isolate and remedy malfunctions after they have occurred.

Some examples are: articles on how to maintain video security systems, passenger elevators, printing machines, high-pressure pumps, textile looms, and cooling systems.

Questions and Answers

Here you have a series of questions, each of which is followed immediately by its answer. The questions may cover any subject. Features of a good question-and-answer article are:

A. Short, direct questions
B. Specific answers
C. Coverage of one segment of subject
D. Clear beginning, middle, and end

This kind of article is excellent where the reader will have a future use for the material, as in examinations. It's also good where the writer wishes to concentrate attention on one segment of a subject. By choosing questions dealing only with that segment, the writer can rid the discussion of useless side issues.

Some examples are: questions and answers on the cost of conserved energy, the reduction of health hazards in the workplace, common problems in fuel-injector installation, and application of standard specifications in roadway design. See Figure 4.2 for an excerpt from a typical question-and-answer article.

Management Techniques

This kind of article can cover any of a variety of management or business procedures. If often features:

---

1. What is meant by cetane number of fuel oil?

A. Cetane scale is applied to fuel oils, gasoline, diesel fuel, etc., to show oil's ignition quality. This quality means oil's ability to ignite in diesel cylinder—or under similar conditions. Fuel with good ignition quality will auto-ignite at low temperatures and is, therefore, preferred. This quality affects engine starting, smoking, and knocking.

Hydrocarbon family ranges from simple combinations, such as methane gas, through various solids, liquids, and gases. Some have complicated chemical structures. All have various ignition characteristics. Formula for cetane is $C_{16}H_{34}$.

Cetane is an excellent diesel fuel in its pure form, but isn't easy to separate completely from other hydrocarbons. Cetane content for various crudes differs greatly. $C_{10}H_7CH_3$ is another hydrocarbon, but it has poor ignition qualities. Cetane number is based on proportions of these two hydrocarbons.

Cetane number of 100 means ignition quality equal to pure cetane, while zero means one equal to that of alpha-methyl-naphthalene ($C_{10}H_7CH_3$). Ignition number of 60 means mixture equivalent to one composed of 60% cetane and 40% $C_{10}H_7CH_3$.

2. What is meant by right- and left-hand engines? Buyers ordering two engines for same plant often specify one each. Is this good practice?

A. Engine is classified as right or left when flywheel is on right or left sides, viewing engine from operating side. Reason one engine of each is specified is because owners believe such an arrangement is easier to attend if controls are on same aisle between engines.

Diesel Engine Manufacturers Association says there may have been reason for such an arrangement years ago before the days of remote governor control, but not with today's common use of such control and the little attention modern engines need. Such specifications require one engine to be nonstandard. Disadvantage to owner is loss of interchangeability, which is so important to avoid carrying large stocks of spare parts. Then, again, any advantage is lost if a third diesel is installed.

3. If engine fails to start, what would you look for?

A. Engine rarely fails to start, but such difficulty can be divided into two groups: (1) Starting mechanism fails to turn engine through an operating cycle. (2) Engine turns through operating cycle but fails to fire. If starting mechanism doesn't work, check starting air pressure or look for run-down or weak battery—depending on starting system.

If air-starting pressure is high enough, air-starting valves may not open fully or at right time. Or by sticking, leaky exhaust or inlet valves may prevent engine from quickly raising compression high enough to fire. Leakage from stuck rings or badly grooved cylinders may have same effect. In cold weather, heat of compression may be too low unless cooling water and (in extreme cases) even lube oil are heated.

At times, bearings are too tight (after overhaul) or there is no lubrication on cylinder walls, etc. All this puts extra load on starting mechanism.

FIG. 4.2 Typical questions and answers for a published article. (*Power*.)

A. One or more case studies
B. Strong personnel slant
C. Specific how-to procedures
D. Emphasis on positive accomplishment
E. Highly practical methods

Management articles cover a wide range of topics—from the simplest business procedures to advanced methods in difficult management problems. The exact level you aim at—anywhere from beginners in business to company presidents—varies with the publication and the subject matter of the article. Thus an article for shop supervisors is written in a different tone from that for general managers.

For many years now two highly popular types of stories in management articles have been: (1) the personality profile and (2) the self-improvement piece. In the personality profile, the author's portrayal of an official or outstanding member of a firm forms the core around which the story of the firm (and the individual) is told. These profiles are almost always written by a member of the magazine's staff. In the self-improvement piece, the emphasis is on easy reading of hard-hitting material that the reader can test and apply in day-to-day practice. Self-improvement articles typically support their suggestions with actual case stories and statistics showing the results of on-the-job application.

Illustrations used in the management type of article today include:

A. "Atmosphere" photos
B. Equipment photos and detail drawings
C. Drawings of processes
D. Cartoons or symbolic representations of management procedures
E. Photos of prominent individuals
F. Reproductions of business forms

Some examples are: how to get along with shop supervisors, how to set up pay schemes for executives, how to increase manufacturing productivity in ten steps, and how John Smith sparked the ABC Computer Co. to its greatest sales record ever.

Departmental Features

Many technical and scientific magazines regularly publish material in one or more "departments." A *department* of a magazine is a section devoted to short items of interest to the readers. Some departments regularly occupy less than a full page; others run three, four, or more pages. Many departments feature practical tips for design, operation, maintenance, production, etc. Generally, the amount of space given to each item depends on its importance to the magazine's readers. Typical departments in which you can take part are:

A. Letters to the editor
B. Readers' problems
C. Shortcuts and kinks
D. Mistakes on the job

The exact names of these departments vary from one publication to another. Thus, *Heating/Piping/Air Conditioning* calls item C "HPAC Data Sheet"; *Hydrocarbon Processing* calls it "Management Guidelines"; *Chemical Engineering* calls item C "The Plant Notebook." All, however, carry similar items—ideas, products, or ways to save time, money, and effort.

Contributing short items to a department of a magazine is an excellent way for you to start a writing career. The editor gets to know your material. If it's good, he or she will be happy to consider any longer pieces you do. Also, you can learn a lot by comparing your original copy with the edited version that runs in the magazine.

Before you contribute anything, study the entire magazine, particularly the department, for several issues. This will give you a feel for the level, writing style, length, and general approach the editor prefers.

For best results when submitting departmental material, try to follow these four pointers:

1. Keep the text short, and to the point.
2. Use a photo, sketch, or graph if possible.
3. Keep your explanations simple.
4. Give step-by-step procedures whenever you can.

Some examples are: reducing the load by using air doors on a heating system, six steps in drawing a fan scroll casing, easy ways to clean printer elements, ways to avoid static charges from electronic equipment, and computer programming shortcuts.

News and Equipment Releases

Many magazines divide news items into three categories:

1. Industry news
2. New-equipment items
3. Personnel changes

Industry news items: These articles cover the big and small stories of interest to people in the field covered by the magazine. The test question for such items is, "Will they interest a large segment of the people in the industry?" If the answer is yes, an item has a good chance of being published. Some examples are: changes in laws governing the industry operations, new licensing requirements or provisions, late data on prices of raw materials, important technical developments, new practices used abroad, and upcoming conventions and other industry events.

New-equipment items: These items are run for the benefit of all the readers of a magazine. To get best results from the new-equipment item:

1. Keep it short and concise.
2. Furnish a good photo or drawing with the text.
3. Always give the capacity, size, materials, and other pertinent details of the product.
4. Never knock a competitor's product.

5. Don't submit the item unless the unit is really new.
6. Omit all unnecessary information.

Let's examine these points quickly.

1. Do not submit new-equipment items longer than one double-spaced typewritten page. Unless your unit will revolutionize the field, items longer than one page are often a waste of time. Study the magazines. Note that almost all the items run about 15 lines on a 12- or 13-pica type measure. So be brief. Editors will love your for it.

2. Try to submit a good photo or drawing with every release. You tell a better story, and the reader gets the point sooner. Compare the items in Figure 4.3. Note how well the illustrated item tells its story. Your eye flicks from the illustration to the text and back again, giving you the complete idea.

Unless your product is extraordinary, don't submit more than two or three illustrations with it. Supplying more is wasteful because rarely is more than one illustration used to show the unit. The purpose of supplying two or three is to give the editor a choice of illustrations without overburdening your budget.

3. Be certain to give the capacity, size, applications, materials, and other pertinent data about the unit. Remember, many readers of new-equipment items are looking for units to solve a specific problem. Unless you give readers the major details about a device, they cannot tell if it is suitable. Experience shows that many busy people will not stop to inquire further unless the data show that there's a good chance that the device is suitable.

4. Never knock competitors' products in your release. It takes up space you could use for your own product. No good editor wants to see the other manufacturer criticized, regardless of personal feelings or beliefs about the product.

5. New-equipment departments are for new products. If you distribute an existing product in a new color, don't write a release about it. The product must be either a completely new design or an updated design involving extensive changes.

6. If the president of your firm introduces a new product at a picnic with every salesperson and distributor present, fine—but don't include this detail in your release. Such information is unnecessary. Space in new-equipment columns is limited, and you'll have enough trouble cutting the pertinent data down to size. Stick to the product and its important characteristics in your releases. Address your release to the new-products editor and include the full name and address of your firm. We will discuss equipment releases later in this handbook.

Personnel changes: These releases (including obituaries) should be kept short and to the point. Where your release concerns a promotion, transfer, or change in assignment, give both the new and the old position and title of the individual. Do not give the person's complete career history unless he or she is being advanced to a really important position—chairman of the board, member of the board, president, executive vice president, or vice president. For other positions, there generally is not enough space to give complete career details. But to be sure, study the magazine. There is some variation from one publication to another.

In obituaries, give the person's age, date and place of death, title and position, and a short summary of his or her career. If the person made significant contributions to the field, describe these concisely. Survivors are usually listed last. If one or more survivors are active in the same field as the deceased, give a short summary of their activities.

Electronic grommets

Rubber and vinyl grommets for use in electronics applications are offered in 26 and five sizes, respectively. Rubber grommet

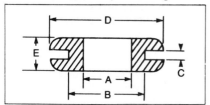

range of dimensions includes ⅛ to ⁴⁹⁄₆₄, ³⁄₁₆ to 1³⁄₃₂, 1¹⁄₃₂ to ¼, ⁵⁄₁₆ to 1⅝, and ⁵⁄₃₂ to ½. Vinyl grommet range includes ⅛ to ½, ¼ to ⅝, ¹⁄₁₆, ¹¹⁄₃₂ to ⅝, and ³⁄₁₆ to ⁹⁄₃₂. **Essex Electronic Supply Inc.**, 19 Baltimore St., Nutley, NJ 07110

Push-in fasteners

Six series of push-in fasteners can be used in such applications as joining plastic, light sheet metal, and insulating material, as well as serving as push-in guides and bumpers.

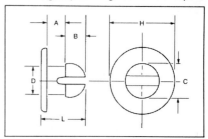

Head diameters ("C") range from 0.105 through 0.3. The "H" diameter is 9/32 to 19/32, depending on series. Drill hole sizes range from 0.096 to 0.265. **Essex Electronic Supply Inc.**, 19 Baltimore St., Nutley, NJ 07110

Oscilloscope package is flexible

Snapshot Storage Scope package turns IBM PC/XT/AT or compatible computer into a digital oscilloscope and data acquisition system. Upgraded version 2.5 can acquire up to 16 channels,

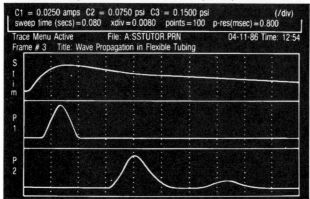

displays any eight channels, defines units and labels for each channel, and averages trigger-synchronized waveforms. Unit also plots X-Y, specifies title for each frame of data, and reads/writes data files using one-fifth the storage and time of standard files. Package is menu driven and requires no programming. **HEM Data Corp.**, 17025 Crescent Dr., Southfield, MI 48076

FIG. 4.3 Typical new-equipment items from a technical magazine. (*Courtesy Machine Design Magazine.*)

EXAMPLES OF ARTICLE TYPES

Listed below are 52 magazines that feature good examples of each of the 13 types of articles we discussed above. Note several things about this list: The magazines listed are by no means the only publications running representative articles of the various types. They do, however, run excellent examples of one or more types of articles. Actual article titles and page numbers are not given, as they would soon be outdated. Styles in magazines, like fashions in clothes, change constantly. The vogue today may be ignored tomorrow. When you examine actual publications in your field, be careful to note that some technical and scientific magazines use little or no contributed material. However, many of the magazines listed here use predominantly contributed material.

Use the list carefully. You can find copies of the magazines in most large libraries, or you can purchase them from the main or branch offices of the publisher, from large news dealers, or from back-issue stores. If one of the magazines strongly appeals to you, subscribe to it for a year or more. There is no better way to get to know the preferences of an editorial staff than reading every issue for a year or more.

For typical examples of the 13 types of articles discussed earlier, refer to the following magazines:

1. System and Plant Descriptions

    Plant System and Equipment
    Chemical Engineering
    Energy Engineering
    Power

2. Process Descriptions

    Hydrocarbon Processing
    Food Engineering
    Textile World
    Chemical Process Engineering

3. Program Descriptions

    Software
    Data Management
    Computer Aided Design Report
    InfoWorld

4. Design Procedures

    Machine Design
    Heating/Piping/Air Conditioning
    Engineering News-Record
    Engineering and Mining Journal

5. Product Functions

    Fleet Owner
    American Machinist
    Marine Engineering
    Heating & Plumbing Product News

6. Calculation Methods

    Electronics

Control Engineering
Electronics Design News
Specifying Engineer

7. Graphical Solutions

    Chemical Engineering
    Machine Design
    Air Conditioning, Heating and Ventilating
    Civil Engineering

8. Operating Procedures

    World Oil
    Personal Computing
    Coal Age
    The Welding Engineer

9. Maintenance Procedures

    Industrial Maintenance and Plant Operation
    Electrical Construction and Maintenance
    Butane-Propane News
    InfoWorld

10. Questions and Answers

    Refrigerated Transporter
    Industrial Distribution
    Electrical Merchandising
    Public Works Magazine

11. Management Techniques

    Microsystems Management
    Managing
    Electrical World
    Journal of Applied Management

12. Department Features

    Power Engineering
    Oil and Gas Journal
    Machinery
    Power Transmission Design

13. News Releases

    Purchasing World
    Aviation Week and Space Technology
    Electrical Engineering
    Petrochemical News

HOW TO FIND AND DEVELOP ARTICLE IDEAS

The published article is the result of a process. The process of writing for publication moves through several overlapping stages, including study of typical magazines and articles in your field, which we discussed above.

Crucial to the success of any article writing process is the effort to find and develop ideas. Look around you today. Article ideas are everywhere. To latch onto a few, train yourself to think in the right terms. Such thinking is not a matter of following a pat formula or set of simple rules. If you force your thoughts into the straitjacket of a tried-and-true scheme, you may limit your ideas and your ability to apply them. On the other hand, there are general principals of thinking that can help you find interesting ideas and develop them into excellent technical and scientific articles. We'll see how writers use these principles here.

In finding and developing article ideas, scientific personnel, engineers, and technicians are particularly fortunate. For them, article ideas can be found so quickly they may never have time to put them all to use.

Look at Your Job

Try to remember the day you started on your present job. Remember how new everything seemed? When your supervisor or an associate explained some procedure, you probably followed it step by step, classifying each task in your mind. Now those procedures seem second nature. You probably know them so well that you hardly have to think about them.

But what about a newcomer to your position? He or she would see your duties in much the same light as you did during the first few days. Focus your thoughts on those early days and list the steps in some of the procedures you learned. Show such a list to a new person and watch that person's eyes glow. Why? You have relieved the newcomer of thinking through the steps. You have saved the person time and energy and have made his or her life a bit easier. This is the intent of almost every article run in technical and trade magazines today. You must help or inform the reader in some way—if you don't, forget your article.

So look at your job and duties. There is a wealth of article material in both. But don't stop there. If you are taking a course or studying some subject on your own, stand back and take another look. What phase was especially tough? Now that you understand it, can you give a better explanation? If so, you probably have a good idea for an article.

List Your Ideas

If your job initially presented no new problems, then use another approach. Write your job title on a piece of paper. What is it? Design engineer, vice president for systems analysis, staff scientist? In these three titles there should be at least 18 article ideas. Let's find a few of them.

Assume you are a design engineer. What do you design? We'll say it's industrial air purifiers. How do you design them? Stop right there. That's your first article idea: how to design industrial air purifiers.

Now think further. What happens when you design a purification unit? You run into problems—glitches, bugs—whatever you call them, you probably get hundreds of problems. The story of how any of these are solved would probably be a good article. So include some of your toughest problems in the list.

Once the unit is designed, it must be built. Then what happens? More problems. The same is true for operation and maintenance. Now let's see how many article ideas we have found. List them like this:

1. Designing industrial air purifiers
2. Filtration problems in industrial air purifiers

3. Location problems in industrial air purifiers
4. Particle buildup problems in industrial air purifiers
5. Assembly procedures for industrial air purifiers
6. Operation and maintenance of industrial air purifiers

This list is just a beginning. Item 1, design, could include the design of every one of the many parts of the purifier as well as of the unit as a whole. Alternatively, you could compare and contrast the design requirements of several types of industrial air purifiers. The construction and installation problems—items 2, 3, and 4—could be expanded almost indefinitely. Ask any design engineer. The person will talk for hours.

Item 5, manufacture, could deal first with the actual purification unit, or separate parts of the unit. After you had written a few articles on this phase, you could switch to the production equipment and techniques for assembling the units. This subject should provide material for at least two articles. Much the same is true for item 6—operation and maintenance.

Thus, if you really want to write technical or scientific articles, start an idea file. List every idea that comes to you. Take pencil in hand a few times and build ideas from a job title, a special responsibility, problems you've encountered, or any other subject. Soon ideas will be coming to you when you least expect them. Jot them down immediately and toss the list into a folder. Work this scheme for a few weeks and you will have a sizable file.

Tested Methods for Producing Ideas

Below are nine ways to produce ideas. All have worked for writers of technical articles. Try each method. Choose those that work best for you and concentrate on them in the future.

1. Ask your associates what their technical problems are, and how they solve them.
2. Study the articles and ads in the magazines in your field.
3. Expand your know-how of a subject.
4. Survey a field, and summarize its literature (i.e., papers, reports, monographs, books, etc.).
5. Check with your firm's public relations director or agency.
6. Study the handbooks in your field.
7. Look for photo and sketch subjects.
8. Watch for graphical ways to solve problems.
9. Ask your associates what subjects they think need discussion in articles.

Let's take a quick look at each method to see how to use it. Remember that you can alter any technique to suit your needs.

Ask Your Associates. Concentrate on people in the same field as your own. Then you will have a good understanding of most of their problems. First ask, "What is your worst problem?"

Listen carefully to the answer. If the method used to solve the problem is different in one or more aspects from usual procedures, you probably have an article idea. Your associate, in most cases, will be willing to give you the complete details for use in an article, or he or she may volunteer to be your coauthor.

Study the Magazine. Read the outstanding technical, scientific, and business publications in your field. Go through the magazines from cover to cover. Study both the articles and the ads. Some articles may tell only part of a story that you know well. If so, check with the editor about doing another piece from a different viewpoint. Ask the editor if he or she plans any issues with special themes or year-end features. If you can obtain a copy of the editorial schedule, you can write to suit special issues.

Other articles may suggest new ideas to you. An article on electronic security systems may make you wonder if anything has been done on circuitry problems in certain kinds of systems. Check with the editor. If nothing has been written, you're the right person to do the circuitry article.

Ads are a fine source of article ideas. Almost every advertised product provides a clue to the interests of the magazine's readers. The more ads about a given type of product, the greater the interest. Reading the ads also broadens your knowledge of the firms in the field, their latest products, and the information available about them. All this is valuable background for your writing.

Expand Your Know-how. Is there a phase of your work you should know better? If so, why not take some courses or do some specialized study of this phase? Courses and research, coupled with your previous education and experience, can give you a valuable knowledge of your field. They may even give you enough know-how to write some articles. However, when using this method, watch out for the trap of writing on the basis of too little knowledge. Editors spot shallow articles almost immediately. So get a good, solid background of information before you sit down to write. Be sure you can obtain all the data you will need to write the article, and try to make it timely. Then you will be almost certain of finding a publisher for your material.

Survey a Field. If you're interested in a field related to your own, make a careful survey of it. Review its major statistics, the firms doing business in it, the literature available, the outstanding authorities, and its history. A survey should give you many article ideas. Even if it does not, you can almost always use the literature summary as an article. Be sure your summary is up to date and that it includes the important contributions to the field.

Check with Public Relations. If you work for a large firm, chances are there is a public relations (PR) officer on the payroll. Get to know this person. He or she is probably eager to meet personnel interested in writing technical material. The PR officer may have so many ideas for articles that he or she can keep you busy for years. Follow your PR person's recommendations. Most PR officers have had long experience with the communications media; they know what editors need.

Smaller firms, and some large ones, may use a public relations agency, or an ad agency having a PR department. If this applies to your firm, contact the agency. The person handling your account is certain to have some article ideas and to know what editors want, if the agency is worth its fee. Work with PR or ad agencies in the same way that you would with your own PR department.

Study Handbooks. Well-written engineering, scientific, and business handbooks are excellent idea sources for alert writers. Pick up a handbook on your field and see for yourself. Read a few of the sections. Note how each summarizes its subject.

If your mind is receptive to article ideas, you will find that certain key words suggest topics, especially in the most up-to-date handbooks. In a section covering nuclear reactor containment, for example, the details about materials may suggest a product-function article surveying the various kinds of containment materials. Or a section on purchasing may give you a lead to an operating-procedure article on the purchase of specialized equipment—stress-monitoring devices, control consoles, floodlights, microcomputers, business or engineering software, etc.

Search for Picture Stories. Get yourself a good camera, or line up a few good industrial photographers. Then search your job and plant for picture stories. Look for any step-by-step operation that's important in your business. This is almost certain to be of interest to readers. But before taking any photos, check your idea with some editors. See our discussion below on tips for working with editors.

If you don't feel you're a good photographer or if the story does not lend itself to photos, try sketches. Get a pad of unruled paper and make a simple, clear sketch of each step. Write a caption for each sketch, then perhaps a short, general statement about the problem or process. Write a query letter to editors.

Watch for Graphical Solutions. Many firms prepare charts, computer programs, schedule boards, and other devices to make routine jobs easier. Some of these are excellent subjects for graphical-solution articles or short departmental features.

Make a list of the graphical solutions that are available. Find out if you would be allowed to publicize them. If so, contact an editor. When you develop a graphical solution yourself, you need not get company permission to publish it unless the firm has specific rules regarding writing. But play it safe and get permission anyway.

Ask Your Friends. Check with your associates for topics they'd like to see written about. Ask them, "What article do you think would do the most good for our field today?" You'll be amazed to see how many excellent article ideas pop up. You probably won't be able to do an article on every idea suggested, but there will be plenty that you are qualified to do.

In addition to asking your associates, you can often obtain ideas by scanning your daily newspaper and general-interest news magazines for articles relating to your field. This scheme has the advantage of timeliness. The ideas you acquire are those that are important at the moment. For example, if an elevated walkway collapses at a new hotel, the news media will carry stories on it; the stories will draw widespread attention to the need to prevent such accidents in the future. If you can write an article on one or more safety measures to prevent similar problems, the article will almost certainly be published. You will make an important contribution to your field. And every astute magazine editor loves timely ideas.

Classify Your Ideas

Once your ideas begin to roll in, they will give you trouble unless you classify them. Classifying your ideas is one of the first steps toward successful technical and scientific writing. For unless you organize by classifying, your growing batch of ideas will float around your mind uselessly. Follow these four steps to classify your ideas:

1. Write the idea on paper as soon as it hits you.
2. Toss the sheet of paper into a general-idea folder.
3. When you have time, review your ideas.
4. Put each idea into a special folder or file.

These steps make it easier for anyone to get a good start on classifying his or her article ideas.

Step 1. You can't classify ideas unless you can remember them. And if you are like many technical and scientific writers, you will remember mainly your poor ideas. The good ones flash through your mind and disappear, unless you write them down. So play safe. Write down every idea as soon as it hits you. This is the only sure way to hook the big ideas and place them on paper permanently. Besides, once you've written down one idea, you can forget it temporarily and turn to looking for new ones.

Step 2. Toss your ideas into a general-idea folder. Once you begin to write and get your material published, you'll start to think of this folder as a bank. For it will contain the principal on which you will draw for much of your future writing.

Allowing an idea to rest in a general folder for a few days or weeks has advantages. Ideas seem to season or grow stale, depending on whether they're good or bad. Coming upon an idea "cold" allows you to judge it better. This way you are more likely to scrap the poor ones and save the good ones.

Step 3. Take time out to review your ideas at regular intervals. Close your office door and tell your secretary or associates you need some quiet. Go over the ideas quickly. Toss out those that sound trite or would require more effort than you can expend. Don't keep poor ideas around—they have a way of spoiling your good ones. Not only that, you waste time whenever you review a poor idea after you've decided it is no good. Put your good ideas into a neat pile before classifying them.

Step 4. Set up a 13-folder file. Label each folder with one of the article types listed earlier in this section. Now take your article ideas, and with only one to a slip of paper, toss each into the correct folder. If you are in doubt about an idea because it could fit into more than one category, put it into the folder in which it seems to have the greatest use.

If you write only a few articles a year, say six or less, file folders are good for classifying your ideas. The same classifying scheme can be used on a computer; merely substitute data files for paper folders. But if you write a large number of articles, it is better to organize your materials more extensively. Here's one way that works well for many technical writers.

Idea Classification Form

Figure 4.4 shows a typical form useful for organizing article ideas after they have been classified. Prepare the form on 8½- by 11-inch paper for each of the 13 ar-

Article Type: System Descriptions						
(1) Idea	(2) Date entered	(3) Work needed before writing	(4) Publication possibilities	(5) Date writing started	(6) Date writing finished	(7) Publication details
Local area network database processing	4/15/xx	Get 2 illustrations: 1. Photo of office network 2. System flowchart	Interface Age Data Management Computer Decisions Journal of Systems Management	5/1/xx	5/28/xx	Published in _____ 12/xx Article length: 5 pages Reprints ordered
New systems for construction, engineers: using PERT, CPM, dedicated features	5/16/xx	Get manufacturers' descriptive materials and specs; write software company for applications details, etc.	Construction Specifier (for Dec. issue) Construction Equipment Dixie Contractor	8/4/xx	9/2/xx	Returned by _____ 9/30/xx; sent to _____ 10/15/xx Accepted 11/1/xx to be published 12/15/xx

FIG. 4.4 Form for recording article ideas. You can alter this form to suit your writing needs.

ticle types listed earlier. Use punched paper so that the forms can be kept in a ring binder. If you use a word processor or computer, you can store the forms in the system's memory. However, it is always wise to retain hard copies of your material on paper too.

Divide the paper into seven vertical columns and enter the headings shown for each, from left to right: (1) idea, (2) date entered, (3) work needed before writing, (4) publication possibilities, (5) date writing started, (6) date writing finished, and (7) publication details.

Under heading 1, list the idea as taken from the folder. Be sure to give enough data so that you can recall the exact article in the future. Enter the date the idea is listed on the form. This will be close to the date you originally got the idea if you review your ideas regularly. If you do not, then put the date on the paper when you first make a note of the article idea.

Be sure to keep the various dates relating to an article—when you first had the idea for it (column 2), when you started writing, and when you finished. Why? Well, if you write regularly, you'll begin to spot a pattern in the development of an article. Once you reach this stage in your writing, the ideas will come more easily, the words will flow smoothly, and you will probably produce at least twice as much as when you started. Dates also help you see when an article has been around too long, and when you need a little prodding to reduce your backlog of ideas.

Column 3 tells you what to do before starting to write. If photos or drawings are needed, make a note in this column. Get all these materials before you start the text. There is nothing quite so disappointing or wasteful as to write an article and then find that the key illustrations are not available.

In the order of preference, list in column 4 the magazines for which you think the article is suited. Then you can direct your writing better.

Insert your starting and completion dates in columns 5 and 6. Enter publication details—the name of the magazine and the date on which the article appeared—in column 7.

The form in Figure 4.4 is not the only one you can use, of course. Alter it to suit your particular needs. Avoid the use of complex forms, however. The longer it takes to fill out a form, the less time you have for writing. Complex forms just give you one more excuse for not writing. If you are like most authors and would-be writers, you have enough of these excuses already. So why invent another?

Make Your Ideas Blossom

Though it sounds old-fashioned, every article must have a beginning, a middle, and an end. These parts correspond to the introduction, body, and conclusion in a report or scholarly paper. This form applies whether the piece is long or short. Once you sense the need for these three elements, your articles will become better balanced and more professional in their makeup.

It takes skill and experience to develop your ideas so that they have a beginning, middle, and end. One method that works well for many technical and scientific writers resembles brainstorming. (See Section 2 on report outlines.) You might call this method "randomizing." To use it, do the following:

1. Write the article idea at the top of a piece of paper.
2. Under the idea, list every word or short group of words related to the idea that you can think of.
3. Don't attempt to edit or arrange the words in any specific order; simply list them as they occur to you.
4. Think of the article idea from as many different angles as possible—from the viewpoint of the general reader, or a mechanic working on the equipment, or of an engineer designing it, etc.; continue listing words that occur to you.
5. Read as many related articles and books on your subject as you can; continue listing words.
6. Review your list of words and phrases and arrange them in a rough outline form—under "Beginning," "Middle," and "End."
7. Mark those ideas that will be illustrated, noting the type of illustration to be used—photo or drawing.

Now let's take a look at how this method might be applied to a specific article idea. To make the example as useful as possible, we will choose a fairly common type of article—a description of a new energy management system. Make your idea blossom by first listing words at random like this:

AB Company's New Energy Management System

Size of building
Type of work handled
Cost of system
Hours of operation
What energy equipment preceded new system

Heating equipment
Ventilating equipment
Air-conditioning equipment
Amount of building occupied
Reasons for new system
Personnel
Designers and builders
Future projections
Where located
Electricity usage
Load control methods
Construction troubles
When started
When finished
Cost before installation
Payback period
Return on investment
Type of building
Special aims in design

There are 24 random ideas here. They've all grown from the simple article idea under which they are listed. One interesting and important aspect of this method is that when you list your ideas, they're all related to a single article. Other methods sometimes allow unrelated ideas to creep in, complicating your job. Now let's rearrange these ideas so they fall roughly into a beginning, middle, and end.

Beginning

Type of building
Where located
Energy savings from new system
Size of building
Type of work handled
Personnel
Hours of operation
Amount of building occupied
Reasons for new system

Middle

Designers and builders
Special aims in design

Load control methods
What energy equipment preceded new system
Heating equipment
Ventilating equipment
Air-conditioning equipment
Electricity usage
Cost before installation
Cost of new system
Construction troubles

End

When started
When finished
Payback period
Future savings projections

With this general arrangement of your article, you are now ready to prepare a formal outline. But before sitting down to do the outline, make certain you've taken care of one other detail—permission to do the story.

Get Permission

If your article is based on data from your firm, check with your supervisors to see that they approve the use of the material. There is no point to writing an article and then having it killed by management. When using data relating to government defense work, be certain to secure a written release from the appropriate agency or department.

Some firms are extremely cautious when it comes to releasing material concerning their work. So be careful. Always get permission to use the material and illustrations that will go into the article. Then there is relatively little danger that you will have to scrap the article later. Killing an article after it is in type is an expensive and painful ordeal for the author, the author's firm, and the editors. The best way to avoid this kind of unpleasantness is to get permission from all the proper authorities connected with the data you intend to include in your article. Always get permission—start now.

HOW TO OUTLINE AN ARTICLE

Use an outline for every article you write. Outlining takes time, but the time you spend on the outline reduces your writing time. For in writing any article, you must make a number of decisions—what form to give your story, what data to present in the beginning, middle, and end, what illustrations to use, etc. You will make these decisions while outlining the article.

Steps in Outlining Articles

There are three basic steps in outlining any technical or scientific article:

1. Assemble all your facts.
2. Collect your illustrations.
3. Classify your facts and illustrations.

Let's see how you can best perform each step.

1. Assemble Your Facts. For your first few outlines, and for difficult articles, use the methods discussed earlier in this section for developing ideas and making them blossom. Once you have gained some experience in writing for publication, you can work more directly. To do this, set up three columns on a sheet of paper. As in our discussion above, the first column should list items for the beginning of the article, the second for the middle, the third for the end.

Under each heading, insert key words for the important topics you intend to cover. If you have your order firmly in mind, you can insert these key words in the exact sequence you expect to use them in the article. If not, list the key words randomly, and then reorder them according to their relative importance. Then you will be ready to transfer the key words from this rough outline to your comprehensive outline.

2. Assemble Your Illustrations. Secure all the illustrations you intend to use in your article. Don't start the outline until you have every illustration you need. Key drawings or photos can make or break a technical or scientific article. If you cannot obtain good illustrations, you might as well not write the article, unless the editor plans to supply them.

Some writers start to secure their illustrations immediately after deciding to write an article. This is often long before they start producing copy. So while working on other stories, they keep tabs on the material rolling in for future articles. This makes sense because it usually takes from several weeks to a month or more to obtain illustrations from various sources. Don't wait until a week before your deadline to obtain illustrations; the whole process of approving, producing, and delivering illustrations takes time and can have unexpected delays.

3. Classify Facts and Illustrations. Put your facts where they belong in the story. Don't drop them in at random as you write. Smooth writing comes from good outlining. But what goes where in your outline and story? Let's look at the basic parts of a technical or scientific article. Then we will examine some typical outlines you can use.

The Lead

Most editors and writers refer to the beginning of an article as the *lead*. Your lead for a technical or scientific article should contain answers to three questions:

What is this article about? Why should it be read? How will the reader benefit from it?

The answers may not always be given directly—some parts may be implied. But by answering all parts of these questions, you summarize the article for the reader. The summary type of lead is preferred by many editors today, even though you see other types of leads used.

Summarizing your article in the lead paragraph helps readers decide if they must read your entire story. Not only do you save their time, but you may encourage them to read further. A good lead thus increases the readability of an article by arousing readers' interest and curiosity. So remember—what, why, how. See our discussion on writing effective leads later in this section.

The Middle

Here is where you give the solid facts to your reader. You answer, in a detailed way, the three questions posed in your lead. Tell your reader what the equipment, procedures, or other items do, give, or produce, or how they help. Don't skimp on facts. But don't go back to the beginning of time to start your explanations.

Know what kind of a reader you are writing for. Keep this person in mind. If the typical reader is a computer programmer, see that your ideas stick to his or her needs—programming languages, dialects, applications, writing and testing procedures for the programs, documentation, etc. But if your reader is a computer engineer or systems analyst, direct your words at equipment, design, performance, data management procedures, and the like.

The End

A good writer knows not only how to begin an article but when to stop. The basic guideline for ending most successful technical and scientific articles is: Summarize your results, then stop. Show what your story means for the future, and end. Don't drag out the ending. You will lose your reader if you do.

One paragraph is often enough for ending a technical article. Some editors avoid any formal ending. When the story finishes presenting all the pertinent facts, it ends. Reading this type of article gives you the feeling that its tail was chopped off. But some people use it because it saves space. Generally, use a one-paragraph ending for your articles. It is easier for editors to delete a paragraph than to write an ending themselves.

TYPICAL OUTLINES FOR TECHNICAL AND SCIENTIFIC ARTICLES

Below you'll find a typical outline for each of the 13 kinds of articles we discussed earlier in this section. Use these outlines as a guide when preparing your own articles. The sample outlines will save you time and energy. Generally, they will get you easier admission to editorial offices when you submit a query. One word of caution—use these outlines only as guides. Fill in your own data for the particular story you are considering.

System and Plant Descriptions

I. Lead
   A. Name, owner, and location of the system or plant
   B. Type of system or plant
   C. Outstanding features of the system or plant design
   D. What is expected from these features
II. Details of the System or Plant (Middle)
   A. Location and its effect on equipment
      1. Special conditions of building structure or locality
      2. Effect of these conditions on the design
      3. Steps taken to cope with unusual conditions
   B. Equipment schemes considered
      1. Discussion of usable schemes or configurations
      2. Reasons for choosing the one used
      3. Advantages and disadvantages of the one used
   C. Principal components and peripheral devices
      1. Type
      2. Manufacturer
      3. Capacity
      4. Power (electric or other)
      5. Operating characteristics (speed, flexibility, etc.)
      6. Unusual features
      7. Tabulation of principal units and their features
   D. Flow details
      1. Flowchart
      2. Paths of major and minor processes, materials, data
      3. Flow characteristics (e.g., rates, pressures)
      4. Flow variations with load changes
   E. Operating details
      1. Control systems
      2. Automatic controls
      3. Manual controls
      4. Personnel requirements
      5. Operating cycles
      6. Personnel work schedules
      7. Materials or data supply
      8. Storage media
      9. Safety and security features
      10. Utilities
      11. Other details
   F. Personnel facilities
      1. Work stations
      2. Dining area
      3. Conference rooms
      4. Parking
      5. Other facilities

       G. Design advantages
          1. Initial-cost savings
          2. Operating-cost savings
          3. Worker-hour savings
          4. Time savings
          6. Other savings
       H. Expected results
          1. Data or product output
          2. Efficiency
          3. Operating life
          4. Overall advantages
          5. Other expectations
III. What This System or Plant Means to the Industry [End]
       A. Today
       B. Tomorrow (several years distant)
       C. Design or building trends to watch

Process Descriptions

  I. Lead
       A. Name, licensor, and application of the process
       B. Name, owner, and location of the system or plant
       C. Outstanding features of the process
 II. Process Detail [Middle]
       A. Product
          1. Rated output
          2. Characteristics (e.g., capacity and specific gravity)
          3. Other properties
       B. Raw materials
          1. Name
          2. Quality
          3. Consumption
          4. Catalysts
          5. Chemicals
          6. Other materials
       C. Flow details
          1. Flowchart
          2. Paths of major and minor materials, fluids, etc.
          3. Flow rates, pressures, temperatures
          4. Flow variations with output changes
          5. Catalyst consumption and regeneration
          6. Chemical reactions
       D. Process equipment
          1. Type
          2. Manufacturer
          3. Capacity
          4. Operating characteristics
          5. Utilities
          6. Unusual features
          7. Tabulation of principal units and their features
       E. Operating details

        1. Control systems
        2. Automatic controls
        3. Manual controls
        4. Personnel requirements
        5. Operating cycles
        6. Personnel work schedules
        7. Raw materials supply
        8. Materials storage
        9. Safety and security features
     10. Utilities
     11. Other details
    F. Process advantages
        1. Initial-cost savings
        2. Operating-cost savings
        3. Other savings
    G. Expected yields
        1. Product output
        2. Raw material or charge input
        3. Operating life
        4. Overall advantages
        5. Other expectations
III. Effect of This Process on the Industry [End]
    A. Other installations (location, product, etc.)
    B. Future installations
    C. Design or building trends of the future

Program Descriptions

I. Lead
    A. Name, owner and/or designer, and general applications of program
    B. Type of system used to run the program
    C. Outstanding features of the program
II. Details of the Program [Middle]
    A. Basic classification (application, utility, processing)
    B. Specific applications: List and describe
        1. Input requirements
        2. Data manipulation
        3. Output
        4. Examples
    C. Advantages (compared to programs with similar applications)
        1. Speed
        2. Capacity
        3. User requirements
        4. Ease of implementation
        5. User interface
        6. Adding and deleting records
        7. Error messages and menus used
        8. Other advantages
    D. Design specifications
        1. Problem solutions
        2. Data organization

3. Displays
4. User access
5. Other specifications
- E. Coding specifications
    1. Name and dialect of programming language used
    2. Coding rules
    3. Sort fields
    4. Other specifications
- F. Program flowcharts
- G. Testing
    1. Selected test cases
    2. Results
- H. Documentation
    1. Level of user for which it is written
    2. Readability
    3. Comprehensiveness
    4. Organizational scheme
    5. Clarity and usefulness

III. What This Program Means for the Industry
- A. Specific field in which it is applied
- B. Computer programming in general
- C. Other programs and systems
- D. Future applications
- E. Programming trends of the future

Design Procedures

I. Lead
- A. Name, size, and type of item to be designed
- B. Principal design problems
- C. How this procedure overcomes these problems

II. Procedure Details [Middle]
- A. Design requirements
    1. Output, capacity, allowable load, rating, etc.
    2. Allowable atmospheric conditions: temperature, humidity, etc.
    3. Allowable operating conditions
    4. Size and weight
    5. Materials
    6. Appearance, color, finish, packaging
- B. Alternatives meeting the design requirements
    1. Description of each alternative
    2. Advantages of each
    3. Disadvantages of each
    4. Reasons for choosing the alternative selected
- C. Design problem
    1. Concise statement of the problem and its parameters
    2. Description of the methods for solving it
    3. Reasons for choosing the solution used
- D. Steps to follow
    1. In setting up the problem

        2. In mathematical, graphical, estimating, or other procedures presented
        3. In using tabular, graphical, or other data presented
        4. When solution is obtained
    E. Verifying results
        1. Reasons for checks
        2. Precautions in making and evaluating checks
        3. Checking intermediate results
        4. Checking final results
        5. Comparing results with those from other methods
    F. Applying results
        1. To a single unit
        2. To a series of units
        3. To work in your plant
        4. To work done in other plants
        5. To other situations
    G. Other considerations
        1. Materials
        2. Finish
        3. Color
        4. Packaging
        5. Shipping
        6. Price, as related to design
III. Future Designs [End]
    A. Relation of this procedure to future ones
    B. Probable trends in design methods
    C. Other considerations

Product Functions

I. Lead
    A. Name, type, and size of product
    B. Typical applications of the product
    C. Problems met in product use
    D. Problems solved by product use
II. How the Product Works [Middle]
    A. Product description
        1. Moving parts
        2. Stationary parts
        3. Other parts
        4. Materials used for parts
        5. Similar types of products
        6. How this product compares with others
        7. Needed accessory equipment
    B. Product working cycle
        1. Steps in a cycle, from beginning to end
        2. Pressures, temperatures, flow rates, etc.
        3. Flow pattern in or through the unit
        4. Allowable working ranges of the product
    C. Product operation

       1. Starting procedures
       2. Routine operating procedures
       3. Stopping procedures
       4. Operating pointers
    D. Product maintenance
       1. Routine maintenance
       2. Special overhaul and repair procedures
       3. Tool and other equipment for maintenance
    E. Product selection
       1. Determining size or capacity required
       2. Choosing the most suitable unit
       3. Checking the selection
    F. Product application
       1. Installation
       2. Initial start-up
       3. Operating the new unit with existing ones
       4. Other considerations
III. Results This Product Gives [End]
    A. Output, efficiency, savings, etc.
    B. Summary of major advantages

Calculation Methods

I. Lead
    A. Item to be computed
    B. Difficulties met in the computation
    C. How this method reduces these difficulties
II. Calculation Procedure [Middle]
    A. Data needed
       1. Variables, constants, and other facts
       2. Where to obtain these data
       3. Precautions in assembling data
       4. Other considerations
    B. Equations or computational methods used
       1. Basis of equations or methods
       2. How equations or methods were developed
       3. Symbols used, and their units
    C. Illustrative example
       1. Problem statement
       2. Problem solution shown in a series of steps
       3. Explanation of reasoning for each step
       4. Major exceptions likely to occur
       5. Other considerations
       6. One or more additional illustrative problems, as needed to explain the method
III. Limitations of the Method [End]
    A. Where the method is suitable for use
    B. Where it is unsuitable for use
    C. Special precautions

Graphical Solutions

I. Lead
   A. Problem to be solved
   B. Advantages of a graphical solution
II. Solution Details [Middle]
   A. Description of the graphs or charts
      1. Equations or data on which charts are based
      2. Symbols, and their units
      3. Limits and other ranges used in the charts
      4. Calculations for which the charts are suitable
      5. Where the charts cannot be used
   B. Illustrative example
      1. Problem statement
      2. Problem solution, shown using the charts
      3. Explanation of each step in the solution
      4. Major exceptions likely to occur
      5. Other considerations
      6. One or more additional illustrative problems and solutions, as needed to explain the method
III. Advantages of This Method [End]
   A. Time
   B. Labor
   C. Expense

Operating Procedures

I. Lead
   A. Why careful operating procedures are needed
   B. Advantages of using these procedures
II. Recommended Operating Methods [Middle]
   A. Steps in start-up
      1. Before the item is running
      2. Factors to check
      3. Putting the item into motion
      4. Checks during and immediately after starting
      5. Other precautions
   B. Routine operation
      1. Items to check at regular intervals
      2. Recording the items checked
      3. Routine maintenance (servicing, lubrication, etc.)
   C. Stopping procedures
      1. Reducing load, speed, capacity, etc.
      2. Cutting off power
      3. Stopping operation or motion
      4. Checks to make after completion
      5. Other procedures
III. How to Obtain Best Operating Results [End]
   A. Correct procedures
   B. Records

C. Personnel training
D. Regular maintenance

Maintenance Procedures

I. Lead
   A. Name, type, size, and service conditions of unit
   B. Problems usually met in maintenance
   C. How correct procedures overcome maintenance problems
II. Step-by-Step Procedures [Middle]
   A. Detecting trouble
      1. From operating records
      2. From performance tests
      3. How to make performance tests
      4. Trouble warnings: error messages, noise, vibration, output reduction, etc.
      5. Other ways to detect trouble
   B. Disassembly methods
      1. Unit shutdown
      2. Removal of working power, fluid, data, etc.
      3. Opening, removing, and inspecting casing, shell, or cover
      4. Precautions during disassembly (tagging of parts, use of special tools, etc.)
      5. Safety considerations
      6. Other considerations
   C. Overhaul and repair methods
      1. Removal of working parts
      2. Inspection of working parts
      3. Repair of worn parts
      4. When to replace old parts with new ones
      5. Other overhaul and repair methods
   D. Reassembly and reactivation
      1. Precautions before starting reassembly
      2. Steps in reassembly
      3. Testing the reassembled unit
      4. Other considerations in reassembly
III. Getting More from Maintenance [End]
   A. Records
   B. Personnel training
   C. Correct operating methods

Question and Answer

I. Lead
   A. Basic question whose answer defines the field to be covered by the article
   B. One or more additional questions amplifying the first
II. Specific Coverage [Middle]

A. Questions and answers selected to cover the particular areas chosen for the article
B. Additional related questions that add to the reader's knowledge about the subject matter
III. Other Coverage [End]
A. Questions devoted to future expectations
B. Exam questions or other specialized items readers should know

Management Techniques

I. Lead
A. Reasons why the article material is needed
B. Who needs and can use the material
C. What problems it will solve for the reader
II. The Problem and Its Solution [Middle]
A. Problem statement
1. Where it occurs
2. Usual difficulties it presents
3. Other characteristics of the problem
B. Usual solutions
1. Common methods
2. Other methods
3. Why they fail, or are only partly effective
4. Why better solutions are needed
C. The new or better solution
1. How it is superior to older solutions
2. What its advantages are
3. Other reasons for its use
D. Applying the new solution
1. Under usual conditions
2. Under unusual conditions
3. Precautions in using the method
4. Other considerations in using the method
III. Getting the Most from the New Solution [End]
A. Personnel factors
B. Financial factors
C. Other factors

Departmental Features

I. Lead
A. Short, concise problem statement
B. Agreement or disagreement with existing methods, opinions, or other data
II. Main Thought of the Feature [Middle]
A. How, why, where, or when the method, idea, data, etc., can be used, changed, corrected, etc.
B. Main reasons for recommending or criticizing the item
III. Action Urged [End]

A. To secure better results
　　B. To give continued success

News Items

　I. Lead: What, Who, When, Where, Why, and How
　II. Detail of the Story [Middle]
　　A. Technical data
　　B. Business data
　　C. Personnel details
　　D. Consultants, engineers, architects, designers, and others connected with the story
　III. Future Expectations [End]

New-Equipment Items

　I. Lead
　　A. Name, type, capacity, and other general details of the unit
　　B. Principal advantages
　II. Construction Data [Middle]
　　A. Materials, model number, dimensions, and any other significant details of the unit
　　B. Performance characteristics
　III. Where to Obtain Additional Information [End]: name, address, and phone number, including a key letter or number, if desired

HOW TO USE THESE OUTLINES

There are three steps in the successful use of the outlines in this section: (1) Choose the outline that most closely fits your planned article. (2) Prepare a rough listing of items for the beginning, middle, and end of your article, as described above. (3) Insert items from your listing in the appropriate places in the sample outline.

After choosing the correct outline, you will probably encounter one of three situations: The outline may suit your article exactly, it may contain too many items, or it may contain too few. The first situation will rarely occur. If the outline has too many items for your article, simply insert all you have available. Then delete the extra ones from the outline. The extra items are not an indispensable part of the outline—they are suggestions for the elements of a typical article of one kind. These sample outlines have been prepared to cover as wide a range as possible.

Where the outline does not contain enough items to provide space for all those in your listing, try the following procedure. Insert all the items from the list that you can. Then, carefully, take the remaining items and insert them where they belong in the overall development of your article. No matter how you develop the story—through a logical progression of details, chronologically, or otherwise—you should wind up with a good outline, if you give it thought.

WORKING WITH EDITORS

Half the reward of technical writing is working with editors. As a group, magazine editors are alert, perceptive, informed. You are likely to acquire useful ideas from every editor you meet. But editors are busy. Unless you know how to work effectively with them, you waste time and miss valuable information. Let's see how you can improve your dealings with all technical and scientific editors.

Pick the Right Publication

One of the major complaints of editors today is that they receive too much unsuitable material. Many articles are submitted carelessly. And editors lose valuable time reading and rejecting poor material. So your first rule for making and keeping editorial friends is: Pick the right magazine before submitting any material. How? Here are three useful hints.

1. Study the Magazines. Learn what they are publishing. Get the last six issues of the technical, scientific, or business publications that run the type of article you plan to write. Study each issue, while your article is still in outline form. If you have the time and are interested, set up an analysis chart as shown in Figure 4.5. Enter the various items in the proper columns as you study an issue. List the number of articles in the issue, the number of pages, words, and illustrations used per article, the types of illustrations, the writing styles, the kinds of headlines, the types of articles used, the readers they are intended for, and the kinds of subheadings used.

	Magazine Analysis Form				
Name of magazine	Usual length of articles (Words) (Pages)	No. of illus. used/ Type of illus. used	Types of articles used	Publication frequency	Author occupation
Machine Design	800 — 1000 1000 — 1500 1500 — 2000 2000 — 2500 1,2,3, and 4 pages	2 — 3 3 — 4 4 — 6 6 — 8 Line drawings, photos; charts	Design procedures Product functions Calculation methods Graphical solutions Management	Weekly	Engineers, designers, stress analysts, college professors
Power	200 — 800 800 — 1000 1000 — 1500 1500 — 2000 2000 — 2500 ½, 1, 2, 3, and 4 or more pages	1 — 2 2 — 3 3 — 4 4 — 6 6 — 8 Photos; charts; line drawings	Plant stories Design procedures Operating procedures Maintenance procedures	Monthly	Plant and power engineers; designers; college professors

FIG. 4.5 Form for analyzing articles in technical and scientific magazines.

When you have done this for six issues of a magazine, compare the outline of your article with the tabulation. If your article is a type used by the magazine and your illustrations are the kind that appear in the issues, you are on the right track. The other items—style, length, and headline—can be tailored to suit the particular magazine. This is why you should pick the magazine while your article is still in outline form. You can alter the handling of your story more readily while it is in outline form than when all the copy is written and illustrations chosen.

2. Check for Other Magazines in the Field. If you know of only a few magazines in a field, don't assume there aren't any others. Research. Where? One excellent source of information about technical magazines is the *Directory of Technical Magazines and Directories*, published by Fairmont Press. Others are: *Government Scientific and Technical Periodicals*, published by Scarecrow, Inc.; *Ayer Directory of Publications*, by Ayer Press; *Writer's Market*, by Writer's Digest Books; and *The Gebbie Press All-in-One Directory*, by B. Klein Publications.

Once you have learned the names of all the magazines in the field covered by your article, get copies of them. Study and analyze the articles according to the plan in item 1 above.

3. Study the Magazine's Ads. Editors of most technical and scientific magazines usually have little to do with the advertising in their publications. But the editorial material they choose attracts certain kinds of advertisers. Thus, in studying the ads, you learn the extent of the field covered by the magazine.

If your article covers a type of product or service advertised in the magazine, there is an excellent chance of reader interest in your story. It is also likely that the editor will be interested.

What if no ads cover your product, service, or analysis? Don't be discouraged. There is almost as good a chance of your story being used if your material fits. For if your article covers part of the field served by a magazine, there will be readers for it.

Once you are reasonably certain the editor can use your article, contact the editor. But be certain your article is right for the magazine. If there is any doubt in your mind, wait a few weeks and think it over. Or get six more issues of the magazine and study them. You are almost certain to find definite preferences, as shown by the articles the magazine publishes.

Now you may ask two questions: "Why spend so much time studying the magazine before contacting the editor? Isn't there an easier way of learning the editor's requirements than studying six issues?"

The answer to the first question is that every minute you spend studying the magazine before contacting the editor is worthwhile because you learn a great deal about the editor and the publication. Few technical writers bother to study the magazines before writing. So you'll be a jump ahead. Your copy will more nearly meet the editor's needs, and you won't waste his or her time with unsuitable material.

As for the second question, there are easier ways to learn an editor's needs—but there is no better way than studying the magazine. Market data, published for professional writers, give a short summary of editors' needs. But such data are almost useless when compared to study of the actual magazine.

In the actual magazine you see what the editor publishes. You observe the length of story, the types of illustrations, the writing style the editor prefers, and the readers the articles are aimed at. Study the *masthead*—the listing of editorial personnel that appears near the front of the magazine—and you will know the names of all the editors on the magazine, and in some cases, their specialities. Also check the author's bylines on each article to get an idea of the writers' credentials. You never waste time when studying the magazines you'd like to write for.

How to Contact an Editor

There are five ways of contacting an editor: (1) Write. (2) Telephone. (3) Have your public relations expert call or write. (4) Ask your firm's advertising agency to call or write. (5) Speak to the magazine's advertising space salesperson, asking him or her to give your material to the editor. Of these five only the first three should normally be used. If at all possible, steer clear of dealing through an advertising agency or space salesperson. Now let's take a closer look at each of these approaches.

Write the Editor. Check the magazine's masthead. See if an editor is listed for the specialty covering your article. If one is, address your letter to him or her. When there is no editor listed for your specialty or if the specialties of the editors are not listed, address your query to the chief editor.

Generally, avoid querying the chief editor, unless the magazine's editorial staff is extremely small (three persons or less). In a larger staff the chief editor will turn your letter over to an associate or assistant editor handling your subject. By going directly to an associate or assistant editor, you can save time and have a better chance of having your article accepted. The reason for this is that the associate editor is usually closer to the field and is more actively concerned with actually securing manuscripts than the chief editor. Many chief editors today are more concerned with policy decisions and other administrative duties than they are with manuscript solicitation.

Make your letter short, concise. Give the editor: (a) the subject of your article, (b) the facts on why the magazine's readers would be interested in your story, (c) your name, and (d) your company affiliation and job title.

Concentrate your effort on the reader interest your article has. Tell the editor exactly why his or her audience will want to read the piece. Point out the new features you cover or the old problems you solve. Remember: every good editor has readers in mind at all times. If you have a story that appeals to the readers, the editor will spot it instantly.

Do not try to sell your article forcibly. You will only build resentment if you try to pressure an editor. Let your story sell itself. To do this, review the story carefully before contacting the editor. Pick out the points of major interest. Tell the editor what these are. If the ideas merit publication, the editor will tell you what steps to take in submitting your article. Follow the steps exactly.

Here is an example of a good query letter:

Dear Editor:

In the last 30 months, 16 commercial office buildings in Illinois have converted to evaporative cooling. Why? Because they're saving anywhere from 10 to 45 percent on their fuel bills. How? By smart engineering of the installation, intelligent operation, and careful maintenance.

Complete statistics for these buildings are available. These statistics include installation, operating, and maintenance costs. Photos and drawings of the installations are also available. The data and illustrations tell a forceful and informative story that will make engineers in the cooling and air-conditioning fields stop to examine their thinking.

Would you be interested in this story? If so, what length would you prefer? How many and what kinds of illustrations would be best?

I am a design engineer with the ABC Engineering Company and have published a number of articles and papers in various technical journals.

Please call or write me at your earliest convenience.

Call the Editor. Check the masthead as in preparing to write a query letter. If you find an editor handling your specialty, call and ask for that editor. Otherwise, call the editorial department and ask for the editor handling the subject matter of your article.

When you have the right editor on the phone, tell him or her concisely and quickly the essential facts needed to judge your idea. These facts are the same as those listed above for query letters.

Have Your Public Relations Representative Contact the Editor. This can save you time and energy. But be certain to supply the PR person with complete, accurate details. If the PR person does not have an extensive technical background, present the facts about your story as simply as possible. Avoid big words, long explanations, and complicated questions. Instead, summarize your story as in item 1 or 2 above. A smart editor will spot a good story within moments after a PR representative begins to talk.

Have your firm's advertising agency contact the editor. Use this method only when you cannot use one of the first three. Give the agency a concise summary of your story. Direct them to have the editor contact you if he or she wants more information.

This scheme can work well if the agency person contacting the editor forgets advertising completely for the time being. Editors will not buy a story just because there is an ad tie-in. Many editors will be extremely wary of material from an ad agency. But editors also recognize that having the agency make the initial contact can help you. So most editors will listen, if all mention of advertising is kept out of the conversation.

Speak to the Magazine's Space Salesperson. If a space salesperson calls on you or someone in your firm, get to know the salesperson. While space salespeople usually have nothing to do with editorial matters, they know the editors personally. Most space salespeople can tell you the specialties of their editors. You can use this information when calling or writing an editor.

Tell the space salesperson about your article. If it sounds interesting, he or she may send it to an editor on the magazine. This does not mean your article has been accepted, of course; the editor is the only one with authority to accept an article. But it does mean that your article will go directly to the editor concerned, probably with a short note from the salesperson.

What to Show an Editor

Suppose, after a call or a query letter, an editor asks for some sample material for your article. What should you show the editor? There is no simple answer to this. But most editors will be satisfied with (a) your story outline, (b) a few of your illustrations, tables, or charts, (c) a short (one- or two-page) excerpt of the article, or (d) any other related material.

However, take care to see you don't overpower the editor with material.

Choose a few of the best items you have. All the editor is trying to get is a quick idea of how you work and the quality of your material. If your presentation shows forethought and consideration, your story will almost always sell. But always keep in mind that editors are busy and they think of their readers first.

Getting Along with Editors

Here are seven valuable rules for dealing with magazine editors. Keep the rules in mind and you will seldom go wrong.

1. Think Only of the Editor's Readers. Check every article idea for reader interest. And make sure to tell the editor why the magazine's audience will want to read your story. Explain why the idea will appeal to readers, what advantages they will get from reading it, etc. Use statistics and other facts, where possible, to show your article's importance.

2. Give an Editor an Exclusive for Every Feature Article. When you decide which magazine is best for your story, go all out to sell the editor on its worth. But don't try to place the story in more than one magazine at a time. If you do, you may antagonize an editor or two. And the next time you come around with a story, you'll find that the welcome mat has been mislaid.

Therefore, give the editor an exclusive on your story; that is, don't submit your story anywhere else until the editor gives you a definite turndown. Generally, if the editor accepts your article, don't submit it elsewhere, before or after the magazine publishes it. Give the editor sole use of your story, unless he or she does not want the sole right to use it. When giving an editor the exclusive use of your story, you assure him or her of scooping the competition. Every active editor loves to do this, particularly with an outstanding article.

There are some exceptions to this rule, however. You may sometimes submit a manuscript to two magazines simultaneously, but only if the two publications have different audiences, and you inform the editors beforehand. For example, a magazine whose sole circulation is among pollution control engineers may accept a system-description article on a waste disposal facility in northern Idaho, even when an Idaho community newspaper or local general-interest magazine is publishing the same piece. The editors will only accept the article if their publications accept "simultaneous submission" rights and when they are sure that publishing the article will not cause competitive damage of some kind. For the writer, the simpler and easier procedure is to avoid simultaneous submission. If you do consider this path, always tell the editors about it.

Note that the specific rights that you give to an editor will determine what you can and cannot do with your article once the editor accepts it. For example, if a magazine buys "all rights" to your article, you forfeit the right to someday offer the material to a different publisher in book form. On the other hand, you can offer an editor "first serial rights" upon acceptance of the article. This approach basically grants the magazine the right to publish your article the first time; following the article's first publication, other rights—such as sale of reprints and books—belong to you.

During the initial stages of your dealings with editors, play it safe. Give the editor an exclusive, as noted above. As you build your reputation and editors learn to appreciate your value as a technical writer, you may want greater flexibility in publication rights. Whatever rights you seek, be sure to get a written copy of any formal agreement you make with the publisher. Consult one of the magazine directories listed earlier in this section for up-to-date information on the specific rights that editors of various technical and scientific magazines seek.

3. Supply Good Photos. Hire a professional industrial photographer, or a newspaper or magazine photographer, to take photos for your article. The results are worth every penny you pay. And when sending the photos to the editor, pack them so they cannot be bent or folded. Never write directly on the photos; you may, however, mark an identifying number on the back of the photo with a felt-tip pen. Type the caption on a sheet of paper and attach it to the back of the photo with rubber cement, or make a separate list of captions.

4. Make Clean, Neat Drawings. Most large-circulation magazines employ professional artists to prepare finished drawings for publication. But you generally should provide the drawings from which they develop the finished artwork. Put each drawing on a separate sheet. Have all the lettering legible; don't crowd the drawing with it. Number each drawing, and either type or glue its caption on.

5. Type Your Text Neatly. Don't think about the form in which your published article will appear—just type everything in your story double-spaced. This includes text, captions, footnotes, tables, quoted material, bibliography, equations, and mathematical material. A copy editor will mark your manuscript to specify publishing details like type size, column width, etc.

Use plain white bond paper. Never use fancy colored paper for articles; editors like white—it's the easiest to read. If you cannot type well, pay a typist to prepare a neat, clean typescript of your article. Keep one copy for yourself. If the editor wants two copies of your article, send the original and one copy.

6. Take the Editor's Advice. Most editors will not know your technical field as well as you do, but they know what their readers want. If an editor is kind enough to tell you how to revise or rewrite your article, listen carefully. And do as the editor says. You will probably improve your article and writing skills by doing so. Remember: Article writing is not a far-out, artistic profession. It is a workmanlike occupation in which you give a reader useful, important facts in an engaging and easily understood way. So listen to the editor—he or she knows how to put ideas across.

7. Be Friendly. Treat editors as fellow professionals. Every editor is concerned with getting the best articles possible for his or her readers. If you can supply good articles, editors will keep you busy day and night.

Use good taste in dealing with editors. You can treat an editor to lunch if the publishing firm allows it. But never offer an editor a gift of any kind. He or she will probably refuse it, embarrassing both of you. The only gift an editor wants from a writer is a good story. Many publications show their appreciation for a good article by paying you for it. Check the magazine directories for honorarium and pay rate information.

How to Irritate Editors

You can learn as much from what irritates editors as you can from how to get along with them. So here are seven ways to irritate an editor. Be certain that you are never guilty of any of these!

1. Threaten to Cut Ad Space if Your Article Is Rejected. As noted, most editors have nothing to do with the advertising in their magazines. Many writers don't understand this, so if their article is slighted in some way, they scream that their company will cancel its ads in the magazine.

Don't try this hoax. It won't work. Why? The editor will probably tell you to go right ahead and cancel every ad for the next hundred years. This still won't change the editor's opinion of the article. An article either stands on its own merits, or it doesn't. Buying every ad page in a magazine will not change the usefulness of an article for the readers.

2. Crumple Your Photos, Drawings, or Copy. Mangled illustrations or copy that is full of typographical errors can ruin an editor's faith in people. If you spend a week or more writing an article, take some time to make sure your material is neat, clear, and wrapped properly for delivery.

Protect all photos with corrugated board or double-thick cardboard. Your postage bill will be a few cents higher, but your photos, drawings, and typescript will arrive safely. And to be sure your material is returned if it is unusable, enclose a stamped, self-addressed envelope.

3. Pester the Editor for a Publication Date. When an editor accepts a good story, he or she plans to run it as soon as the magazine has space for it. So calling the editor every few days or writing nasty letters won't help much.

Every magazine editor has problems. Issue schedules, in which the content of the magazine is planned, are a major problem. Some articles may be too long. Others are too short. Or there may be three articles on the same subject. The schedule must be juggled to make the best possible issue for the readers.

So if publication of your article is delayed, be patient. You will make lasting friends by showing a little understanding of the other person's problems. And the next time this editor accepts an article from you, watch how soon it runs. Patience pays off in article writing.

4. Scream about the Rewrite of Your Golden Words. Some business papers rewrite every contributed article they publish. Reason? To get the story across to the reader faster and more easily.

Don't feel discouraged, then, if your article is rewritten. The published version is probably better than what you submitted. Even if the editor makes a factual error, be gentle in pointing it out. Remember: When agreeing to publish your article, every editor reserves the right to rewrite the material as necessary. If you don't want your material rewritten, submit it to a publication that doesn't rewrite.

5. Bother the Editor for Copies of the Magazine. Many new writers bother editors incessantly for copies of the issue in which their article appears. They start three weeks before publication and don't miss a day until the magazine is out. Again, be patient.

If possible, learn who handles tear sheets and reprints in the editorial office. This person is often an editorial assistant or administrative aide. By being considerate, you can usually arrange to get a copy of the issue as soon as it comes out. And you won't be bothering the editor.

6. Hound the Editor for Payment. Some magazines pay authors on acceptance of contributed material. Others wait until publication. Find out which method "your" magazine uses. Then be patient.

Article payments are seldom delayed. But occasionally the accounting department will be overloaded. Your payment may then be delayed a week or two, perhaps longer. Give the check at least two months, after publication, to arrive. If it doesn't come by then, call the magazine. Be tactful in any inquiry you make; everyone makes a mistake now and then.

7. Make Extensive Changes in Your Reprints. This, your last way of annoying editors, can make editors wish they had never learned to read. Generally, don't make changes in reprints other than correcting typographical or factual errors.

Typos can't always be helped. The best of printers and typesetters make them. Editors will usually not object to correcting these, free of charge. Factual errors that are the editor's fault can also be corrected. But these are usually few.

Errors that you as author are responsible for make editors unhappy. And if you want to make text changes for personal reasons or reasons of company policy, you're almost certain to run into resistance. The only time you can make substantial alterations in reprints is when you own all the rights to the article once the magazine has published it. This means that you would need to have written documentation that the magazine had not bought all the rights to the article.

What Editors Can Do to Your Articles

When you submit an article to an editor you should be ready to allow the editor to alter it to suit the magazine's readers. An editor can: (1) rewrite to suit the magazine's needs, (2) delete parts of the text or illustrations and tables, (3) ask detailed questions about the material, (4) publish your material without clearance from you (few do this, but they can, if they wish), (5) ask for exclusive rights to your story, and (6) reject your material.

Therefore, be sure your story says what you mean, before you submit it. Then you will run into fewer problems with rewrites or cutting. Generally, you will find that the more you work with editors, the fewer the changes they will make in your articles. Your writing will improve, and your choice of article length will become more accurate. But be ready to accept the changes the editor recommends.

Article Length

When you contact an editor, he or she may tell you the number of words your story should contain. However, many editors today do not specify article length in words. Instead, they speak of "a one-pager," "a two-pager," "a three-pager," etc. The reason for using these expressions is that editors continually think in terms of the published article. With the large number of illustrations used in modern articles, it is difficult to set a word limit.

From studying back issues and making notes in a tabulation like Figure 4.5, you will get a good idea of how many words there are in articles of various published lengths. As a guide, you can also use the tabulation in Figure 4.6. But remember: It is only a rough guide. The number of words or illustrations may vary as much as 50 percent, plus or minus.

Try to decide approximately how many words, illustrations, and tables you will have before you start to write. Count each table as equivalent to an illustration. Count the number of words in the article after you finish writing. If you have twice as many words as you planned, cut until you are closer to the original estimate. But if you are only a few hundred words over, do not bother to cut, particularly in a long article. The editor will do this if necessary.

Number of Published Pages	Approximate Number of Words	Number of Illustrations
½	300-400	1 or 2
1	800	1 or 2
2	1,500-2,000	1-6
3	2,500	1-6
4	3,000-3,500	1-8
6	5,000	1-12
8	5,500-6,500	1-15

FIG. 4.6 Table for estimating article length. (Based on a magazine having an 8.5- by 11-inch trim size and three columns of text per page.)

HOW TO CHOOSE SUCCESSFUL ARTICLE TITLES AND LEADS

Since the editor starts reading at the beginning of your article, its title and lead must be good. These vital parts of your article must also be good if they are to stop hurried readers as they flip through the magazine. Remember, your article is competing for attention with every other article in the issue. So learn now how to write effective titles and leads.

Working Titles Win

Article titles (or "heads," as they are sometimes called) should (1) convince readers that they must read your article and (2) help them move into the text smoothly, with minimum effort. The title must "work" for you.

If you stop to analyze good article titles, you will find that they are usually brief, accurate, and often in the active voice. Don't depend on the subject matter of the title to give the punch needed. Besides what the title says, it's important how the title says it.

Good titles use sentences more often than phrases. They emphasize newness, or "nowness." The good title combines clarity, detail, and vividness. Benefits mentioned in a good title are specific—the reader doesn't have to hunt for the benefits in the text. Many good titles tell a story, or promise one.

And many good titles are the W type—they use why, when, where, who, or what. Or they imply the use of or the answer to one or more of these words.

How to Write Winning Titles

Use a label type of head when first thinking of your article. Thus, if you are doing an article on plywood adhesives, use these two words to identify the article in your mind. They'll speed your note making and help organize ideas logically. But they won't make a very effective title. Why not? Because they are just a label. They do not tell the reader anything more than the subject of the article. There is no incentive for reading the story, unless plywood adhesives happen to be extremely important to the reader.

To improve this title, let's try putting it in the active voice. You can do this before you write your article or after it's finished. Some technical writers prefer to wait until they finish the article. Then if there has been any change in the slant of the article, the title can be altered to suit.

But if you start with a good working title, it can help slant your thinking during writing. Live with the title while you are writing your article, and you may find a number of new ways to improve the title. So take your choice. We will try to start with a good title.

Now we know that the elements of this title are the two words "plywood adhesives." Next, as you have seen, you should try to get some newness or timeliness in the title. Also, the title should be as nearly a sentence as possible.

Let's say the adhesives you will discuss in the article are stronger, have longer life, and are cheaper than others. That immediately suggests this title:

NEW PLYWOOD ADHESIVES ARE STRONGER, LONG-LIVED, CHEAPER

Working on this title you might come up with:

SAVE WITH NEW, STRONGER LONG-LIFE PLYWOOD ADHESIVES

Or you could phrase it:

LONG-LIFE PLYWOOD ADHESIVES OFFER EXTRA STRENGTH, SAVINGS

You could find a number of other variations. The important point to see here is that any of these three is better than the label type of title you started with. These three promise readers something for their time. And if a reader does no more than read the article title, he or she learns something.

Once you have worked up three or more titles for your article, make a list of them. Then try these titles on your friends. Keep a record of their votes. Put the titles aside while you write the article. When you come back to the titles, see if your opinion agrees with the votes. If it does, use the winning title; if it doesn't, try to rework the titles. Use the one that appeals most to you.

Examples of Good Titles

Here are 14 examples of good article titles. Note how many of them incorporate the principles we discussed above.

New Communications Modules Boost Accuracy, Productivity
How to Install Laminated Glass for Optimum Protection
Want Increased Data Security?
Nine Key Factors Round Out Apartment Energy Efficiency

How a Five-Person Team Brought an Old Mill Up to Date
Modified Gassification Process Reduces Radiation Hazards
Who Says Technicians Can't Sell?
Trouble-Free Switches Test Old Electrical Traditions
IDEAS: Where Do You Find Them?
Shop-Built Equipment Repairs Tire Damages
Today's Space Station Uses: From Astronomy to Defense
Test Your Design IQ
Research Conferees Set New Directions in Management Policy
Use These Investment Statistics to Calculate Energy Conservation

You can write good titles for all your articles. But it takes time and practice. From now on, study the title of every article you read. Note how good titles "hook" you, encouraging you to read the article text.

Don't be afraid of using subheads in your article. Editors sometimes refer to these as *kickers* because they can do much to alert readers to the value of an article. For instance:

> Mile-a-minute typists find that...
> New Keyboards Up Productivity

Or use a short kicker:

> A case study of...
> Automated Energy Management

Here's another good one:

> New Jersey Refinery will be first plant where...
> Remote Electronic Tank Gaging Drives System Controls

Another tool that magazines often use to catch reader interest is the blurb. *Blurbs* are usually longer than subheads, and their purpose is roughly equivalent to that of an abstract in a report or paper: blurbs summarize the text. Most technical and scientific magazines place a two-, three-, or four-line blurb at the top of each feature article. Generally, you need not be concerned about writing your own blurbs for articles, since most editors write them themselves. Here are two examples of effective blurbs:

> With skyrocketing telephone rates, it is time to take a look at a communications technique that has been overlooked for many years.

> To specify the right cooling system for your building, you need to know the basic differences between various types. Here's a review of the generic varieties, their limits and their range of usefulness.

Good titles, subheads, and blurbs never come easily—not even to experienced writers and editors. So if you sweat over a title for 30 minutes, don't become discouraged. Others do it all the time.

After the Title Comes the Lead

Like the title, the lead paragraph must attract and hold the reader. If it doesn't, the reader will flip the pages to the next article. If you want to be read, be certain to make the lead paragraph worth the reader's time. How? Let's see.

While you can devise an infinite number of leads for your story, you will find that most trade magazines use one of nine types for many of their articles. These types are: (1) summary, (2) descriptive, (3) problem solution, (4) comparison, (5) question, (6) purpose, or objective, (7) interpretive, (8) news, and (9) historical.

No matter which type of lead you choose, try to give the reader the key facts about your story in the first paragraph. If this makes the first paragraph too long, use two paragraphs to present your facts. Some leads are as short as one sentence; others run as long as several paragraphs. Leads in most technical and scientific articles are fairly short. Ordinarily, don't use more than two paragraphs because your reader may lose interest. Now let's look more closely at the various leads.

Summary. As we noted earlier, the summary is one of the most popular types of article leads; it is also one of the simplest to write. To write it, list all the pertinent facts in your story in the order of their importance. You will probably have too many facts for one or two paragraphs. So pick the three or four facts that are most vital to the main point of your story. In a summary lead, sentences are typically short, yet are loaded with information. The reader can decide from the first few sentences whether reading further will be worthwhile.

Descriptive. Here you can pack the main thought of the article into one sentence. This is often the first sentence in the article. Typically, the descriptive lead paints a picture in the reader's mind, thus laying the foundation for what follows. You can use a definition or a statement of fact to arrest your reader's attention. Or you can give tersely the details of a scene, process, product, trend, a prediction of future events, or any other subject. Your descriptive lead can dazzle readers with "color," and grab them with facts. Be sure your description is clearly pertinent to the point of your text.

Problem Solution. This lead is ideal where your article is concerned with the solution of a basic problem. You will find it popular in articles about new systems or structures, or the modernization of old ones. The problem-solution lead may state a problem directly or indirectly. Almost invariably, it notes the chief advantages secured from a specific solution, and it does this within the first paragraph of the article.

Comparison. To use this type of lead you must have a story in which there is a choice between two or more methods, devices, or schemes. Often you will compare the familiar with the unfamiliar. This sets the scene for your readers, helping them move easily into your story.

Question. Use this lead when you can summarize your article in one or two short, simple questions. By keeping the question(s) simple, you make the lead easier for readers to follow. And if you make the questions appeal to your readers' interests, they will be eager to read further as you explain the answers.

Purpose or Objective. Here you tell the reader what your article is about. This type of lead sometimes gives writers trouble when they prepare it. Why? Because if you have not thought the story through thoroughly before writing, it is hard to say exactly what the story is about. And not every writer takes the proper steps in preparing to write. To avoid difficulty with this type of lead, make certain you have pinned down what your aim is in writing the story. Do this while you list your article ideas and develop your outline.

The typical purpose or objective lead is concise. It simply states what the article is about and may begin "This article describes..." or use a similar direct phrase. Along with the article title, the lead tells readers exactly what to expect from the story.

Interpretive. Here you orient the reader's mind to your story. You may be covering a new development in his or her industry. Or you may be presenting conclusions based on a study, or predicting the future based on existing facts. You can also use this lead to connect a seemingly unrelated subject to a current problem. The interpretive lead is particularly well-suited to an article in which you want to state your opinion or make an observation about the item that you concentrate on throughout the text. This type of lead is usually longer than a purpose or objective lead, often requiring more than three sentences to make its point and give the facts upon which the point is based.

News. The news lead resembles the typical leads you see in many newspaper articles. It makes full use of the W's—why, when, where, who, and what. Use it to tell something that the industry in general did not know before. This lead is excellent for significant first installations and major engineering, scientific, or technical achievements.

Historical. As its name suggests, this type of lead introduces the history of the item discussed. The historical lead is sometimes highly effective. But be careful when choosing this type. You could use it for almost every story, if you wanted to. But most readers are not interested in a historical lead because they seek results—not bygone events. Further, the historical lead tends to read slowly. So choose the historical lead mainly for feature articles that tell of the evolution of a device, system, firm, individual, or idea.

Good Leads Score High

Whenever you write, aim at leads that have interest-getting ideas. Use short, accurate sentences in your lead. If you do, you will catch many a reader, as well as the editor. More importantly, you will write your article better. Why? Let's see.

Good leads (1) gear your mind to your article, (2) give you a picture of where you are going, (3) speed your writing, and (4) give you more momentum to finish your article.

The accurately written lead gears your attention to the essentials of your story. It tells you exactly how to write your piece, because in most articles the body merely fills out the lead. This is why many professional writers spend a proportionately greater amount of time writing the lead than they do writing the body of the article. Remember: There is no room for wasted words in today's technical article. A good lead gets directly to the point of the story. Further, it sets your thinking in the right direction at the start, so your writing goes faster. And it is less painful along the way.

Just as important is the stored energy or momentum that an effective lead builds in your mind. With the essence of your article neatly stated in the lead, you can dive into the detailed part of your story with complete enthusiasm. Don't ever overlook the importance of an enthusiastic belief in your article. It gives an urgency to your writing that readers recognize and like. Your story moves swiftly and accurately from the lead to the body and to its logical end.

KEEPING THE MIDDLE LEAN

An engaging lead gets your reader into your article. But you will not keep the reader there unless the rest of your story is interesting. To hold today's reader, you must present your story quickly and succinctly. So you must keep your article lean in the middle. Here's how.

Five Rules for Keeping It Lean

There are five rules for making the body of your article engaging. These are: (1) Be concise. (2) Stick to your outline. (3) Keep your story in mind. (4) Write a little each day. (5) Finish soon.

1. Be Concise. The modern technical or scientific reader wants facts in a hurry. He or she is usually a busy person and cannot waste time on an article that wanders. So move directly from your lead into the most important part of your story. Make certain your reader will understand clearly what the article is doing. Don't let anything remain a mystery, or the reader will become frustrated and stop. To get the most clarity from your paragraphs, use good transitional devices. These are described shortly.

2. Stick to Your Outline. If you start to make major changes in your outline after you have written the lead, you may later have to rewrite the lead. If you have used one of the outlines in this section, you can be reasonably sure that it will need only minor changes. Follow your outline. It will help your writing, and your story will flow more smoothly.

3. Keep Your Story in Mind. Some writers refer to this as being "in focus." As long as you keep your story in mind there is no danger of wandering from your subject. It is wandering that causes a story to slow down and bore the reader.

You may find it helpful to write a short summary of the main points of your story. Keep this in front of you while you write. Read it frequently, particularly before you begin writing each day. With a little mental discipline you should be able to stay in focus at all times while you are writing your article.

4. Write a Little Each Day. Never try to write an article unless you have enough time to finish the writing within about four weeks. Then you will be certain to keep to your story and follow your outline. If you allow several days to pass between writing sessions, you will find that it takes longer to get started again. This wastes your time and energy. By writing a little each day, even only a page, you keep your article in your mind and you can write it better. Also, by writing each day, you build good mental habits of concentration and willingness to work. After some experience you learn that no matter what you are writing, the best way to finish is to work on the project each day. Your thoughts will flow better, and the finished job will have more unity.

5. *Finish Soon.* Don't drag out your writing job. By finishing quickly, you give your writing a certain urgency that readers recognize. Your readers will realize that you are trying to tell the complete story in as few words as possible. They will follow the article with interest because they know that they will obtain enough information to repay them for the time they spend in reading it.

Use Body Builders

Your lead must catch your readers' interest. The body must build from the lead and give the readers all the facts the lead promises. To do this, the body of your article must be strong and full of useful facts. If it isn't, the readers will feel they wasted their time reading your article. The next time they see your byline, they may skip the article. So never shortchange your readers. They will resent it more than you think. Some readers may even write the editor to complain. This can seriously damage your future relations with that editor.

If you write the body of your article well, your audience will read it faster, remember it longer, understand the subject better, and enjoy the article more. Let's see how you can improve the body of your article.

To write the best article body possible, you should: (1) know your audience and purpose, (2) develop your subject from the lead, (3) cite cases and give examples, (4) use illustrations to tell part of your story, (5) compare the new with the old, and (6) cite advantages. You may not be able to use all of these in one article. But if you use three or more, you can be certain that your readers will benefit. Let's take a closer look at each of these items.

1. *Know Your Audience and Purpose.* As with other kinds of technical and scientific writing, your first thought before you start to write the body of your article should be, "Who will read what I write?" Remember at all times that you are writing for individuals—engineers, scientists, technicians, students, or other people in your field. Decide exactly for whom you are writing—try to form a mental image of your typical reader. Keep this person in mind while you write. Use the words he or she does; choose illustrations that use the same symbols he or she uses; tell your story in the typical reader's terms. Then you will be assured of wide readership and appreciation.

Write with a purpose. What are you trying to do in the body of your article? If the answer to this question is not clear in your mind, you can bet the reader will not understand the article or its aim. So write your purpose on a slip of paper, in as few words as possible. Keep this paper in front of you while writing. Make every paragraph of the body of your article a step to helping the reader understand the purpose in your writing.

2. *Develop Your Subject from the Lead.* Bridge the gap from the lead of your article to the body with a good transition. The simplest transition is a word; but a good transitional sentence can be highly effective, especially when moving from one paragraph to the next. Typical useful transitional words are *and, but, so, in-*

stead, then, for, thus, and *however.* Study examples of articles in your field for additional transitional words.

Other transitional devices include a contrast, an exception, a contradiction, a review of the previous text, steps in a procedure, a chronological progression, and a series of related ideas.

3. *Cite Cases and Use Examples.* Empty words are the surest way to bore your readers. You can write 10 pages telling them how a device saves money. But these 10 pages will not be anywhere nearly as convincing as one paragraph describing an actual installation of the device and stating the number of dollars it saves in a given time. Specifics like these build reader interest by instilling your article with substance and color. Therefore, show your readers by actual cases what you are trying to prove to them; don't merely tell. If you have the facts available for two or three cases, cite each case. The more cases you can cite, within reason, the better your chances of convincing your readers. Generally, do not cite more than about five similar cases. If you do, you may overwhelm readers and lose them.

Try to choose cases that are as different from one another as possible. Then your readers will see that the principles you are discussing are usable in a number of varying circumstances.

Suppose you do not have even one case to cite. What then? Explain to the reader that no actual cases are currently available, and go on to tell them what the results would be if an actual case were available. Set up a typical case for a mythical company, plant, city, etc. Then cite the savings, advantages, or other assets you are trying to explain. This method is almost as convincing as actual cases.

Along with cases, or instead of them, you can use worked-out examples, showing readers how to calculate results, how to calculate factors in a problem, or any of a number of other items you are discussing.

4. *Use Illustrations to Tell Your Story.* More and more, today, people shy away from words. So use illustrations instead of words wherever you can. Your reader will be more interested in the article, will give it greater attention, and will be more satisfied by the appearance of the article.

While you can use an illustration or two as part of your article lead, the bulk of your illustrations will relate to the body of your article. Therefore be sure to refer to your illustrations in the body of the article. Try to write the article so that you refer to an illustration every few paragraphs. This helps you carry your readers' interest and adds unity to your article.

5. *Compare the Old and the New.* Readers understand contrast. Show them how the old system or plant was inefficient and wasteful as compared with the new one, for example. Every reader will understand immediately why you prefer the new to the old.

6. *Cite Advantages.* Today's readers want to know how they will profit by using the methods or devices you suggest. While you can spend many pages telling them what the advantages are, a few examples will be far more convincing. So cite advantages and then use actual cases or examples to prove your statements. The body of your article will be much stronger if you use these methods to convince your readers. By citing several cases or examples, you can often move the reader to investigate further. And this, after all, is the best proof that your article has helped its readers.

Write Clearly and Effectively

Up to now we have discussed only the factual aspects of the body of the article. The reason we did this is that your facts must be clear and presented to advantage if your reader is to profit from reading your text. But all the facts in the world will not convince your readers unless your presentation is clear and effective. Let's see how you can improve the readability of the article body.

Here are seven basic rules for making the body of your article easier to read: (1) Weed out abstract words. Instead, use concrete and familiar words. (2) Beware of pomposity. (3) Slash wordiness. (4) Interlock your ideas for unity and coherence. (5) Choose action verbs. (6) Give the right emphasis to your ideas. (7) Vary your style.

1. Avoid Abstract Words. Here are eight abstract words: *alleviation, attribution, collusive, equivocation, inadmissible, modicum, nonpareil,* and *subservience.*

Try to define each of these words. You will probably find that it is difficult to define most of these words in terms of concrete symbols. Abstract words like these bore a reader. Your reader must stop and define each of these words mentally while reading it. On the other hand, the body of your article will move much faster and the reader will acquire more information if you use words that are familiar and easy to understand. Words that represent visible, tangible materials help you express your ideas clearly and quickly.

Of course, there may be times when you cannot use a concrete word instead of an abstract one. But try to avoid the abstract word as much as possible.

Here are some typical examples of concrete words: *plutonium, turbine, planet, meter, transistor,* and *antenna.* Note how each of these words has a definite meaning when used in the body of your article. So be specific; use as many concrete words as possible, to help your readers.

2. Shun Pomposity. Technical and scientific articles, by their very nature, tend toward using long words. But you can still avoid pomposity and make your article easy to read. How? By eliminating all unnecessary technical terms and phrases. If you want to impress your readers, the best way is to express your ideas so clearly that readers understand them on the first contact. This means using only those technical terms that are necessary to convey your message. Readers will finish your article with the feeling that its author really knows how to express the subject in written words.

Pomposity can turn a reader against an author and the author's subject. If the reader does not feel at ease, he or she may give up in the middle of your article. And since this is where most of your important facts are located, you lose the full effect of your writing.

3. Slash Wordiness. Be succinct. Technical writing demands careful construction, for the same reasons that a car or boat demands proper design. Careless design can make a car or boat too heavy to carry passengers economically. Many passengers would refrain from traveling, rather than pay the extra cost. Likewise, readers of technical and scientific articles won't stay with a story that overloads them with useless words.

Make every word count when you write the body of your article, just as you do when you send a telegram. And don't be afraid of using too few words. For if you find that the body of your article is too much like a bare chassis, you can always add illustrative examples and expand your text. This is much easier to do than cutting out words.

4. *Interlock Your Ideas.* Give unity and coherence to your article by relating one idea to another. By using subheads, you can often bridge the gap between two or more ideas. Illustrations and tables are also useful in relating your ideas to one another, as we have shown. Examples of how you can interlock ideas with subheads appear in this section and in articles in the technical magazines cited earlier.

5. *Choose Action Verbs.* Words like *turn, press, touch, connect, lock, grind,* and *switch* suggest action and movement. But words like *maintain, exist, remain,* and *seem* are static. Rarely if ever do they stir readers or interest them.

So whenever possible, choose action verbs. Your writing will move faster, and you will need fewer words to tell your story. This is extremely important in the body of your article because it is here that the reader may begin to tire.

6. *Emphasize Important Ideas.* You can put key points in your article lead, but never leave your most important ideas out of the body of the article. Why? For this reason: Your reader may never read beyond the start of the body of your article. He or she may tire, be interrupted, or have the habit of not reading beyond the first page of an article. If you leave your most important ideas for the end, such a reader will never see them.

Further, in the rush to meet a publication deadline, the editor may need to cut part of your article to fit in a given space in the magazine. If you leave vital facts until the end, there is more of a chance that they will be cut accidentally since editors expect the most important data to be toward the front of the story.

As with technical papers and reports, try to arrange your ideas in an order of descending importance. Make your most significant idea the first part of the body of your article. Follow this with the second most significant idea. Continue this way until you reach the least important idea. This should be near the end of the body of your article.

This arrangement may seem to be extra work, but it gives you a better chance of getting your most important ideas read. And that, after all, is why you are writing your article.

7. *Vary Your Style.* Keep your readers awake. Vary the sentence and paragraph length in the body of your article. If all your sentences and paragraphs are about the same length, the words have a lulling effect on the reader. He or she may begin to drowse.

When you use a long sentence—say 20 words or more—try to follow it with a short sentence. Do the same for paragraphs. Generally, try to have at least three short sentences and one or two short paragraphs on each double-spaced typed page. But avoid choppiness. Sentences and paragraphs become choppy when one short sentence or paragraph follows immediately after another. Here is an example of a group of choppy sentences.

> First, take the plug from its socket. Measure the gap. Check for carbon deposits. Observe the wear of the plug points. Then clean it. Finally, replace it. Connect the lead wire. Work with each plug in this manner.

Note the stop-go effect of this writing. If you want to express thoughts in a staccato way, list them as a series of steps with a number or letter in front of each. But if you do not want to make a list, then rewrite your sentences. The above instructions might be rewritten as:

To make sure that the cylinders are getting a spark, you should remove the plugs and examine each one closely; then you should clean them, set the gap, and replace them. But take one at a time. Otherwise, you may switch them and possibly delay the job of finding the weak cylinder.

Another way of varying your style is by personalizing your writing. Get your reader into the act. To do this, use personal and active words. Typical personal words are *you, your, hold, turn,* and *write.* Words like these imply that the reader actually takes part in the work or procedure you are describing. He or she can follow it much more easily when it is expressed in these terms.

Make Your Conclusions Clear

Most technical magazines today do not give much space to conclusions at the end of their articles. The reason for this is that many editors prefer to state the essence of the article in the blurb or lead paragraph. But in your writing, it is best to use a conclusion for your article. The conclusion rounds out your thinking. If the editor wants to transfer your conclusions to a blurb, it is easy to do so. Here are some hints on writing your conclusion.

Write your conclusion so it summarizes the body of your article. In a short article one sentence is enough as a conclusion. In longer articles you may need a paragraph or more. But no matter what length you choose, don't get windy. The place for details is in the article body.

So state your conclusions and stop. Be succinct, specific, straightforward. Readers will appreciate the effort you devote to making your conclusion short and direct.

If you wish to include a bibliography or references for your article, insert them immediately after your conclusion. Then your readers can use this material or ignore it, depending on their needs.

Since the end of your article is the last chance you will have of influencing your readers, write the conclusion so that your strongest point will be freshest in the readers' minds when they finish your piece. You will often find that the best conclusion is simply a restatement of the first or second sentence of your lead. In many articles you will be able to restate the ideas from your lead in fewer words than you used in the lead paragraph.

In a long article—say six or eight published pages—your conclusion should be comprehensive. Summarize the main details of the article first. Then show how these details relate to one another. Your purpose here is to bring the entire article into clear focus. If the reader finishes with only a hazy notion of what you have said, your article has failed. But if your reader finishes the article with a clear idea of what you have said, your article has succeeded. A well-written conclusion can help you achieve this kind of clarity.

POLISHING AND SUBMITTING YOUR ARTICLE

Technical and scientific personnel meet two major troubles when writing. Either they (1) write too slowly, trying to revise as they write or (2) do not spend enough time revising their material after it is written. Both these troubles can cause many disappointments. Can they be avoided? Yes, they can. Let's see how.

Concentrate on Technical Facts

When you write your article, don't be overanxious about grammar and the mechanics of writing. If you must worry, then worry about the technical facts in your article. Once you have your facts on paper, it is easy to revise. In your revision you can apply forcefully the suggestions given in this book for making your writing as effective as possible. Moreover, your writing will improve with practice.

Write quickly to keep your ideas flowing. If something new occurs to you while writing, include it. If the data is not pertinent, you can easily delete it later. And if a sentence or paragraph does not seem to be as good as you think it should be, don't stop; keep writing. Make a short note in the margin and revise the material later.

Now let's summarize. Don't be overconcerned with punctuation, grammar, or the other mechanics of writing while you are putting words on paper. Instead, focus on getting a smooth flow of ideas. Write quickly and finish the article as soon as possible. If you do this, your article will be stronger and there will be little chance of overlooking good ideas because you are too concerned with the mechanics of writing.

How to Polish

As soon as you finish your article, put it away and forget it for one week. Start to work on another article or some project which is not in any way related to the article you have finished. Don't review the finished article in your mind. Try to get as far away from the article as you possibly can.

Take the article out at a time when your mind is clear and free of problems. Read the article as though you had never seen it before. Read as quickly as you can. While you are reading, note in the margin any new ideas that occur to you. If a sentence or paragraph is not clear, mark it. If there is a sudden jump from one subject to another without a transition, mark it. Be critical of your own work. Nothing but improvement can result.

If you find that a rewrite is necessary, don't hesitate—do it at once. If you put off the rewrite, you may forget what was wrong. Many writing experts consider rewriting so vital to success that they feel, "There's no such thing as good writing, only good rewriting."

Faults to Check

Here are eight faults to check: (1) mile-long leads, (2) yard-long sentences, (3) big words where shorter ones will do, (4) overly long paragraphs, (5) too many illustrations or tables, (6) unrelated material, (7) sloppy manuscript or illustrations, and (8) too much material for the length of the article you plan.

1. Mile-Long Leads. In your early article attempts you may be so enthusiastic over your story that you try to tell it all in the lead. Check your lead first. For most articles, it should not be more than about five sentences. If it is longer, break the lead into two or more paragraphs.

Also consider the idea content of your lead. Don't try to cram all the ideas of your article into the lead. If you do, you will give your readers an acute headache. Instead, try to limit the idea content of your lead to the main idea of your article and possibly one or two other related ideas. See our discussion of the different leads for help.

2. Yard-Long Sentences. When you write quickly, you will probably tend to use shorter sentences. But some of your ideas may be so complex that your sentences expand. So check for unusually long sentences.

There are few rules for ideal sentence length. In general, the average length of sentences in your article should not exceed about 20 words. You can have much longer sentences if you need them, but be sure to follow a long sentence with a short one—between 2 and 6 words. This technique helps increase reader comprehension.

If you type your article yourself, watch for sentences that are longer than two typewritten lines. With 1-inch margins the average typewritten line contains approximately 10 to 12 words. So a two-line sentence has about 20 to 24 words. See to it that your article does not have many sentences longer than two typewritten lines.

3. Long Words. Long words are usually abstract words. As we noted earlier in this section, the use of too many abstract words can bore your readers. So be on guard—if you find three or four long words in one sentence, ask yourself if they are necessary. Get rid of any needless words by using smaller, more concrete terms.

4. Overly Long Paragraphs. A paragraph is like a milestone. Your readers feel a sense of accomplishment when they finish a paragraph. But if you make your paragraphs too long your readers get discouraged.

For most modern technical articles you should have between two and four paragraphs per typewritten page. There are few subjects today that rate a paragraph that runs a full typewritten page. You can easily break an overly long paragraph into two or more. When you spot a long paragraph, mark the place where you want to split it with a special symbol. The proofreader's mark for the beginning of a new paragraph looks like a backward p. See the symbols chart in Section 14 for further details.

5. Too Many Illustrations or Tables. Check a few technical magazines. You will find that most of them use about two illustrations or tables per page. Some articles may have more and some less; but this is the average.

Choose your most important illustrations and supply these to the editor with your article. If there is a choice of illustrations, supply the alternates but mark them properly. Note this fact in your letter to the editor.

Generally, use as few tables as possible. Tables are expensive to set in type. Most readers today do not expect a magazine to carry detailed tables that properly belong in a catalog or handbook. Confine tables that you do use to a few important values. Your readers can still learn how to use the table. But you will spare them the glut of useless tabular data.

6. Unrelated Material. If you are writing an article on piping, stick to pipes. Don't let your text wander. If you do, the article will become weak. Since most engineers and scientists have special interests, they look for the narrow article that covers a particular area of their field.

7. Sloppy Manuscript or Illustrations. Editors remember an author first for his or her technical skill and second for the condition of the manuscript the author submits. *Remember:* The editor must work with your manuscript after receiving it. If your manuscript is in poor shape, the editor must spend extra time on it. This will not endear you to busy editors.

Have your manuscript typed by a professional typist, if possible. It may cost you a few dollars extra, but the result will be worth it. The editor will then be able to process your manuscript with the least effort.

8. Excess Material. Try to gage how much text and how many illustrations you need for the article. If you plan a two-page article, you should have up to about six illustrations and about eight typewritten, double-spaced pages. Note that all manuscripts or articles should be typed double-spaced. If you plan a longer article—say, four published pages—the amount of typewritten material should be about double, but you should have only six or eight illustrations.

Submit Your Article

When your article is ready—after careful checking and flawless preparation—send it to the editor. If you have talked with the editor about the article, you can deliver it by hand; but you don't gain much by this. No editor will decide for or against the publication of your article while you sit and wait.

If your article contains photos and drawings, or you just want to give it extra protection, send it by registered mail. Use the replacement cost, or article payment you expect, whichever is greater, as the value for which you register the material.

IN RETROSPECT

We've discussed the requirements for successful writing of many types of technical and scientific articles in this section. If you've read each requirement and the representative examples, you should now have a good knowledge of many of today's article writing practices. Of course, we couldn't reproduce examples of all the hundreds of varieties of articles being published today. For these you must study the current issues of magazines in your field. Study an article by noting the various items discussed in this section, such as title, subheads, lead, article body, and end. With these key parts of an article as guides, you will find it easy to evaluate the effectiveness of each author's presentation.

But if you are to write good articles, there comes a time when you must stop studying and begin writing. Use the suggestions given earlier in this section and the checklist below. Your writing will soon improve and take less time.

CHECKLIST FOR ARTICLE WRITING

Before You Write

1. Do you know what kind of article you will write? Use the list at the beginning of this section.

2. Have you prepared a comprehensive outline of the article? See the sample outlines in this section.
3. Have you chosen the necessary illustrations and tables? Do you know the target audience for your article?
4. Have you obtained clearance for the article from your supervisor, your firm, and any sources of classified data?

Title

5. Is the title specific? Does it name its main topic?
6. Does the title use as few words as possible?
7. Have you used the active voice where it will help?
8. Will the title arouse interest? Does it emphasize newness, savings, help for the reader? See the discussion of article titles in this section.

Lead

9. Does your first paragraph give a concise statement of the problem the article solves?
10. Have you briefly described the other important aspects of the problem or possible solutions?
11. Can you use numbered lists or other arrangements of data to speed reading time?
12. Is the lead (a) brief, (b) clear, (c) interesting?
13. Does the lead provide a smooth transition to the article body?

Article Body

14. Are pertinent facts presented in their order of importance?
15. Do you define new concepts and terms as soon as they are introduced?
16. Does the text explain each illustration and table? Or does each illustration and table have a clear caption?
17. Have you, where possible, referred to the illustrations and tables in the same order as they are numbered?
18. Are illustrative examples used where they will help the reader?
19. Can you make the divisions of your subject clearer by using subheadings?
20. Is the text straightforward, succinct, engaging?
21. Are conclusions needed in the article?
22. If conclusions are needed, have you stated them clearly, at the end of the article, where readers will see them quickly?
23. Have you provided a short list of useful references, if the article has used any?

Before Submitting the Article

24. Have you read the article and checked it for technical accuracy?
25. Has the article been checked for grammar and usage, spelling, conciseness, and clarity?
26. Last, is your manuscript typed double-spaced on good bond paper? Do you have at least one duplicate?

SECTION 5
CREATING CONTRACT-WINNING PROPOSALS

Each year the U.S. government spends more than $100 billion to buy goods and services from private-sector firms. State and local government offices and thousands of corporate buyers also participate in the unending search for assistance from qualified companies and individuals. These government agencies and corporate buyers seek help for a limitless procession of needs—from truck repair to deep-sea ballistics, from prison architecture to windmill research, from space station plumbing to urban planning.

How do they decide who fills the needs? Selection criteria vary as widely as do needs, but most contracting of this kind comes about through some form of competitive procurement process. Federal and state laws mandate specific procedures for government agencies, and some corporations have adopted similar rules for their own purchasing departments.

THE NEED FOR PROPOSAL EXPERTS

In practice, these procurement methods mean that the government or corporate buyer asks firms to write and deliver proposals for the buyer to evaluate. The buyer then invites the sources of superior proposals to engage in negotiations regarding cost and other special concerns. Finally, the buyer awards the contract to the source that—through the proposal and negotiations—shows itself best-suited for the job.

Superior Proposals Are Essential

Many organizations in technical, engineering, and scientific fields live—or die—by proposal writing. Because government requires continual private-sector help in technical and scientific areas, superior proposals are essential—both to the success of government operations and to the survival of private firms specializing in such help.

More than 300,000 companies in the United States obtain at least part of their work through government contracts. Many get all their jobs through public agencies. In addition, some do business with private corporations that—like government customers—award contracts on the basis of proposal evaluation or competitive bidding.

Clearly, proposal writing is big business. In some engineering and scientific firms, it is the key to multimillion-dollar projects. Many newly formed companies depend on a superior proposal to get their first government contract. Hence, a skilled proposal writer is a highly valued professional. His or her expertise can command a huge salary. But even more precious is the engineer, scientist, or technician who creates contract-winning proposals. Such individuals can often gain top positions on the technical or management rosters of their organizations. After all, proposal writing is among the most vital editorial functions in many technical and scientific organizations.

Proposal Writing and You

If you do business in a field involving technical or scientific projects—regardless of whether you currently work in government or private industry or in a small or large firm—you will benefit by learning how to create superior proposals. Why? Because effective proposal writing involves all the principles of writing a superior technical report, paper, or article, and much more.

To create a contract-winning proposal, you must know how to develop a comprehensive strategy for selling your services, products, or ideas. That's what proposal writing is all about: You write a proposal to sell. But about two-thirds of all proposals written today fail to pass even initial evaluations by government agencies.

Competition is keen. Clearly, you must do more than learn to write good proposals: You must learn to create winning ones. Once you can do this effectively, you will be better-equipped not only to win contract awards but also to evaluate other people's proposals, to satisfy the requirements of your clients and supervisors, to write advertisements and promotional literature, and to perform other writing tasks with confidence.

Proposal writing is a challenge. Few, if any, other technical writing tasks require you to bring more skills together at one time. Creating a superior proposal can be a hectic and complicated task. But it's not hopelessly difficult.

In creating any proposal, you should be aware of certain features of the procurement process. Let's examine these features now. We'll also draw some fundamental distinctions between different proposals. Then we'll cover the common proposal elements and discuss specific steps that you can take to create superior, contract-winning presentations.

TYPES OF PROPOSALS

A proposal can be characterized in many different ways. For example, we can discuss a specific proposal in terms of the subject it covers, how long the document is, what elements it contains, what format the proposal uses, or many other factors. These terms are helpful when writing a proposal because they offer insights into the general characteristics of the presentation. But creating a unique proposal involves much more than thoughtful writing.

Because you create a proposal to sell your goods or services, preparing it requires special planning, research, and organizational work. As a result, proposal writers typically devote as much as three-fourths of their time on any one project to preparation, and they spend only about a fourth of the time doing the actual writing.

To prepare properly, the proposal writer must first know certain basic facts about the type of proposal and the origin of the writing project. Based on origin, proposals can be divided into two main types. These are: solicited and unsolicited.

Solicited Proposals

Stated simply, federal procurement regulations and the laws of many states and municipalities require government agencies and departments to solicit proposals from a number of different sources when the government needs outside help meeting its requirements for goods and services. The resulting presentations are known as solicited proposals.

Most agencies or departments appoint a group known as a *source selection board* to choose firms that it considers potentially qualified to do the required work. Once the board makes this preliminary selection, the government office sends a solicitation package to each of the prospective proposers.

The source selection board may consist of several members of the government office who are familiar with the needs or problem being considered. Usually, some of these board members help to prepare the solicitation package. Under the direction of a contracting officer, board members may also write parts of the solicitation, such as the evaluation criteria and request for proposal (RFP).

Soliciting proposals in this manner has certain advantages over other procurement methods. In contrast with bidding procedures that base contract awards mainly on price, the solicited proposal allows contractors to offer distinct solutions to a given problem or need. The written presentation thus serves as the basis for a comprehensive evaluation, instead of one limited to economic factors. Procurement authorities feel this helps them to get the best results possible and to ensure fair, consistent, and timely solutions.

Equity in the treatment of all proposers is one function of proposal solicitation. Accordingly, contracting experts and watchdog groups generally agree that the majority of contract awards in government today are ethical.

Formal advertisement, which frequently concentrates on cost rather than technical aspects of a project, is similar to the purchasing procedures followed in many private-sector organizations when they award contracts to other commercial firms. In these contracts, the competing firms may not necessarily be required to submit technical proposals for the job. Instead, the contract goes to the firm with the lowest bid.

Most federal agencies today have abandoned formal advertisement in favor of negotiated procurement and proposal solicitation. Following World War II, government contractors recognized that some contracts demanded a much more thorough appraisal than competitive bidding alone could provide. Today the federal government accomplishes about 85 percent of its purchasing through solicited proposals. State and local governments and some corporations prefer this method too. Hence, solicited proposals are standard today.

Parts of the Solicitation Package. The typical solicitation package contains a number of standard elements. These include: (1) the RFP, (2) evaluation criteria, and (3) statement of work (SOW). The RFP makes the formal request and outlines particular points that the proposal should address. The evaluation criteria, on the other hand, show the factors upon which the proposal will be judged. The SOW is the heart of the package and explains the problem the requestor needs to have solved.

Depending on the size and nature of a given problem, the elements of the solicitation package can describe a project in great detail or in extremely general terms. For example, in soliciting a proposal for a supply of machine bolts, the Department of Defense might issue an RFP prescribing an exact length for the proposal, with evaluation criteria showing the numerical scheme evaluators will use to score the presentation, and a SOW listing technical specifications. But in another solicitation package—say, for a proposal to design a space-based missile defense system—the need might be so broad and complex that the solicitation package can describe the project only in general terms. Thus, the RFP for the project might allow any length, the evaluation criteria might list several major categories and no rating scheme, and the SOW might give everything from overall requirements to a partial list of specifications. Figure 5.1 is a typical listing of evaluation criteria from a recent RFP.

The specific characteristics of any given solicitation package are important insofar as they could affect your ability to respond to the requestor's needs. As we'll see, you must analyze your customer's concerns carefully and demonstrate clear understanding of all needs if you want your proposal to win a contract. Fortunately, procurement officials recognize the difficulty that vaguely written solicitation packages pose for contractors, and some agencies are working actively to improve the solicitation process.

Many government agencies now use standard forms to request proposals. Figure 5.2 shows Standard Form 33, an RFP used by some federal offices. Other forms, including Standard Form 18, are sometimes used by federal entities, including the Department of Defense, Department of Transportation, Department of Energy, and their respective agencies. In addition, agencies may notify contractors of an RFP issuance through the *Commerce Business Daily,* as shown in Figure 5.3.

Source Selection. Typically, customers follow one of three general patterns in deciding what firms to solicit proposals from.

In the first, the contracting officer and source selection board draw the names of potentially qualified contractors from the contracting files of the customer (i.e., the agency, department, or organization requesting the work). Most government offices maintain such files, which contain the names of contractors that the agency or department has used previously, or which the officers otherwise know to be a source of goods and services in its field.

In the second pattern, the requestor designates its own list of possibilities. But in either of these two cases, the basic procedure is for the contracting officer to solicit proposals from potentially qualified firms, following an initial screening by the board.

The third pattern differs substantially from the two above. Here the contracting officer places a notice directly in *Commerce Business Daily,* inviting responses from interested parties.

The process of soliciting proposals is a complex one, but space limitations prevent us from discussing the procurement process in further detail here. As we've noted, you should be aware of the specific practices relating to any proposal you intend to create. Issues that are peculiar to any one contract—such as how the

Evaluation Criteria	Weight (%)
1. Technical Approach	
A. The ability of the offeror to review and evaluate the plant designs in a timely manner with emphasis on the structures, systems, and components important to safety.	25
B. Extent to which the proposal presents a clear understanding of the objectives set forth in the "Statement of Work"	10
C. Extent to which potential problem areas are identified, and the adequacy of the proposed solutions	
Subtotal	40
2. Management	
A. The roles and dedication of the project manager and key personnel to perform the work	
B. Adequacy of the Program Plan and management controls utilized to control costs, schedules, and technical quality	10
C. Extent to which the offeror can provide the type of computer facilities and programs, experience, and knowledge necessary to achieve the project management objectives	5
Subtotal	25
3. Related Past Experience	
A. Experience of the offeror in plant construction, design, engineering, and evaluation of factors regarding structural integrity	15
B. Experience of key personnel in the necessary disciplines such as architectural engineering, mechanical engineering, electrical engineering, and safety engineering	15
C. Experience of key personnel in geological engineering, meteorology, and environmental sciences	5
Subtotal	35
Total possible score	100

FIG. 5.1 Typical listing of evaluation criteria from a recent RFP. *(RFP No. RS-NMS-82-030, U.S. Nuclear Regulatory Commission.)*

request originated, what types of contracts are available, or how the contract will be budgeted and administered—may affect your pursuit of a given contract award. But here we're concerned chiefly with learning to develop better proposals. You can learn more about the procurement process by studying what goes into the creation of a contract-winning proposal.

SOLICITATION, OFFER AND AWARD

3 CERTIFIED FOR NATIONAL DEFENSE UNDER BDSA REG 2 AND/OR DMS REG 1 — RATING | 4 PAGE 1 OF

1 CONTRACT (Proc Inst Ident) NO. | 2 SOLICITATION NO. | 5 DATE ISSUED | 6 REQUISITION/PURCHASE REQUEST NO.

☐ ADVERTISED (IFB) ☐ NEGOTIATED (RFB)

7 ISSUED BY CODE | 8 ADDRESS OFFER TO (If other than block 7)

In advertised procurement "offer" and "offeror" shall be construed to mean "bid" and "bidder"

SOLICITATION

9. Sealed offers in original and ______ copies for furnishing the supplies or services in the Schedule will be received at the place specified in block 8, or if handcarried, in the depository located in ______ until ______ local time ______ (Hour) (Date)

If this is an advertised solicitation, offers will be publicly opened at that time.

CAUTION – LATE OFFERS: See pars. 7 and 8 of Solicitation Instructions and Conditions.

All offers are subject to the following:

1. The Solicitation Instructions and Conditions, SF 33-A, ______ edition which is attached or incorporated herein by reference.
2. The General Provisions, SF 32, ______ edition, which is attached or incorporated herein by reference.
3. The Schedule included herein and/or attached hereto.
4. Such other provisions, representations, certifications, and specifications as are attached or incorporated herein by reference. (Attachments are listed in schedule.)

FOR INFORMATION CALL (Name & telephone no.) (No collect calls) ▶

SCHEDULE

10 ITEM NO	11 SUPPLIES/SERVICES	12 QUANTITY	13 UNIT	14 UNIT PRICE	15 AMOUNT

See continuation of schedule on page 4

OFFER (pages 2 and 3 must also be fully completed by offeror)

In compliance with the above, the undersigned agrees, if this offer is accepted within ______ calendar days (60 calendar days unless a different period is inserted by the offeror) from the date for receipt of offers specified above, to furnish any or all items upon which prices are offered at the price set opposite each item, delivered at the designated point(s), within the time specified in the schedule.

16 DISCOUNT FOR PROMPT PAYMENT (See par. 9 SF 33-A)

____ % 10 CALENDAR DAYS ____ % 20 CALENDAR DAYS ____ % 30 CALENDAR DAYS ____ % ____ CALENDAR DAYS

17. OFFEROR CODE | FACILITY CODE | 18 NAME AND TITLE OF PERSON AUTHORIZED TO SIGN OFFER (Type or print)

NAME AND ADDRESS (Street, city, county, State and ZIP code)

AREA CODE AND TELEPHONE NO. ▶

19 SIGNATURE | 20 OFFER DATE

☐ Check if remittance address is different from above – enter such address in Schedule

AWARD (To be completed by Government)

21 ACCEPTED AS TO ITEMS NUMBERED | 22 AMOUNT | 23 ACCOUNTING AND APPROPRIATION DATA

24 SUBMIT INVOICES (4 copies unless otherwise specified) TO ADDRESS SHOWN IN BLOCK ______ | 25 NEGOTIATED PURSUANT TO ☐ 10 U.S.C. 2304(a)( ) ☐ 41 U.S.C. 252(c)( )

26 ADMINISTERED BY (If other than block 7) CODE | 27 PAYMENT WILL BE MADE BY CODE

28 NAME OF CONTRACTING OFFICER (Type or print) | 29. UNITED STATES OF AMERICA BY (Signature of contracting officer) | 30 AWARD DATE

Award will be made on this form, or on Standard Form 26, or by other official written notice

33-130

Standard Form 33 Page 1 (REV. 3-77)
Prescribed by GSA, FPR (41 CFR) 1-16.101

FIG. 5.2 (*a*) Standard Form 33 used by federal agencies to request proposals.

Unsolicited Proposals

Technical writers refer to the second type of proposal as *unsolicited,* but some find this term misleading. In theory, an unsolicited proposal is a highly appealing way of getting government contracts: There is no formal announcement of a prospective project, no solicitation package to pore over, no competition. All you

REPRESENTATIONS, CERTIFICATIONS AND ACKNOWLEDGMENTS

REPRESENTATIONS *(Check or complete all applicable boxes or blocks.)*
The offeror represents as part of his offer that:

1. **SMALL BUSINESS** *(See par. 14 on SF 33-A.)*
   He ☐ is, ☐ is not, a small business concern. If offeror is a small business concern and is not the manufacturer of the supplies offered, he also represents that all supplies to be furnished hereunder ☐ will, ☐ will not, be manufacturered or produced by a small business concern in the United States, its possessions, or Puerto Rico.

2. **MINORITY BUSINESS ENTERPRISE**
   He ☐ is, ☐ is not, a minority business enterprise. A minority business enterprise is defined as a "business, at least 50 percent of which is owned by minority group members or, in case of publicly owned businesses, at least 51 percent of the stock of which is owned by minority group members." For the purpose of this definition, minority group members are Negroes, Spanish-speaking American persons, American-Orientals, American-Indians, American Eskimos, and American-Aleuts.

3. **REGULAR DEALER — MANUFACTURER** *(Applicable only to supply contracts exceeding $10,000.)*
   He is a ☐ regular dealer in ☐ manufacturer of, the supplies offered.

4. **CONTINGENT FEE** *(See par. 15 on SF 33-A.)*
   (a) He ☐ has, ☐ has not, employed or retained any company or persons *(other than a full-time bona fide employee working solely for the offeror)* to solicit or secure this contract; and (b) he ☐ has, ☐ has not, paid or agreed to pay any company or person *(other than a full-time bona fide employee working solely for the offeror)* any fee, commission, percentage, or brokerage fee contingent upon or resulting from the award of this contract; and agrees to furnish information relating to (a) and (b) above, as requested by the Contracting Officer. *(Interpretation of the representation, including the term "bona fide employee," see Code of Federal Regulations, Title 41, Subpart 1-1.5.)*

5. **TYPE OF BUSINESS ORGANIZATION**
   He operates as ☐ an individual, ☐ a partnership, ☐ a nonprofit organization, ☐ a corporation, incorporated under the laws of the State of _____

6. **AFFILIATION AND IDENTIFYING DATA** *(Applicable only to advertised solicitations.)*
   Each offeror shall complete (a) and (b) if applicable, and (c) below:

   (a) He ☐ is, ☐ is not, owned or controlled by a parent company. *(See par. 16 on SF 33-A.)*
   (b) If the offeror is owned or controlled by a parent company, he shall enter in the blocks below the name and main office address of the parent company:

   _____

NAME OF PARENT COMPANY AND MAIN OFFICE ADDRESS *(Include ZIP code)*		
(C) EMPLOYER'S IDENTIFICATION NUMBER *(SEE PAR. 17 on SF 33-A)*	OFFEROR'S E.I. NO.	PARENT COMPANY'S E.I. NO.

7. **EQUAL OPPORTUNITY**
   (a) He ☐ has, ☐ has not, participated in a previous contract or subcontract subject either to the Equal Opportunity clause herein or the clause originally contained in section 301 of Executive Order No. 10925, or the clause contained in Section 201 of Executive Order No. 11114; that he ☐ has, ☐ has not, filed all required compliance reports; and that representations indicating submission of required compliance reports, signed by proposed subcontractors, will be obtained prior to subcontract awards. (The above representation need not be submitted in connection with contracts or subcontracts which are exempt from the equal opportunity clause.)
   (b) The bidder (or offeror) represents that (1) he ☐ has developed and has on file, ☐ has not developed and does not have on file, at each establishment affirmative action programs as required by the rules and regulations of the Secretary of Labor (41 CFR 60-1 and 60-2) or (2) he ☐ has not previously had contracts subject to the written affirmative action programs requirement of the rules and regulations of the Secretary of Labor. *(The above representation shall be completed by each bidder (or offeror) whose bid (offer) is $50,000 or more and who has 50 or more employees.)*

CERTIFICATIONS *(Check or complete all applicable boxes or blocks)*

1. **BUY AMERICAN CERTIFICATE**
   The offeror certifies as part of his offer, that: each end product, except the end products listed below, is a domestic end product (as defined in the *clause* entitled "Buy American Act"); and that components of unknown origin have been considered to have been mined, produced, or manufactured outside the United States.

EXCLUDED END PRODUCTS	COUNTRY OF ORIGIN

Standard Form 33 Page 2 (REV. 3-77)

(b)

FIG. 5.2 (*b*) Standard Form 33 (reverse) showing checkoff items.

need do is determine how to solve an existing problem or fulfill a government need, and submit a proposal. If you've made a fairly good offer and have the right credentials to support your claims, you may have a customer. In reality, this almost never happens. More often than not, the unsolicited proposal develops in a manner very much different from that described above. Typically, an unsolicited proposal receives its initial spark not so much through the spontaneous innova-

Dept of Environmental Quality, POB 94381, Baton Rouge, LA 70804. Attn: Annette Sharp, Undersecretary
U - PHASE IV - OUTREACH OPERATOR TRAINING PROGRAM The Louisiana Dept of Environmental Quality (the Department) gives public notice that it is requesting proposals from interested, qualified persons or firms to improve, through operator training, the performance of municipal wastewater treatment plants in the five million gallons/day effluent category or less that were funded with P.L. 92-500 monies. Copies of the Sol for Proposals (SFPs) No. 24011-88-01 are available at the Dept of Environmental Quality, 7th Floor, State Land and Natural Resources Bldg, 625 North Fourth St, Baton Rouge, LA. They also may be obtained by writing to Louisiana Dept of Environmental Quality, PO Box 94381, Baton Rouge, LA 70804. Attn: Laura McDonald, or by calling 504/342-1211. Proposals may be sent to the above address, or delivered to the Department at the 7th Floor, State Land and Natural Resources Building. Proposals must be received NLT 15 Oct 87 (closing date) after which time and date they will not be accepted. Proposals will be evaluated according to criteria defined in the SFP, and all proposers will be notified by mail of any contract awarded after such evaluation within (30) days of the closing date. (254)

15 – PLATE, FAIRING, LANDI NSN 1560-01-253-0360, P/N 369H2551-4, FSCM 02731 IAW McDonnell Douglas Helicopter Co, Dwg. Appl to the OH-6 Acft. Sol DAAJ09-87-Q-4392 (P23/JA) closing date o/a 26 Oct 87 POC: Joi R Fowler, Buyer, AMSAV-PSRG, 314/263-1026. FOB Origin with del 95 ea to Texarkana TX 75507-5000. Proposed del schedule is 95 ea 180 days ADC. See Notes 22, 26, 36, 56, 73 and 81. All responsible sources may submit a proposal which shall be considered. This procurement required prequalification. Contrs who desire to be pre-qualified msut submit qualification documentation IAW procedures contained in the clause entitled "Sources Eligible for Contr Award Consideration". This clause can be obtained from the buyer identified above. This proposed contr action is for supplies for which the Govt does not own specs and dwgs to permit full and open competition. Firms that recognize and can produce the required items described above are encouraged to identify themselves to the buyer identified above and provide supporting evidence to the buyer which will permit their participation in the current or future suls. Intended source is McDonnell Douglas Helicopter Co. Diane C Dedert, Contr Officer.

Directorate of Contracting & Manufacturing, Oklahoma City, ALC, Tinker AFB, OK 73145
★ 53 -- SPACER, RING NSN 5365-01-183-5351JF, mfr WMZ Mfg Co Inc, Morgan Products Inc and Cylectron Corp, P/N 9502M43P01, shape; round. Matl hardness rating 24.0 Rockwell C min and 35.0 Rockwell C max case hardened, silver plated. Hole shape: round. Dia 5.815 in min and 5.820 in max outside dia; 6.116 in min and 6.118 in max. Outside thkns 0.191 in min and 0.196 in max. One outside corner chamferred 0.04 porm 0.06 in at 45.0 deg; diagonal inside corner chamferred to dia 5.962 porm 5.972 in. Matl; iron alloy, qty 87 Oct, 140 ea; 89 Sep, 140 ea. Appl to F101. Dels to Tinker AFB Base, OK. Sol will be issued to WMZ Mfg Inc, 359 Burnham St East Hartford, CT 06108; Cylectron Corp 24027 Ryan Rd Warren, MI 48091; Morgan Products Inc 4459 13th St Wyandotte, MI 48192. All responsible sources may submit a bid, proposal, or quotation which will be considered. RFP 88-45011. To issue o/a 2 Oct 87, closing date o/a 2 Nov 87. For sol contact OC-ALC/LIN McNeill/PMZMC/405/739-4292. See Notes 27, 33, 81, 73, 40.
★ 53 -- SEAL, WASHER, NSN 5330-01-183-4934JF, Mfr Sundstrand Aviation, Parker Hannifin Corp, P/N (Sundstrand) 5903512 and (Parker-Hannifin) 828005, washer fits btwn the head of a machine bolt and the inlet housing; the function of a washer is to distribute the pressure applied to a specific surface through torqing action. Matl not avail, qty 88 Apr, 345 ea; 88 May, 345 ea; 88 Jun, 5410 ea and 88 Jul, 5415 ea. Appl to F101. Dels to Tinker AFB OK. Sol will be issued to Sundstrand Aviation 4747 Harrison Ave Rockford, IL 61101, Parker-Hannifin Corp O-Seal Div 10567 Jefferson Blvd Culver City, CA 90230. All responsible sources may submit a bid, proposal, or quotation which will be considered. RFP 87-32479. To issue o/a oct 2, closing date o/a 87 Nov 2. For sol info contact OC-ALC/D.D. Pilate/PMZMC/405/739-4294. See Notes 27, 33, 81, 73, 40. Ref Note 33. For qualification procedures, contact OCALC/CRS.

FIG. 5.3 Examples of RFP notices appearing in *Commerce Business Daily*.

tions of the proposer as through a series of contracts and discussions between a contractor and a particular government office.

For instance, a project engineer working on construction of docking facilities for a maritime police station in a coastal city might get to know the city inspector assigned to review the project's progress. Then one day the inspector mentions that his office is having structural problems with a dry-dock facility elsewhere in the city.

Let's say the engineering firm has recently developed a proprietary design to remedy dry-dock problems of this type. Recognizing the opportunity for an unsolicited proposal for the dry-dock work, the project engineer checks with her supervisors to discuss the idea of offering a solution to the city's problem. The contractors study the idea further, attempting to determine whether they could do the work profitably, on time, and to the customer's full satisfaction. If the company is interested in pursuing a contract, the supervisors or managers would then check with the city's contracting officials to be certain the city would welcome an unsolicited proposal. The contractors might hold additional discussions about the availability of budget money and other details. Then the contractors might submit a "white paper" or executive summary to the contracting officials, and, finally, a formal proposal.

If all goes well, the unsolicited proposal may be approved. But not all projects proceed so smoothly. Some contracting officers—worried that other contractors could lodge a formal complaint against the agency for awarding a contract on a sole-source, noncompetitive basis—might insist on issuing an RFP to other firms. Thus, what began as a truly unsolicited proposal might not end up as one.

Unsolicited proposals are not always subject to skepticism, of course. Take a look at nearly any *Commerce Business Daily* and you will see notices of contracts or negotiations that are planned "pursuant to an unsolicited proposal." Many engineering and scientific firms—both large and small—win a substantial number of projects in this way.

The chief requirements of unsolicited proposals are:

1. You must analyze your customer's problems and concerns and know the customers' needs and limitations. Usually, this requires holding discussions with the customer before submitting your unsolicited proposal.

2. Since the recipient of an unsolicited proposal cannot issue a solicitation package, you must make an extra effort to make the details of your unsolicited proposal fit your customer's situation. This, too, requires extended preparation and discussion. It may also mean submitting a white paper or summary to test your ideas with the customer prior to submitting the final proposal. In one sense, you substitute your discussions with the customer for the material that would otherwise be contained in a solicitation package.

3. You must have something unique to offer. In other words, if you want your unsolicited proposal to succeed, your firm must clearly be the original source of the project idea. Make certain that you have a way to substantiate this. By maintaining strict control over project materials—such as work records or a proprietary design—you should be prepared to prove that your proposal is truly an unsolicited one.

If you know your way around a particular government agency, department, or other organization and can fulfill the requirements listed above, there's a good possibility that you'll be able to enjoy the advantages of writing unsolicited proposals for at least a portion of your business. You'll have less competition for certain awards, and you'll have greater opportunity to negotiate a price that satisfies both your company and your customer.

TECHNICAL AND COST PROPOSALS

Besides distinguishing proposals according to their origin, we categorize them on the basis of content. Proposals in engineering and scientific fields range from simple one-volume documents covering general aspects of a project to long and complicated documents containing numerous volumes, each volume devoted to a separate topic. A single proposal may contain the work of one writer, or it may contain information put together by an entire team of specialists.

Technical Proposals

Virtually all proposals in engineering and scientific fields—regardless of whether they are solicited or unsolicited—include a section or volume devoted exclusively to engineering solutions and technical specifications. This is known as the technical proposal.

The technical proposal presents the engineering or scientific aspects of a project, independent of the cost. It is the heart of the presentation that you use to sell your goods and services to the customer. Generally, the technical proposal is the first thing evaluators look at when they receive your written document. Why? Because evaluators can often take one look at a technical proposal and tell whether or not the source deserves further consideration for the contract. Rarely will a customer be concerned with nontechnical issues—such as cost and contractual details—before evaluators can verify that your proposal shows technical merit.

A well-written technical proposal demonstrates clear thinking on the part of the individual or team assigned to prepare it. Further, it follows all the rules of good writing and utilizes principles of effective selling. In short, the technical proposal represents your opportunity to show the customer that you can do the required work—competently, methodically, and reliably.

In writing any proposal, make certain you respond to every item of information your customer has requested. In developing your technical proposal, this means you must analyze carefully the RFP, evaluation criteria, and SOW to determine exactly what the customer is looking for. Then you write the proposal, giving concrete evidence to prove that you know your customer's concerns and you can meet all requirements. Simple as these steps may sound, many proposers—through poor preparation or simple neglect—ignore them. Thus it's not surprising that two-thirds of all proposals written do not pass beyond the initial evaluations of government officials.

What do evaluators look for? Typically, a superior technical proposal demonstrates certain crucial facts about the proposer. Sometimes a customer will remind proposers—in the written evaluation criteria or in formal discussions—of the need to cover particular areas. But not all customers make their needs explicit. For example, a customer may simply forget to mention certain criteria. Or a customer may fail to explain all concerns clearly. Since nearly all evaluators pay close attention to certain key factors in every technical proposal, you must always remember to cover these items.

The five key factors that you should cover in developing every technical proposal are:

1. Understanding. Your customer will want to be certain that you fully understand his or her needs and purposes. That is, you must show that you know the goals of the customer's agency, department, or organization. Further, you must prove that you comprehend clearly the specific problems or needs for which solutions have been requested. To prove your understanding, be careful not to restate material contained in the RFP or SOW; instead, translate the problem into your own words, expanding and modifying the customer's information as you build your technical arguments.

2. Analysis. In developing your understanding of the customer's situation, you must conduct a painstaking and comprehensive analysis of the specific problems in question. Your proposal should prove this. Tell the customer what techniques you have used to examine the problem—such as graphic analysis, critical-path monitoring (CPM), drawing-task system (DTS) by computer, accomplishment monitoring (ACCOM) by computer, etc. But don't stop here—show the customer how you've analyzed the problem, giving references to specific parts of the solicitation package, engineering specifications, project scheduling, design criteria, etc.

3. Planning. Make certain that your proposal shows you have developed a comprehensive plan to meet the customer's technical needs. Here your main evidence is the general quality of your proposal. But you can also build the evaluators' confidence by summarizing and outlining the main points of your overall approach at strategic spots in the document. Remember to discuss potential difficulties or problems in the work and to explain what contingency plans you have developed. Also be sure to include important dates, worker hours, organizational charts, management procedures, and other planning data.

4. Practicality. Regardless of how much importance your customer attaches to innovation and creativity, you must show that your plan is technically viable. Even the most unusual technical solutions must be practical and have a high probability of success—or they fail as solutions. Substantiate your claims with references to design criteria, research findings, prior work, similar projects completed by other firms, standard specifications, performance statistics, test results, etc.

5. Reliability. If you want to convince evaluators that your firm is the one best-qualified to undertake the engagement, your proposal must prove that you and your staff are fully reliable. Here the key is to show your qualifications: Résumés of proposed staff members and summaries of corporate experience are among the most vital elements of superior technical proposals.

But it is usually not enough to pull a few résumés from a file and insert them in your proposal. Demonstrating your reliability is a task that demands special effort. It is usually worthwhile to spend some extra time tailoring each résumé to best reflect the technical needs of the specific project and the customer, almost as if your staff members were applying for positions within the customer's own organization. In fact, some customers tend to examine the résumés in proposals as diligently as though they were considering résumés of their own prospective staff members.

In addition to your corporate and individual qualifications, there are other areas in which you must prove your reliability. For instance: You must present clear, realistic schedules, specify your deliverables, and support your technical commitments as firmly as possible. And remember to respond to all items of information your customer has requested. If evaluators find they can rely on your written proposal to address the customer's concerns, they will be more likely to give your overall presentation a high rating for reliability.

For example, if the RFP suggests a tentative deadline for project completion, make certain you address the idea clearly in your proposal. You may decide to accept the suggested date, to offer a new deadline, or to create a schedule that is totally different from the one your customer envisions. But you must respond explicitly to your customer's ideas if you want your proposal to build confidence in your firm's reliability.

Choosing an Approach for the Proposal

How you choose to address the five key points listed above will depend on a number of considerations—including specific project requirements, additional data your firm may have gathered through special contacts, and the personal preferences of your managers and associates on a proposal writing team. There are many, many variables involved in the preparation of any technical proposal. As a proposal writer, it is your responsibility to develop a strategy that suits the technical requirements of each project and each customer.

For example, the material you use to demonstrate your understanding of the customer's requirements in one project might overlap the material you use to detail your analysis and project planning. In another proposal, you might address these factors separately. For small contracts, you might write the bulk of the technical proposal yourself; for large contracts, you might devote your efforts to a single technical consideration, while other authors and technical managers cover their own areas of expertise. Further, depending on the customer's stated requirements, your technical proposal may or may not include a discussion of management plans and procedures. And while you might present detailed organizational charts and other illustrations in some proposals, in others you might use few of these items, if any.

No two technical proposals are exactly alike. But nearly all winning proposals share this one feature in common: Instead of merely telling evaluators about the competence of the proposers and the excellence of the proposed solution, the winning proposal shows the evaluators exactly what makes the offeror and its solution superior.

Almost anyone can write a technical proposal that restates material from the solicitation package. And anyone can fill a proposal with superlatives and statements like, "We will provide the fastest, most effective solution available." In fact, that is the kind of approach that many inexperienced proposal writers use. But the experienced proposers—those who consistently win lucrative contract awards—use a different strategy. Instead of repeating statements from the RFP or telling how great their proposed solution is, the winners prove their points. How? By supporting their statements, whenever possible, with strong evidence.

Generally, experienced proposal writers use a somewhat indirect approach in covering the key issues we've examined. In other words, a well-written technical proposal does not normally contain sections labeled "Understanding," "Planning," etc. Nor does it make broad contentions like, "Our team of qualified technicians has studied the request and developed a thorough understanding of the customer's requirements." Instead, the superior proposal employs a selling strategy. For example, it might present a brief analysis of the customer's needs and technical goals, along with a description of how the firm perceives its technical task. The proposal might then proceed through a series of logical arguments which help the reader appreciate the advantages of the proposed technical approach. If the technical proposal is truly comprehensive, it presents a detailed plan, leaving little to the imagination. A careful selection of graphics, matrices, résumés, and other exhibits increases the general effectiveness of the presentation. In sum, the well-written technical proposal addresses all the customer's concerns, offering a distinct set of benefits.

Cost Proposals

A proposal for a technical engagement would be incomplete if it covered only the engineering or scientific aspects of a project. Financial considerations must also be covered. Therefore, the technical proposal is typically accompanied by a separate volume or section that gives a cost projection for the proposed technical engagement. This is the cost proposal.

Compared to most technical presentations, the cost proposal represents a relatively simple task for the proposal writer. A typical cost proposal covers the total cost of a project and breaks that total down into separate components. Where the difficulty lies is in developing an accurate estimate for the proposed project.

The basic purpose of the cost proposal is twofold: First, it shows your customer what price you propose to charge for your goods and services. Second, it outlines the approach that you've used to develop that projection.

Usually the customer tells proposers exactly what details to include in the cost proposal. Guidelines normally appear in the RFP and evaluation criteria; but many customers provide proposers with a preprinted form that specifies the items that the customer wants. While you can list the required information on your own form, most contractors use a standard form whenever the customer supplies one. Using a standard cost form can sometimes simplify the task of creating the cost proposal.

Regardless of whether or not your customer supplies a preprinted form, most cost proposals contain essentially the same information. However, some customers may ask you to combine your cost presentation with a description of some other aspect of the project, such as your proposed management scheme. Hence you'll find RFPs that ask for either a "cost and management proposal," a "contractual and cost proposal," or simply a "cost proposal." Since cost is the bottom line for these kinds of proposals, we'll concentrate on cost projections here.

What Goes into the Cost Proposal? Typically, a cost proposal will contain these basic items: (1) estimated direct labor costs, (2) other direct costs, (3) general and administrative cost, (4) overhead rate, and (5) fee or profit margin. Let's take a quick look at each.

1. Direct Labor Costs. This estimate is derived from the hourly labor rate that you pay on your payroll, combined with the number of worker hours that your firm considers necessary to satisfy the requirements of the contract. Direct labor includes only that labor which is directly applicable to the contract. Therefore, if the technical proposal has shown an electrical engineer putting two months of time into the project, the cost proposal must show the same two months.

2. *Other Direct Costs.* These are costs required for you to satisfy the engagement—excluding direct labor, administration, overhead, and profits. Typical costs in this category include subcontractor and consultant fees, travel, special equipment and materials, and other direct costs necessary for you to perform on the contract.

3. *General and Administrative (G&A) Costs.* Here you list your indirect costs, ignoring overhead and the direct costs of the proposed project. G&A costs cover that part of the day-to-day operation and administration of your company which is not considered overhead. G&A costs are usually given in terms of a current rate—i.e., the percentage of total costs for your organization. There is no standard rate that is generally applicable to any one type of firm, so G&A rates have some flexibility. But most contractors report G&A rates of less than 20 percent. Many firms have rates running as low as 3 percent.

4. *Overhead Rate.* Overhead is an indirect cost representing that part of your daily requirement for doing business which G&A costs do not cover. Like G&A costs, overhead is usually given as a rate—the percentage of labor costs necessary to pay your operating expenses. Typical costs you may include when calculating your firm's overhead rate are: rent, utilities, taxes, depreciation, insurance, and fringe benefits. Since these may sometimes represent a substantial portion of a contractor's total costs, evaluators and contracting officers tend to scrutinize overhead rates carefully. Thus an RFP may ask you to show how you've calculated your overhead rate, breaking it into precise elements. Here again, there is no standard rate that applies generally. But the overhead rates of many successful engineering and scientific firms range between 70 and 90 percent. Some firms have rates higher than 100 percent. If your rates reach that high, you may have to show some justification in order to win a customer's confidence.

5. *Fee or Profit Margin.* Here is where your cost presentation will be most heavily influenced by the type of contract you are pursuing. For example, if the RFP states that the contract is expected to be of the cost-plus-fixed-fee type, you would give a specific fee calculated on the basis of a percentage of your costs. But if the contract were of the time-and-material variety, you might give your profit percentage, showing how it combines with labor and overhead rates to form your general billing rate.

There are many other contractual possibilities besides these, of course. Make certain that you study every potential engagement carefully—before you plan your proposal strategy—to determine what type of contract will be involved. Once you fully understand the prospective arrangement, you can decide whether you should provide estimates or fixed figures and determine how you should calculate the fee or profit margin.

6. *Other Fiscal Data.* In addition to the five typical components listed above, your cost proposal may include other financial information, depending on the customer's stated requirements. For example, you may be asked to do a substantial amount of research and development (R&D) on a particular project, which might require a separate listing of R&D costs. Or your customer may ask to see a schedule of charges for such items as blueprint reproduction and computer time, especially if the total amount of these costs cannot be determined before project completion. Providing a standard schedule of such charges may help to tip the proposal evaluation in your favor. A schedule may be particularly valu-

able where you can offer competitive rates and when the contract requires such services throughout its lifetime.

Occasionally, customers take the cost evaluation a step further and ask the offeror to provide a complete financial statement. Thus you should be prepared to cover your company's annual revenues, the value and costs of your current contracts, and other fiscal data, as requested.

Nowadays the customer also typically tells proposers about the general requirements or the problem that needs to be solved, leaving most or all of the technical solution to the contractor. The evaluators—the customer, in effect—review each technical proposal on a comparative basis and then examine the technical proposal against the cost proposal. If the presentation survives preliminary evaluation, the evaluators scrutinize it further. Provided that the information contained in the technical proposal is consistent with that of the cost proposal, the document may be subjected to a final evaluation. Then the customer may call the offeror in to resolve any remaining questions and to negotiate the "best and final" details of the contract.

Make your cost and technical data consistent. One of your chief goals in writing any cost proposal should be to maintain consistency between your costs and your technical proposal. If your technical proposal says the project will require a certain amount of electrical engineering work, the cost proposal must cite that labor. Any variations between the cost proposal and the technical proposal are likely to be viewed by evaluators as a sign of poor proposal planning or plain incompetence.

The rule for consistency applies to all statements you make in the technical proposal, regardless of whether you discuss direct or indirect labor, project organization, how much travel you anticipate, or how much time you'll need for R&D. *Remember:* make certain that your cost proposal follows the technical proposal exactly.

Key Points in Cost Proposals. What are the key items that evaluators look for in cost proposals? Basically, there are three. We can sum them up as follows:

1. The extent to which the proposal follows the guidelines set forth in the RFP and evaluation criteria, or in the customer's stated requirements
2. How well the data contained in the cost proposal follows the information given in the technical proposal
3. How the rates and estimates compare with those of the proposer's competitors and with the current rates for other firms in the field

FORMATS AND ELEMENTS

Proposals come in many shapes and sizes. For any given contract, a customer may receive numerous documents, each having a different format. The makeup and organization of one proposal may be totally different from that of another. Evaluators reading for content sometimes ignore these differences. But the format of a proposal may actually determine whether or not the proposal gets your message across and brings in a contract award.

All the elements that make up the format of a proposal do not simply fall into place. To organize your contents effectively, you need a strong structure. Some formats work better than others. Why? Because a proposal is a sales presentation, and as such, it relies on certain principles of selling to achieve maximum impact. One pivotal principle is: Structure your presentation to make it easy for the customer to buy.

You can turn that principle to your advantage in every proposal you create. By selecting a suitable format for each presentation, you may not only improve your selling but also increase the effectiveness of your technical arguments.

But what about mandated formats? Occasionally a customer becomes tired of trying to compare proposals that differ in format. Since varying formats can lead to difficulty in making accurate comparisons, the customer may mandate a specific format. In that case, the mandated format is usually described in the solicitation package. Unfortunately, these descriptions are sometimes poorly written, and the mandated format may leave a lot to be desired in regard to its overall effectiveness.

When a solicitation does prescribe a specific format, try to follow it as closely as possible. Evaluators will usually notice that you've followed instructions carefully, and this may be a point in your favor. You may even be able to turn some of the disadvantages of a restricted format into sales influences. How? By designing every element of your presentation to win. We'll look at some ways to do this when we examine typical proposal elements later in this section.

In practice, many customers are reluctant to mandate a proposal format. Most solicitations outline the kinds of information to include in the proposal or describe a preferred organizational scheme. Even when a customer indicates a preferred format, evaluators generally recognize that it is not always possible to use that format. Thus the RFP may use the term *preferred* instead of *prescribed* or *mandated*.

For most contracts, the customer tries to tell proposers what the problem or the general need is; but recognizing that the proposers are the technical specialists, the customer leaves the responsibility for developing an effective presentation to the proposers themselves. Whenever you are in doubt about what elements to include or about how to organize them in your proposal, follow the RFP and evaluation criteria.

This section contains a description of some of the typical elements of a technical proposal. It also includes a format that applies principles of effective selling to the written presentation. We believe this approach will help you build skill in handling most kinds of proposals.

Possible Elements

The basic format of a technical proposal has three major divisions:

1. Front matter
2. Body
3. Back matter

The format may be made up of few or many elements, depending on the proposal's specific requirements: scope of work, desired length, approach, major concerns of the customer, etc. Figure 5.4 lists the possible elements of a technical proposal. Remember that not all of these are required. Few, if any, proposals would include them all. However, you should know these elements, as they rep-

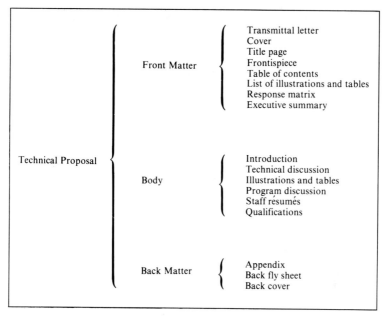

FIG. 5.4 Possible elements of a technical proposal.

resent an important part of the proposal writer's stock-in-trade. Below is a description of each of the possible elements.

Front Matter

Transmittal Letter. The transmittal letter (also called a *cover letter* or *letter of transmittal*) is sometimes not even required by the solicitation. But most proposal writers consider it an essential accompaniment to the submittal of cost and technical proposals. The letter is customarily addressed to the contracting officer who is responsible for the proposed engagement, and it formally states such information as the origin of the proposal and the number of copies included in the transmittal.

The transmittal letter may be written by the person who has directed the proposal project, although some firms prefer to have the transmittal letter written and signed by a division manager or other top executive. In any case, the proposal leader should review the letter to ensure that it agrees with the proposal in both fact and focus.

For some writers, the transmittal letter has special significance. They view it as a way to deal directly with the contracting officer and generate early support for the proposal.

Since contracting officials tend to concentrate on matters of cost and technical viability, it is usually advisable to include the following items in the letter:

1. A few short paragraphs summarizing the features of your proposal which you feel represent the key concerns of the contracting officer—e.g., special as-

pects of the technical approach that increase the reliability and safety of the engineering solution, cost-saving devices or procedures, or records of successful test results
2. Affirmation of the specific period for which the the offer is made, based on the acceptable period stated in the solicitation
3. A statement noting that the offer is firm and binding and that the signer has been authorized to transmit the proposal on behalf of your organization
4. A commitment of total support for the proposal by your organization, with an offer to supply more data or to meet personally with the requestor if desired

The transmittal letter is typically written after the proposal has been finished. Once the final documents have been produced, the letter is delivered as a separate document accompanying the proposal. The letter may be reproduced on a different paper stock than that used for the proposal. Usually it is printed on the offeror's letterhead, with the firm's logo. When the transmittal letter contains more than one page, it is paginated internally, as any separate letter would be, and does not affect the pagination of the proposal itself.

However, many proposal writing experts consider it worthwhile to include a copy of the letter in the proposal document as well. In that case, it may be bound directly inside the cover, before the title page.

Cover. Proposals are bound in many types of covers. Covers range from plain, inexpensive, heavy paper stock, to "perfect" bindings like those used for commercial books. Other possible bindings include simple loose-leaf, metal spiral, or plastic comb. The type of cover chosen will depend on several factors, such as the number of volumes required by the solicitation and the resources of the company producing the proposal.

The proposal cover may represent an opportunity to strike a good first impression with evaluators. Since it is often the first item evaluators see (with the possible exception of the transmittal letter), it can be used to command attention. Successful proposers have enhanced the appearance and impact of their covers in various ways. These include reproducing a procedural flowchart on the cover, using special colors, and printing the customer's logo on the cover. Some experienced proposers prefer a simpler approach. When in doubt, it is best to avoid fancy colors or designs and choose a standard cover.

Basic information carried on the cover includes the RFP number, the name and departmental affiliation of the customer, the name and address of the proposer, and the type of proposal (technical or cost). If the complete proposal comprises more than one volume, the cover should show the number of the volume.

Proposal writers usually do not need detailed knowledge of covers—paper stock, costs, types of materials, etc.—because covers are normally handled by a technical publications group. In a small firm, however, the writers themselves may design the cover. Writers should develop a general knowledge of cover types, and the project manager or editor should check a proof of the cover for correct spelling and information.

Title Page. The title page of a proposal gives essentially the same information as the cover—RFP number, name and departmental affiliation of requestor, and name and address of the offeror. In addition, the title page should note the date of the submittal. If the proposal contains information that is proprietary with the proposer, the title page may carry a short statement noting any special conditions under which the data is being offered.

In some proposals the title page is identical to the cover, and both may be printed from the same plate. The title page does not carry a page number, but it may be considered the first page of the proposal for printing and production purposes.

Frontispiece. A frontispiece is an illustration that faces or immediately precedes the title page of a document. Although the majority of proposals do not include one, a frontispiece may furnish a special opportunity to arouse interest. For example, some experienced proposers and technical publication specialists bind a foldout page into their proposal volumes. As a frontispiece, the foldout greets the reader with informative graphics—line drawings, pictorial flowcharts, bar graphs, etc.—at the same time as the title page. Such an approach may help "pull" the reader into the proposal text.

Table of Contents. The table of contents is a standard element of all types of proposals. It lists the contents of the presentation in sequential order, providing enough detail to enable readers to locate important data easily. Generally, the table of contents gives section headings, subheadings, appendix titles, and page numbers for each of these items.

In multivolume proposals, each volume may carry a table of contents for the entire set of volumes. Or each individual volume may list the main titles of other volumes and detail only its own contents.

Illustrations and tables are also listed in the table of contents—unless the proposal is a large one, has an unusual number of tables and illustrations, or is bound in more than one volume. In these latter cases, a separate list of tables or illustrations may be used.

The format of the table of contents varies with the solicitation requirements and the stylistic preferences of offerors and their publication specialists. Figure 5.5 is a table of contents used for a technical proposal consisting of four sections; here the proposers have followed a format recommended by the customer.

In many proposals the table of contents is the last element of front matter. In that case it immediately precedes the first page of text in the body of the document.

Response Matrix. A well-organized response matrix can be a highly effective complement to the table of contents. The response matrix normally lists all the most important items of the solicitation, such as specific evaluation criteria, and it shows where in the proposal the responses are located. Items may be listed according to various schemes. For instance, the response matrix might provide an alphabetical listing of evaluation criteria items, or it might list the items by their order of importance as designated in the solicitation package. Figure 5.6 is an example of a typical format for a response matrix.

A response matrix should not be confused with an index, which is a less commonly used element of proposals. An index is an alphabetical listing of contents (not responses). When an index is used, it forms part of the back matter.

The response matrix is a unique type of listing. It offers evaluators immediate visual evidence that your proposal responds to specific concerns of the customer. Further, it acts as a sales influence by making it easy for evaluators to refer to the points that interest them most.

```
┌─────────────────────────────────────────────────────────────────────┐
│       REPOSITORY FACILITY DESIGN, ENGINEERING, AND CONSTRUCTION     │
│                             Volume 1 of 2                           │
│                          TECHNICAL PROPOSAL                         │
│                                                                     │
│   Section      Description                                  Page    │
│                                                                     │
│     I.      INTRODUCTION AND SUMMARY                         1-1    │
│                                                                     │
│    II.      TECHNICAL DISCUSSION                             2-1    │
│                                                                     │
│             A. Design, Development, and                      2-2    │
│                Testing of Structures                                │
│             B. Technical Approach to Each Task               2-6    │
│             C. Interpretations, Requirements, and            2-21   │
│                Assumptions                                          │
│             D. Identification of Problems and Solutions      2-22   │
│                                                                     │
│    III.     PROGRAM PROPOSED                                 3-1    │
│                                                                     │
│             A. Staffing and Organization                     3-3    │
│                of Project Team                                      │
│             B. Management: Concept, Plans,                   3-6    │
│                Procedures, and Controls                             │
│             C. Labor-loading Schedule                        3-12   │
│             D. Performance Schedule                          3-15   │
│             E. Deliverables                                  3-17   │
│             F. Quality-Control Program                       3-20   │
│             G. Experience, Education of Key Personnel        3-24   │
│                                                                     │
│    IV.      QUALIFICATIONS                                   4-1    │
│                                                                     │
│             A. Organizational Relationship                   4-2    │
│             B. Experience with Similar Projects              4-4    │
│             C. List of Projects, References,                 4-11   │
│                and Contacts                                         │
│             D. Resources and Facilities                      4-16   │
└─────────────────────────────────────────────────────────────────────┘
```

FIG. 5.5 Table of contents for volume 1 (technical proposal) of a two-volume proposal.

Executive Summary. The executive summary is used more frequently than a response matrix. Its purpose, as its name implies, is to highlight the key points of a proposal for managers and administrators who don't have time to read the whole document. Experienced proposers have found that an executive summary helps to draw wider attention to a proposal's chief appeals. The wider interest arises partly because more people want to read a synopsis labeled "executive" than one labeled merely "summary" or "introduction."

In any case, it is usual to include some kind of summary in a long proposal. In short presentations, the summary may be combined with the introduction to the text. But proposal writers should not overlook the benefits of using a separate executive summary since it can be so easily tailored for maximum sales impact. For example, an executive summary can arouse reviewers' interest in the proposal by stating briefly the technical advantages of the proposed approach and comparing them to a competitor's ideas. It can also create a desire for those ad-

Topic	RFP Reference	Proposal Reference
Review and evaluation of plant designs	Part 1, page 2	Section 2, pages 2-2 through 2-6
Understanding of objectives	Part 1, page 3	Section 1, pages 1-1 through 1-3
Identification of problems and solutions	Part 1, page 5	Section 2, pages 2-22 through 2-28
Management plan and controls	Part 2, page 7	Section 3, pages 3-6 through 3-12
Company experience	Part 3, page 2	Section 3, pages 3-6 through 4-4

FIG. 5.6 Portion of a response matrix, showing the typical format.

vantages, listing a few key facts that show the superior dependability and expertise of your firm.

One approach to writing a summary is to arrange the points as a series of paragraphs, each in bold type, with a bullet or a number for emphasis. Remember to keep the summary brief—about one page in length—unless the solicitation has called for an exceptionally large proposal. For more information on executive summaries, see Section 2.

Body

Introduction. A strong opening is essential to an effective technical presentation. The introduction, as the first part of the proposal body, should set the scene for what is to follow, while generating interest in the proposed project. Depending on the requirements of the solicitation, the introduction may include various kinds of data to put the detailed proposal in proper perspective. But the basic content of the introduction does not differ greatly from one proposal to another.

The introduction serves several fundamental purposes:

1. Cites the solicitation and identifies your organization as a suitable respondent
2. Acquaints the reader with your firm
3. Makes a brief, preliminary statement proving your understanding of the customer's problem and requirements
4. Makes a formal offer for the contract

Beyond the purposes listed above, a well-thought-out introduction can have a number of other important effects. For example, you might use the introduction to cite your organization's successful track record in projects similar to the one proposed. Or you might name a well-known technical expert or other top-level member of the team who will work on the proposed project if the contract is won.

In overall effect, the introduction should convey the message that the proposer is the logical awardee for the contract. In many proposals the introduction presents a one- or two-page overview of the offeror's technical approach, the reasons for pursuing the contract, and the role of the pursuit in the offeror's general program.

When a separate executive summary is used, the introduction should not restate the data covered in the summary. Instead, the introduction should focus on one or more key concepts to pull the reader into reading the technical discussion itself.

One of the most important aims in writing any proposal introduction is to achieve maximum readability. An easy-to-read introduction is a vital tool for generating a favorable impression among reviewers. Commonly, evaluators must read so many proposals for each contract that they get tired and their attention begins to wander. A proposal that uses short sentences and simple words to promote easy reading is often a refreshing change from the pompous and long-winded approach of many technical presentations. If you can combine a high level of readability with a few dramatic points in your introduction, you'll almost certainly draw increased attention to your entire proposal.

The introduction should be as short as is reasonably possible, depending on the total amount of data required by the contract. A small contract may require an introduction of between 1 and 3 pages in length, whereas a very large contract may require as much as 100 pages of introductory material, or more.

Technical Discussion. The second section of the proposal body is the heart of the technical proposal. Following the line of thought begun in the introduction, the technical discussion continues to fulfill certain minimal requirements of the solicitation and to utilize some kind of selling strategy. The technical discussion may contain as many headings and subheadings as are necessary to cover the technical aspects of the project.

At the very minimum, the technical discussion must include these elements:

1. A discussion of the customer's problem and requirements, demonstrating your understanding of all the issues—especially those areas that are of most concern to the customer
2. A statement identifying the customer's real needs (e.g., a discussion of the root causes of a symptom noted in the RFP—not a restatement of the solicitation)
3. An analysis of the customer's needs as a means of exploring the available solutions
4. An elaboration of potential problems involved in the work, including any interpretations or assumptions contained in your analyses; ideas on how to reduce risks and any contingency plans you've developed
5. An expository review of all the options, demonstrating the pros and cons of each
6. A statement of the conclusions drawn from a logical comparison of the available options
7. A detailed presentation of the proposed project.

The overall purpose of the technical discussion is to make your customer feel confident that your proposal is the best approach to the problem. That is, you use this second section to sell your approach through a series of technical arguments. The specific objectives of each proposal vary, of course, and you must tailor the strategy of your arguments to the customer's precise needs.

For example, a proposal to a state agency to supply a simple service such as metal deburring will use a totally different line of reasoning than an offer to design and build an international system of electronic mail for a multinational corporation. The deburring proposal might concentrate on the superior performance and cost-effectiveness of the offeror's work, comparing its services to those of its competitors. The electronic mail proposal would require a more complex treatment but would place greatest emphasis on specific technological issues and the innovativeness of the proposer.

Despite these kinds of differences, the development of the technical discussion generally follows a logical flow similar to the one outlined above—from analysis and understanding of the need to selection and presentation of the solution.

It is difficult to say whether any one part of the technical discussion is more vital to the success of a proposal than another. Each proposal should cover all the aspects noted above. However, there is one area in which many proposers fail to give adequate coverage. That is in the definition of risks and uncertainties.

A proposal that offers a frank discussion of such issues may earn a more favorable evaluation than one that ignores or skims over them. This sometimes occurs even when the latter proposal offers a more innovative technical solution. For instance, evaluators may reason that the latter solution has not been subjected to sufficient field testing and therefore represents a larger risk to the customer. Despite the apparent merits of the innovative solution, the customer may see the risk as too large to be worth taking. As a result, the innovative approach would most likely be rejected.

Similarly, the proposal writer should make a point of explaining any modifications or exceptions that have developed between the original solicitation and the work that is being proposed. *Remember:* Always strive to make your proposal easy to "buy." So if the customer's SOW has discussed a symptom (such as a noisy heating system) instead of the real problem (cracked pipes or a malfunctioning boiler), you must give careful consideration to this in your response. In so doing, you'll rise above competitors who fail to make these necessary distinctions.

At the same time you must be careful to present your argument as clearly as possible. Design every element of the technical proposal to address what the customer wants. Suggest modifications in the work scope, if necessary, but make certain your argument maintains its focus on the customer's requirements, rather than on any prejudices of your own.

By all means, avoid wordiness, jargon, and complex terminology. However, do not try to condense any part of your technical discussion into summary form. Regardless of how well-acquainted you are with the customer (or how familiar the customer is with your technical concepts), you must explain exactly what your approach is, how you plan to implement it, and why it is the best solution available to meet the customer's needs. Otherwise the customer may not be able to justify an award on the basis of your proposal.

In addition to the "what," "why," and "how" aspects of the contract, the technical discussion may include such exhibits as a preliminary design layout, systems flowchart, and other kinds of illustrations and tables. In some proposals the technical discussion covers program and management considerations, but these are normally placed in a third section: the program discussion.

Illustrations and Tables. Graphics may be used to increase the effectiveness of both the technical and the program discussions. As in other types of technical communication, illustrations and tables can help to dramatize the most important points of a proposal.

Another advantage of using graphics is that it helps to ease the burden on those who evaluate proposals. A few good illustrations and tables, carefully integrated with your text, may get your message across without adding to the eyestrain and boredom that tend to arise when evaluators have huge stacks of documents to read. And the faster you convey your message, the more likely it is to leave a lasting, favorable impression with reviewers.

There are many kinds of graphics that you can use—organizational charts, flow diagrams, matrices, performance networks, graphs, timetables, and various combinations. Each is well-suited to particular types of information and has its own pros and cons. An important listing, such as a schedule of deliverables, can easily be converted to matrix form, as Figure 5.7 shows. The matrix makes it simple for evaluators to pick out key items quickly. A flow diagram, on the other hand, can be highly effective for showing the major tasks, procedures, and milestones in a proposed program. Figure 5.8 is a project procedural flowchart that was used in a proposal submitted to the Nuclear Regulatory Commission.

Although some illustrations require the work of a professional artist, most can be done by a writer with the proper tools. But if your company does employ an illustrator or artist, make an effort to discuss your writing task with this person. Often the illustrator can provide creative ideas regarding the use of graphics.

Schedule of Deliverables Matrix

Tasks	0	1	2	3	4	5	6	7	8	9	10	11	12	13	14	15	16	17	18	19
Monthly reports		X	X	X	X	X	X	X	X	X	X	X	X	X	X	X	X	X	X	
Mobilization and organization review	•																			
Quarterly technical review					•															
Draft report, task 1							△													
Semiannual technical-management review								•												
Final report, task 1										▲										
Draft report, task 2												△								
Annual technical-management review														•						
Final report, task 2															▲					
Draft report, task 3																	△			
Quarterly technical review																		•		
Final report, task 3																			▲	
Final contract review																				■

X Required submittal
• Required meeting
△ Draft report submittal
▲ Final report submittal
■ End of contract

FIG. 5.7 Matrix for schedule of deliverables helps evaluators pick out key items quickly.

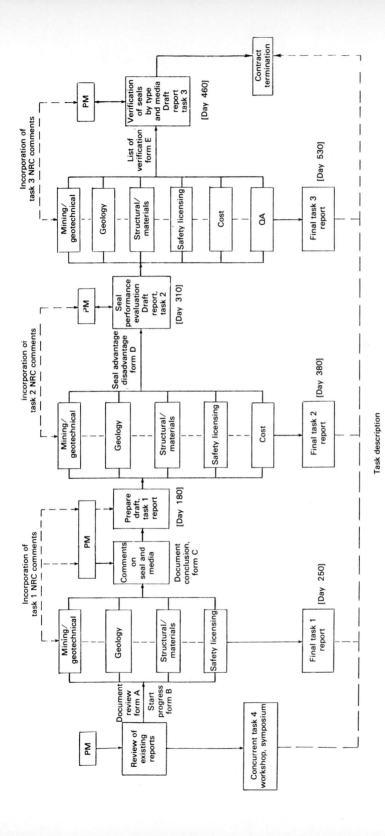

FIG. 5.8 An example of a procedural flow diagram. (This flowchart was used in a proposal submitted to the Nuclear Regulatory Commission.)

Task description

- Task 1: Literature review, technical evaluation
- Task 2: Seal performance evaluation
- Task 3: Seal performance verification
- Task 4: General technical assistance

5.25

Program Discussion. In the technical discussion, the proposer analyzes the customer's requirements and recommends a particular solution. In the program discussion, the proposer explains the plan for implementing that solution.

The program discussion is the third section of the proposal and the next logical step in the overall sales presentation. It is also the section of the proposal that tests the offeror's ideas most strenuously. During the technical discussion, you can make as many promises or commitments as you want. But here is where you must prove that you can and will fulfill all those promises and commitments.

This third section is sometimes the largest part of the entire proposal. The number of subsections or topics that it contains may vary considerably from one proposal to another. However, the following items must be covered in every program discussion, no matter what the contract requirements are:

1. Staffing and organization for the proposed project
2. Résumés of key managers and staff members
3. Delineation of work responsibilities by staff member and by number of work hours involved
4. Performance schedule
5. List of deliverable items and services

In a small proposal, some of these topics may not require lengthy treatment in the text. Certain items might be covered quite simply in flowcharts, tables, or lists. But a large or complicated proposal may require breaking the program discussion into a series of subsections, each devoted to a separate aspect. Within each subsection, you might have a comprehensive examination of a single element of your program, possibly covering everything from the basic engineering principles involved to the policies, procedures, equipment, and forms you plan to use to carry out that aspect of the program.

In any case, the program discussion must quantify each and every item that can possibly be quantified. Deliverables, specifications, time, distance, labor, etc.—all must be presented in terms of specific quantities to enable the customer to evaluate your proposal in detail. If a figure represents an estimate rather than a firm amount, be sure to label it as such.

Failure to specify amounts for these types of items may lead to serious problems. Evaluators may interpret a failure to quantify as an indication of evasiveness on the part of the offeror. Alternatively, the failure may be seen as a sign of incompetence or general lack of responsiveness in the proposal.

If the offeror manages to win a contract award in spite of a failure to quantify, other problems may arise. For example, the customer may neglect to demand more specific figures in the writing of the contract, and contract disputes may later result. When a contract is signed without specifying exact figures or estimates for all quantifiable items, there is a greater risk that the customer will later take exception to the contractor's choice of materials and specifications. Therefore, let this be a principle in writing every proposal: Quantify every item that you can, regardless of whether or not the solicitation requires quantification. In so doing, you will show that your proposal gives careful consideration to details, and you will help to protect your firm against contract disputes.

Another pitfall to avoid in the program discussion involves the performance schedule. Often, solicitations present a suggested or mandatory schedule for proposers to follow; but it is usually a mistake to commit yourself to this schedule without first developing your own version. Why? For several reasons. The customer's schedule may be vague or incompatible with your specific plans. By promising to adhere to it, you could actually jeopardize the outcome of your project. If, on the other hand, you develop your own schedule, you can tailor it to the specific needs of your proposed solution. Further, you can use your own schedule as proof that you have analyzed the problem diligently and planned it according to the customer's precise needs. This may add to your overall credibility in the eyes of the customer.

Staff Résumés. Because the program discussion should cover all specifics of the proposed program, it is usually a good idea to present the résumés of key project personnel in this section. Principal performers on the proposed project—supervisory engineers, managers, and high-level specialists—are important elements of the program itself. General support personnel—secretaries, typists, drafters, and others—are not key elements in most programs; data on their experience generally belong in the section on corporate qualifications. Data on the support staff tend to be more significant in the overall qualifications of the organization than in the program proper.

A mistake that proposers sometimes make here is to use standard résumés, drawn from company files, for the key staff people. These résumés rarely present an up-to-date account of the skills and activities of the professionals they are supposed to represent. Even when the information in a standard "boilerplate" résumé is current, it usually has not been tailored to the needs of a specific project. Such résumés do not present staff members in ways that best reflect their capabilities to satisfy all the requirements of a given contract.

For maximum effectiveness, each résumé in a proposal should highlight the individual's qualifications for the specific project that has been proposed. Thus the résumé should emphasize specialized knowledge and experience, not general expertise. Each résumé should be rewritten for every new proposal. That way the résumé can concentrate on the specialties that are most relevant to individual projects.

Qualifications. The final section of the proposal body covers corporate qualifications and experience, unless the project requires a section for special considerations, such as management structure or quality control. The three items that must be covered in this section are:

1. Evidence that your firm has suitable technical expertise and professional experience to provide the best solution to the customer's problem
2. Background information showing that your organization is reliable and consistent in providing superior results
3. Description of your organization's facilities, equipment, and financial resources, proving that the company has everything needed to make the project a success

There is a tendency among proposers to attempt to offer this information in the form of standard lists or charts that the company has designed for general use. This is a common mistake, similar to that of using boilerplate résumés. Standard lists of qualifications may save time and effort in a firm that confines its work to a single, highly specialized type of project, but for most proposers this is not appropriate.

It is advisable to begin the fourth section with a brief introductory description of your firm's technical and professional qualifications, emphasizing their relevance to the proposed project. Here you can also introduce your firm's proven track record of dependability, plus any special achievements or awards that the staff has earned. Other data that may be pertinent in this opening subsection are:

1. A short summary of the company's history
2. An overview of the organizational structure—including a description of the parent organization, departments, or divisions as they relate to the proposed project

These opening remarks set the scene for a listing of the organization's specific qualifications. Even when the customer has not requested such a listing, it is wise to furnish one. However, try to avoid using standard lists or pages from past proposals.

Qualifications should be organized in a way that best meets the requirements of each proposed project. When the RFP specifies a particular kind of listing, this may simply mean following the customer's suggestions. Generally, make it a rule to arrange your qualifications by placing the most relevant work first. That is, list all pertinent details of the firm's current and past projects—contract numbers, dollar values, names of customers, key personnel, etc.—but tailor the order of the projects to the proposal. Thus, if the contract focuses on some type of geological engineering work, place the current or past project that is most similar to the contract at the top of your list.

Make certain that the data on each project is internally consistent with the others. If the first item begins with a brief description of the type of work required for a recent contract, the other items should begin the same way. Important data to include in a listing of current and past contracts are:

1. Name and type of contract, identifying number, dollar value, and customer's name, address, and telephone number
2. General requirements and outcome of the work
3. Remarks on significant milestones in the contract (especially favorable ones), such as early delivery of items and pilot test results
4. Names of other key references, such as contracting officials and project managers, with addresses and telephone numbers

Another approach that has proven to be effective in showing an organization's experience is to use a corporate experience matrix. A well-constructed matrix has the advantage of providing quick reference to specific technical activities in which the firm has engaged. This is especially valuable when a lengthy listing would otherwise be necessary. In Figure 5.9, the matrix uses a series of numbers (1 through 14) to key a listing of specific technical activities to each of 11 projects. To gain maximum effectiveness from this type of approach, it is advisable to place the matrix and listing on a foldout (8½- by 17-inch) sheet and bind it in the proposal.

Other items that belong in the fourth section are descriptions of the physical facilities and resources of the offeror. An annual report or other type of financial statement may also be placed here, if the customer has called for such information.

In addition, the qualifications section may include résumés of representative support staff who would be involved in the proposed project.

Back Matter

Appendix. An appendix contains supplementary material that does not belong in the body of the proposal. Often this consists of bulky or highly complex matter that would be of interest to only a few readers. Anything that unnecessarily slows down or confuses the progress of the technical argument and that impairs the sales presentation can be considered for placement in an appendix. However, if the material pertains directly to the technical argument, it probably belongs in the proposal body. The appendix should contain material that is not integral to the development of the proposal.

For example, if you refer briefly to a published paper or article in your technical discussion, you might include this article as an appendix—provided that it supports your overall presentation. Or you might use the appendix for testimonial letters from satisfied customers, for examples of calculations, for photos of previous projects, or for other background items.

Depending on the subject matter of this supplementary material, you may choose to use a single appendix or multiple appendixes. Multiple appendixes are normally used to separate different types of materials or data on diverse topics. For instance, you might put background calculations in one appendix, photos of past work in another, and a technical report in a third appendix.

Careful use of appendixes may enable you to trim some of the "fat" off a lengthy or highly complex proposal, without omitting important data. This helps to streamline the overall presentation.

Back Fly Sheet. A blank fly sheet is often used as the final page of a technical or cost proposal (or of each volume). The fly sheet provides additional protection for the proposal's contents. It may carry an imprint of the company logo, if desired. The fly sheet may also be used for reader's notes.

Back Cover. The stock used for the front cover of a proposal should also be used for the back cover. The back cover is usually left blank, unless the proposal involves classified information. In that case, the back cover may carry a statement or stamp noting the classification or the proprietary nature of the contents. This information may be repeated from the front cover.

PROPOSAL DEVELOPMENT STEPS

The effort to create a winning proposal begins when top management makes the decision to pursue a contract. What typically follows that decision, in many companies, is a seemingly endless series of meetings, memoranda, and disagreements regarding the best way to handle the project. In many cases, the proposal is not actually written until a few days before the delivery deadline, resulting in a last-minute frenzy of disjointed activity. Frequently, the resulting product is a conglomerate of canned brochures and boilerplate material—hardly the makings of a superior proposal.

Proposers who know how to develop a good proposal in a short time are the exception rather than the rule. But there are organizations, both large and small, that turn out excellent proposals consistently. These are organizations that shun the canned material in favor of a more creative approach to each contract they

Client: Nuclear Project	Activity													
	1	2	3	4	5	6	7	8	9	10	11	12	13	14
Louisiana Power & Light Company: St. Rosalie, Louisiana	X				X	X		X	X		X	X	X	X
New England Power Company: Charlestown, Rhode Island		X	X	X		X		X	X	X	X	X	X	X
Central Maine Power Company: Sears Island, Maine		X						X	X			X	X	X
Washington Public Power Supply System: Hanford, Washington		X				X			X	X		X	X	X
Public Service Company of New Hampshire: Seabrook, New Hampshire	X		X	X	X	X	X	X	X	X	X	X	X	X
Delmarva Power & Light Company: Summit, Delaware		X			X	X	X	X	X	X	X	X	X	
Los Angeles Department of Water & Power: San Joaquin, California		X				X			X					
New York State Electric & Gas Corporation: Somerset, New York		X												X
New York State Electric & Gas Corporation: Site Selection Study, New York		X						X						
Carolina Power & Light Company: Brunswick Station, North Carolina		X												X
ENEL: Proposed Nuclear Site, Italy		X												

(a)

FIG. 5.9 (a) Matrix showing organizational experience keyed to specific activities.

Description of Key Technical Activities
Performed on Projects Listed

1. Study and Report Preparation of site selection, subsurface investigations, laboratory testing of soils, environmental engineering, foundation and stability analysis
2. Technical Analyses of specific structures
3. Study, Design, and construction supervision of embankments
4. Instrumentation and Monitoring Programs
5. Study, Design, and Construction of tunnels
6. Design and Treatment of soils and rock
7. Construction for blasting foundation excavation
8. Preparation of construction specifications
9. Preparation of safety reports
10. Documentation and testimony of safety reports
11. Design and Evaluation of dewatering
12. Evaluation of groundwater studies
13. Assistance to Quality Assurance on soils and rock treatment
14. Legal Testimony, Licensing Support

(b)

FIG. 5.9 *(b)* Key to activities in organizational experience matrix of Figure 5-9a.

pursue. To do this, they follow a carefully planned strategy designed to turn the goal of a superior proposal into a reality.

Successful proposers know that writing is a major phase in developing any proposal, but they understand that a great deal of effort precedes the actual writing phase. Although the variations in technique are limitless, experienced proposers recognize that the proposal development process involves a number of different skills, including at least these major ones:

1. General ability to determine the essential problem and requirements of the customer
2. Management skills in selecting the team to develop the presentation
3. Marketing input about the customer and the strategy needed to rise above competitors
4. Technical excellence to define the services and approach required for a successful solution
5. Writing and editing, to convey the message clearly
6. Graphics, to present the project in a dramatic and interesting way

7. Coordination and typing of materials, to produce the document

Rarely are all these capabilities found in one individual. While some large organizations employ one or more proposal writing experts, most must assemble an ad-hoc team of managers, technicians, salespersons, and others to create the presentation for each prospective contract. Coordinating such a team can be a major effort, and even a professional technical writer must have some way of managing input from several sources. Hence, a detailed plan is required.

With experience, astute proposal writers tend to develop specific sets of procedures geared to maximize the strengths and minimize the weaknesses of each proposal writing situation. They start with basics—by identifying the most important phases and functions in proposal development. Once these phases are understood to follow some logical order and progression, they can be used as the basis of a comprehensive selling strategy.

To present the proposal development process in a form that can easily be tailored to individual projects, we've broken it down into a series of major steps. These are shown as being sequential. In practice, however, some of these steps overlap one another, or they may be repeated at various points in the course of developing the whole presentation.

These are the major proposal development steps:

1. Define or identify the principal requirement.
2. Choose the project leader and assemble the proposal team.
3. Assign specialists to outline specific elements.
4. Coordinate the outlines to draw up a whole scenario.
5. Draft the proposals (technical and cost).
6. Review the first drafts and identify problems.
7. Revise the drafts, verifying accuracy and approach.
8. Submit the second drafts for management review.
9. Revise the drafts to make final adjustments.
10. Produce the document.

These steps make it easy for anyone to get a good start on writing a proposal.

Step 1. Define or Identify the Principal Requirement

Analyze the solicitation package thoroughly—RFP, statement of work, evaluation criteria, attachments—to determine the customer's essential requirement. Then state this requirement in crystal-clear terms. *Remember:* You can't write a persuasive proposal unless you know where you are going with your offer. Writing is only one phase of proposal development. And what you write can be only as good as the information you gather beforehand. This takes careful research and planning.

In researching a prospective contract, some organizations do not rely solely on formal solicitations for data. Instead, they may use an "inside track." That is, they monitor the customer's activities, concerns, and procurement ideas through a network of personal contacts or other intelligence-gathering methods. In this way, they sometimes can learn about a contract far in advance of the RFP release. Use of the inside track is a time-honored practice, and for firms that can find such a route, it may be a useful approach. But it can also be a trap. Customers who have close contacts with one firm may bend over backward to avoid appearing biased, for example. Or competitors may lodge formal protests against what they see as an unfair selection process.

Regardless of how you collect your data, the surest way to plan an effective proposal is to assemble all your information on the customer's requirement and submit it to a painstaking series of analyses. Read the solicitation package carefully, paying particularly close attention to the issues that seem to be of greatest concern to the customer. Always take note of matters of scheduling, deliverables, technical and cost limitations, and all written instructions.

There are a wide variety of analytical methods to choose from—graphic analysis, systems analysis, brainstorming, outlining, checklists, etc.—but all require you to think through the problem and define the customer's requirements.

If you are working on an unsolicited proposal, of course, you will not be able to depend on an RFP for your data. Therefore, it is advisable to engage in some extended discussion with the customer before you make a formal submittal. Make sure there are no bureaucratic or budgetary obstacles in your way, and make certain that the person you speak with has the authority to get your plan approved and funded.

For solicited proposals, on the other hand, it is normally advisable to avoid making direct inquiries of the customer. Many customers are reluctant to answer questions that may give one offeror an advantage over others. It is thus the offeror's responsibility to define or identify the customer's problem—even when the customer's written solicitation may be unclear or misleading.

It is fairly well accepted that many RFPs are poorly written. Suppose, for example, that the customer lacks knowledge in the field in which services are needed. Let's say the customer, a manager of a small word processing firm, finds that his computer facilities are experiencing sudden and frequent equipment failure. If the customer has little knowledge of the actual hardware being used, he may conclude that the problem lies in the computer system itself. Unable to solve the problem without calling in a contractor, the customer writes an RFP that describes the problem as "repeated equipment failure." But equipment failure is really the primary symptom, not the problem. The real problem may be power surges in the electrical lines, too many operators using the system at one time, improper use of certain computer programs, poor grounding of electrical charges, or any of a number of other problems.

In cases like this, the need for proposers to analyze the customer's requirements becomes all the more apparent. The principal requirement here is for the contractor to determine the problem that is causing the equipment failure, and to solve it. But the RFP has cited only a symptom, not the real problem. So the proposer must study the solicitation in order to identify the true problem.

Not all customers lack the necessary technical knowledge to state a problem clearly, but virtually all potential contracts require study and analysis before a good proposal can be written. That is why we say you must define or identify the principal requirement.

In many organizations, top management handles the responsibility of defining the principal requirement, which is the chief determinant of the bid-don't-bid decision. However, the task of specifying all elements of the customer's need is a staff responsibility. This task generally belongs to the proposal leader and individual members of a proposal development team.

Step 2. Choose the Project Leader and Assemble the Proposal Team

Large proposals are rarely written by a single individual; they are put together by a proposal team. Usually a vice president or other top executive chooses one person to lead the proposal development effort. This person may be a professional technical writer, project manager, or other key staff member. During proposal development, he or she is commonly referred to as the *proposal leader* or *project manager*.

The leader's job is to coordinate the research, writing, and production of the document. Hence the proposal leader may assemble the team, dividing it into separate groups of specialists as necessary. These specialists may include engineers, business managers, quality-control experts, marketing directors, drafters, support staff, subcontractors, and outside consultants such as attorneys and accountants.

These people, many of whom will produce the proposed system if the contract is won, will be the authors of the proposal. Each specialist (or group of specialists) writes the sections of the proposal dealing with his or her area of expertise. Thus a large proposal can have dozens of contributing authors.

It is the proposal leader's responsibility to give each member a specific, clearly understood writing assignment. Then the leader must monitor progress and mold the work of the team into one cohesive, comprehensive presentation.

In a small firm where the principal does almost everything, the team may consist of little more than a project manager, an engineer, and a typist. Or the proposal may be written by a single individual. Many fine proposals have been developed by professionals working alone. Further, two- and three-member "teams" have been used successfully by small organizations and by large organizations for small projects. In any case, the functions and phases in proposal development are essentially the same—regardless of the size of the organization, the project, or the kinds of technical disciplines involved. Here we'll focus on how it's done by a team with several members.

Once the proposal leader has chosen the potential members of the team, a coordination meeting is held. The leader usually furnishes the staff with copies of the solicitation package, plus a brief summary defining the scope of the work, scheduling requirements, the estimated dollar value of the services requested, and other key information. To organize all the aspects of the writing task coherently, the group may draft a detailed outline at this point. The outline should show the prospective format of the proposal, with descriptions of the various elements and the types of information they must cover.

During the initial coordination meeting, the proposal leader may issue specific writing assignments, asking each member to make a formal commitment to the project. On the basis of these commitments, the proposal leader may draw up a chart showing the formal organization of the proposal team. Figure 5.10 shows an organization chart used for a proposed project submitted to the Nuclear Regulatory Commission. Charts of this type may be used both for the benefit of the proposal team itself and as a graphic device in the written proposal. The organization chart shows the working relationships between the team members.

Another important duty of the proposal leader is to set up the proposal development schedule. Usually the schedule is built by working back from the due date. For instance, you might begin to construct a rough schedule as follows:

Delivery: Not later than 8/15/XX
In mail: By 8/10/XX

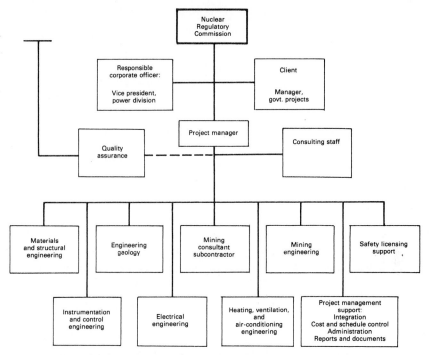

FIG. 5.10 Example of a project organization chart. (This organization chart was used in the development of a proposal submitted to the Nuclear Regulatory Commission.)

Production: Copying and binding 5 sets: 8/8/XX
Final revisions and adjustments: 8/5/XX
Management review: 8/1 to 8/3/XX
Typing of second draft: 7/30 to 7/31/XX

Step 3. Assign Specialists to Outline Specific Elements

A team is useful only if each of its members gains a full understanding of the specific requirements dealing with his or her area of specialization. Each specialist must analyze the RFP, SOW, and evaluation criteria to devise strategies and approaches for the various aspects of the whole presentation. But it is necessary to manage these individuals according to some set of plans or procedures, to prevent confusion and move forward.

There are a number of techniques that have proven effective in actual practice. As we've noted above, the proposal leader and team may conduct brainstorming sessions and use a variety of analytical methods, such as flowcharting for systems and procedures, value analysis, and listing of deliverables. These techniques may be employed in group meetings during the early phases of proposal development. But at some point the specialists must begin to record the results of their analyses on paper.

Typically the proposal leader assigns each team member (or subgroup) to produce one or more outlines for his or her area of responsibility. For a small proposal, this may mean that a group of engineers and other technical people outline the technical proposal, while accounting and management people outline the cost proposal. For a larger proposal, outlines may be required for each of several disciplines, including, for example, such areas as structural engineering, engineering geology, mining, electrical engineering, and cost and schedule control. In addition, subcontractors may be asked to contribute outlines relating to their responsibilities on the proposed project.

Properly coordinated, a group of people can often produce a better final product than can an individual. Each person in the group must have a clearly delineated set of responsibilities, however. It is the proposal leader's job to delineate those responsibilities and to supervise the members to ensure that the team makes progress. To this end, the proposal leader may review the work of each team member to ensure that he or she is producing the required data and organizing it in a way that does not conflict with other parts of the project. The manager may also issue memos and official directives specifying the types of material that must be produced, due dates, modifications in the scope of work, etc.

Either way—whether you have a large team or you are working alone—the proposal will begin to take shape when you produce outlines of the required work. Once the proposal leader reviews the outlines, they should be refined and typed carefully. These outlines may later serve as guides for the table of contents for each volume of the proposal. But first they must be coordinated in a clear, coherent scenario.

Step 4. Coordinate Outlines to Draw Up a Whole Scenario

Gather together the entire proposal team and distribute copies of the outlines. Then arrange the materials according to the order in which they should tentatively appear in the written document. One good way of doing this is to tape the outlines on a wall so the team can easily compare the outlines to the prospective format. This sequence is the essential proposal scenario, from start to finish.

Coordination meetings are necessary to enable the team to monitor its progress and compare notes on the different aspects of the whole project. The cost accounting people, for example, cannot refine their database without knowing the results of work performed by engineers and other technicians. In order to develop an accurate cost proposal, they must know exactly how many work hours will be involved in electrical engineering, what the salary rates will be, what the deliverables are, etc. Meetings of this kind also enable the team to blend the individual efforts into a unified approach.

The meeting provides an overview of all the data to be presented in the proposal. Deficiencies or other problems should be detailed so the relevant specialists can devote extra effort to developing all the project data. The coordination meeting also furnishes an opportunity for the members to discuss the use of graphics for specific parts of the proposal. At this point the team can determine exactly where in the written proposal each bit of information will be placed and how it should be written.

Step 5. Draft the Proposals (Technical and Cost)

Although much of the proposal development process involves preparation rather than writing, the way you and your colleagues write the proposal can mean the difference between acceptance and rejection of the project. The winning proposal is technically accurate, responsive to the customer's requirements, and comprehensive. It must be written so that the key points can be picked out easily by evaluators as they scan the pages. And it must sell.

Follow the tips presented earlier in this section for writing each element of your proposal. In addition, see the guidelines in Section 2 on report writing—many of the guides for effective technical reports are useful when drafting proposals. And follow the proven practices listed below for maximum effectiveness:

- Treat the proposal as a sales presentation.
- Design front matter to stimulate interest.
- Show complete understanding of the customer's need.
- Respond fully to all requirements—especially the customer's areas of greatest concern.
- Emphasize benefits of your proposed project.
- Distinguish your approach from that of competitors.
- Furnish evidence of your organization's total dependability.
- Prove that your plan will work—give firm evidence.
- Use graphics to convey your message quickly and dramatically.
- Reinforce your text with key staff résumés designed to satisfy the customer's chief concerns.
- Incorporate testimonial letters and other references from past or current customers.

Step 6. Review First Drafts and Identify Problems

Once the elements of the proposal have been drafted, they should be assembled and reviewed carefully. If you are the proposal leader, you may wish to review the drafts yourself. For a thorough review, however, copies of the complete draft should be distributed to each member of the proposal team. (Some firms prefer to break the team into groups, with each group's reviewing the parts of volumes of the overall proposal relating to its own area of specialization—technical proposal, cost proposal, management plan, etc.) Usually a group can more easily point out areas of weakness or inconsistency than can a single individual.

Examine the arrangement and contents of the proposal elements. See that they follow the general format specified by the customer in the RFP or in other communications. If no specific format has been mandated, use the general format outlined earlier in this section. If necessary, you can alter the format to better suit the customer's concerns. Of course, if your firm has a standard format for proposals, use it.

When reviewing the proposal draft, be as critical as possible. Every improvement you make before submitting a proposal is one more item in your favor. *Remember:* Once the proposal is submitted, it cannot be done over. Rarely does a proposal effort permit you the luxury of a leisurely schedule. And you cannot sit beside the evaluators—the customer, in effect—and explain the proposal as it is read. The document must stand on its own. And it should be the best your organization can produce.

Step 7. Revise Drafts, Verifying Accuracy and Approach

Although the process of reviewing a proposal should involve the many engineers, technical managers, and others who will perform on the contract if it is won, the job of revision usually belongs to professional technical writers and editors. Their job is to improve the readability and accuracy of the typed document. Thus the proposal leader and editorial staff may mark a copy of the draft to correct errors of grammar or typing and to question any unclear or potentially inaccurate data.

Marketing experts may also provide input at this stage to help enhance the selling aspects of the written proposal. Once all the questions have been answered and the necessary corrections have been made, the final draft should be typed according to the specifications of the RFP or of your firm's publication specialists. The proposal is then ready to be submitted for management review.

Step 8. Submit Second Drafts for Management Review

Submit a copy of the complete document to the appropriate executive(s) for final review. The final draft should be complete in every detail, with each paragraph, illustration, table, and résumé in place.

For most proposals, this is the last step before the whole document is printed and sent to the customer. Thus it may be your last opportunity to catch errors or weaknesses and correct them. Top managers can often point out potential problems that might otherwise pass unnoticed, since their perspective is usually different from that of the specialists working directly on the project.

Managers tend to scrutinize a proposal from a broader perspective. Rather than focus on specific technical considerations, for instance, they may read the document for legal and contractual ramifications and for matters of cost and administration. Hence management may desire to consult with legal and accounting people at this point to clarify any last-minute questions.

Management review of the proposal is especially important because the proposal itself will become the basis for a written contract, if your firm wins the award. However, it is also important to avoid spending too long on this step. Some organizations expedite the process by presenting a synopsis of the document to key managers, instead of having each manager read the entire proposal.

Always make certain that management is alerted to the need for promptness in the final review. Don't let valuable time be spent in a lengthy examination when the time may be needed for some final changes before sending the proposal to the customer. Unless you make everyone involved keenly aware of the deadline for delivery, they may hand you a list of corrections to make on Thursday, for a proposal to be delivered on Friday!

Step 9. Revise Drafts to Make Final Adjustments

Make any final revisions or modifications as quickly as possible. Usually these are limited to major problem spots discovered during management review and to errors of fact, grammar, and spelling. If the entire proposal development team has fulfilled all its responsibilities properly, there should be few, if any, corrections necessary.

Step 10. Produce the Document

The proposal development process concludes with production—typing, printing, collating, and binding. Depending on the size and facilities of the organization, the proposal writers may not be involved in this step.

Most organizations produce the proposal—including as many copies as the customer needs—using their own facilities. A large firm may send the material to its own printing and production department, while a small firm may use one typist and a duplicating machine to do the job. In any event, the aim is to produce a clean, error-free, easy-to-read document for timely delivery to the customer. This goal demands strict attention to quality control and scheduling.

By following the steps and guidelines for proposal development presented in this section, you can improve your proposal writing and make the wonder of a winning proposal happen consistently.

CHECKLIST FOR PROPOSAL WRITING

Preparing to Write

1. Solicited Proposal. Have you obtained and studied all solicitation materials (RFP, evaluation criteria, SOW, attachments) and determined the requestor's immediate objectives? Do you know exactly what the requestor needs? Can you deliver what the requestor wants?

2. Unsolicited Proposal. If necessary, can you provide evidence that your proposal has not involved any type of solicitation? Have you studied the problem thoroughly enough to know that the prospective customer would welcome a proposal? Have you developed a suitable solution to the problem? Do you have reliable contacts within the customer's agency or organization? Are you sure that the necessary funding will be available when and if your proposal wins a contract award?

3. Proposal Strategy. Do you have a comprehensive proposal strategy? Do you know how you will address areas of chief concern to the customer? Have you formed at least a basic set of procedures or an overall plan to:

Win the customer's attention?

Arouse interest in your proposed project?

Create a desire for your solution?

4. Technical Plan and Cost Estimate. Do you have a sound technical plan? Have you estimated the cost of the proposed project? Are your costs reasonable? Do they flow readily from your technical plan?

Writing and Reviewing

1. Identification of the Problem. Does your proposal analyze the customer's needs and identify the true problem? Do you state all requirements in your own words? If your overall argument involves any new interpretations or modifications of the customer's requirements, do you defend them?

2. Format and Elements. Does the entire document conform to the customer's specifications, if any have been stated? If it does not follow the mandated format, do you give firm support for the format used? Does your format follow a logical and effective presentation strategy?

3. Technical Discussion. Does it open strongly, in a professional manner? Does it emphasize benefits or otherwise arouse interest? Does it flow smoothly from the problem definition? Do your technical arguments proceed clearly and logically from the customer's needs? Do you cover the pros and cons of alternative solutions, showing why the approach you've selected is superior? Are all potential problems covered? Does the discussion demonstrate a high probability of success?

4. Program Discussion. Is the proposed project presented in full detail to show organizational structure and management plans and procedures, staffing, and professional qualifications? Are quality-control and contingency plans elaborated? Are all quantifiable items quantified? Are all schedules (deliverables, performance, etc.) presented clearly? Does the discussion provide strong evidence (e.g., key staff résumés) of the excellence of your proposed staff?

5. Qualifications. Are all relevant projects—current and past—described? Are key references given, with the names, addresses, and telephone numbers of contact persons? Is a matrix included to provide easy reference to company experience in activities similar to those involved in the proposed project?

6. Overall Document. Has every section been meticulously proofread and edited? Have all elements been assembled properly—including front matter, back matter, illustrations, and tables? Is every item neatly produced? Does the whole document look professional? Have you produced the number of copies that the customer has requested, plus copies for your organization's library?

SECTION 6
SPECIFICATION WRITING

Engineering and industrial specifications account for a sizable chunk of the total number of words written by technical writers each year. No comprehensive record of technical wordage is available, but many technical writing specialists believe specifications (called *specs* for short) account for millions of worker hours of writing time annually. In industry and government today, specifications must accompany any procurement of goods or services. Specifications are also required for the design, manufacture, testing, purchase, delivery, and use of various items.

To prepare specifications properly, the technical writer must develop a well-structured approach to this unique writing task. As a technical specialist or writer, you can learn to write better specifications by studying how they are produced and applied in engineering and industry today.

SPECIFICATION WRITERS AND USERS

In discussing specifications, you will find that the organizational aspects of the writing task may become quite complex. In some cases a specification will involve data obtained from a number of individuals and sources. For example, several parties may be involved at various points in the process of developing a single specification for, say, a military helicopter. These may include various engineers, design contractors, parts manufacturers, and others. On a complex project the specification may be part of an entire project specification hierarchy, where each party contributes data relating to requirements at one or more levels of the whole project.

Because such a wide variety of factors may affect the writing of nearly any specification, we have structured our approach in this book around the essential information you need to know to begin preparing engineering and industrial specifications successfully. A good place to start is with an overview of the types of individuals who are responsible for specification writing.

Who Writes Specifications?

Four broad groups write specifications today: (1) engineers, (2) architects, (3) specification writers, and (4) marketing departments. The first group includes people from all fields—aerospace, chemical, civil, computer, electrical, mechanical, mining, nuclear, or any of a number of other engineering disciplines. Individuals in the second group—architects—prepare architectural specifications for buildings of all types. Although there is a close parallel between engineering and architectural specifications, we give greater emphasis to the former in this book. The reason why we put more emphasis on engineering specifications is that they constitute a larger portion of the total number of specifications written today. But most of the guidelines that we give for engineering specifications are applicable to architectural specifications.

Specification writers, the third group, are specialists who work for, or with, engineers or architects. Many engineering, architectural, and manufacturing firms employ these writers to prepare specifications on a full-time basis. The specification writers may be graduate engineers or skilled technical writers who command an expert knowledge of the technical field in which they prepare specifications.

In small firms the engineer who designs a product or structure may also write the specifications for it. But in many medium-size and most large firms, the specifications are prepared by a writer who works closely with the design, project, or consulting engineer. This combined effort is often the best way to produce well-written material in the least time. In all the discussions in this section where we mention an "engineer writing a specification," remember that the actual work may be done by a specification writer under the guidance of the engineer.

Items Specified

Specifications must be written for almost every engineering and industrial project undertaken today. If a new power plant is to be built, for instance, a comprehensive specification (or a series of specifications) is required. The same is true of aircraft, weapon systems, truck engines, computer equipment and programs, machinery, ships, and other complex systems. Specifications are also required for each of the components, materials, and products that will make up a system. Thus, specifications are written for a component such as a valve, solenoid, pump, or fuel tank and for supplies like lubricants, cleaning fluids, light bulbs, and paper.

You can safely say that where any structure, device, product, process, or machine is to be designed, built, tested, purchased, or installed, a specification is needed. The specification may be only a few words, as, "Furnish four (4) 10-megabyte hard disks." Or the specification may run to hundreds or thousands of pages, as for a complete weapon system or power plant.

Specification Use

The link between an engineer's concept of a structure, device, or product and the finished item is a specification. For without a specification the person or firm building or supplying the item may not know the exact requirements necessary to achieve the desired results. Specifications thus play a vital role in the procurement of goods or services by government agencies and private clients.

The various requirements determining what a specified item will do and under what conditions it will do it may originate with the ultimate users, sellers, or distributors of the item. Thus, the ultimate user—say, the Department of Energy—

may have its engineers specify the general requirements for a research reactor. But these requirements must be analyzed by the suppliers of each definable item used to make the final product. Based on their analyses, the various suppliers must produce a wide range of specifications covering everything from building materials to design, testing, and construction of the reactor.

The user's requirements are, in effect, the starting point for all the specifications. These requirements flow down from the ultimate user to the parties at various levels in the specification hierarchy, directly or indirectly determining the rest of the specifications. In order for all requirements of the procurement to be met, good specification writing must occur at each point in the overall effort.

An understanding of the relationships between the parties involved in any project is necessary not only for good project management but also for good specification writing. So let's take a quick look at some of the typical parties involved in writing and using a specification. The 10 terms you will probably encounter most frequently are: *engineer, consulting engineer, client, owner, firm, architect, contractor, supplier, manufacturer,* and *subcontractor.*

In a general sense, the term *engineers* and *consulting engineers* refer to people who do the detailed work of designing the product or facility. A *consulting engineer* is usually hired on a fee basis by the *client* or *owner* to do the required work. In some cases the *architect* hires the consulting engineer. Where engineers or architects are on a payroll, of course, they are usually said to be working for a firm. Some industrial companies have permanent employees known as *consulting engineers.*

Construction specifications, which govern work done at the building site, are written mostly by engineering and architectural firms. Engineers and architects may also write a number of other specifications, including ones for equipment and materials purchased and used in the product or facility.

The *contractor* bids on the work covered by the engineer's or the architect's specification. Usually it is the contractor who actually builds the product or facility, although some engineers handle their own construction work. A contractor may employ *subcontractors,* if necessary, to do a portion of the job.

Suppliers furnish components used by the contractor in the construction of a product or facility. These components may vary from tiny silicon chips to huge engines or turbines. The supplier may or may not manufacture the component. But the supplier normally depends upon the engineer or contractor to provide the component specifications. The supplier may also use design and test specifications, supplied by the engineer, in furnishing the required components. In many cases the supplier will furnish a complete system or final product.

The term *manufacturer* refers to the builder of components used in the construction of a product or facility. Some contractors are also manufacturers. Also, some consulting engineers function as contractors; in this case they are termed *construction engineers.*

As we will soon see, the engineer may rely heavily on the contractor for selection of a particular component to be used. In other situations the engineer may limit the contractor's choice of items. But regardless of the method used, specifications are among the most important technical documents being written today. In industry and government, no procurement of goods or services would succeed if engineers did not use written specifications to clarify and amplify their design drawings.

In this section we devote much attention to the function of the consulting engineer in the preparation of specifications. This is because many structures and products are built on the basis of specifications and drawings prepared by consulting engineers. Most of what we say applies equally well, however, to engineers and specification writers in every phase of industry and government.

WHAT IS AN ENGINEERING SPECIFICATION?

An engineering specification is an organized listing of basic requirements directly applicable to the design, construction, purchase, testing, installation, or use of an engineered item, equipment item, or what is generally known as *hardware*. It is a quantitative description of the required characteristics of a hardware item; thus, it defines the desired final product of a procurement.

In one sense, an engineering specification tells the contractors what they are expected to do for the engineer's client. In a broader sense, it defines what is expected of and for the hardware itself. To do these things properly, the specifications should (1) define the scope of the work to be performed; (2) describe the materials of construction, product compositions, and performance required; (3) explain the quality of the workmanship needed, including requirements for durability and reliability; (4) list the governing codes, standards, rules, and regulations under which the work is to be performed; and (5) make clear the responsibilities of suppliers, contractors, and others, regarding additional engineering requirements such as product safety and maintenance.

The specifications, along with project plans, are the visible results of the engineer's studies, investigations, and decisions. They make it possible for the contractor to prepare estimates of cost of labor, materials, and equipment and to tender a bid on the job. They bring together the technical mind of the engineer and the skilled hands of the builder. The specifications should provide the client, within the price range allowed, with the best materials and equipment available, put together in a flawless manner.

CONTRACTS AND SPECIFICATIONS

For any given project, the engineering specifications must be prepared in a way that is compatible with all applicable legal provisions or requirements. These legal matters should, of course, be defined in a contract. Virtually every engineering project involves one or more contracts. Contracts are legal documents; they are drawn up basically to control the cost and delivery of goods and services. Also, contracts state the various conditions and limitations related to a project, such as professional and product liabilities and warranties.

Engineering specifications and contractual requirements go hand-in-hand to outline the responsibilities of the parties involved in the project. Once a procurement has been accomplished, the specifications may be considered a part of the contractual documents. Remember, however, that specifications and contracts constitute two distinct sets of materials. Thus, legal matters belong in the contract itself, not in the engineering specification. Usually, any discussion of legal requirements in the specification would constitute extraneous material, obscuring the technical essence of the written specification.

Further, technical requirements belong in the engineering specification, not in the contract. Since the contract is the controlling document, the legal requirements and provisions contained in the contract may, in effect, impose certain technical specifications on the parties contracted to do the work. And the contract may, if necessary, qualify or alter a specification; but the contents of the two documents usually must be kept separate.

TYPES OF SPECIFICATIONS

Specifications may be classified according to a variety of schemes. This section focuses on the general types of engineering specifications—particularly those that are essential to any effort to develop hardware and make it operational. Regardless of the name applied to a specification type, certain features and steps in preparing it are the same as for any other well-written specification. So if a writer understands the fundamentals of the basic types, he or she can apply this knowledge to any specification writing task.

Origins of Specifications

A specification typically originates in one of three general kinds of projects: (1) government-directed and military work; (2) industrial systems and components; and (3) construction. Certain techniques have been widely used in the preparation of specifications for these three classes of projects and have led to the development of some standardized writing practices.

Government and Military Projects. Specifications for these projects generally have their origin within a procurement effort by a federal agency or department. The specifications may include nearly any goods and services that the industrial, commercial, and professional portions of the private sector can provide. Examples are specifications for research facilities, weapon systems, communications equipment, and office supplies.

The potential procurement usually starts out with the publication of the overall project requirements and design guidelines by the General Services Administration, the military department, or other agencies. Initially, the project may be announced in a government publication such as the Department of Commerce's *Commerce Business Daily;* the actual technical requirements are then published in a solicitation package or formal request for proposal (RFP), which interested parties must send for.

Several government publications cover specification writing practices and standards in wide use. These documents include: Military Standard 490 (MIL-STD-490), entitled *Specification Practices;* a Department of Defense manual (DODD 4120.3) entitled *Standardization Policies, Procedures and Instructions;* and Military Standards 961, 962, and 963.

Industry and Industry-Related Projects. Specifications in this class of projects may originate with any commercial organization, business partnership, or individual involved in the design, manufacture, or testing of an item. This includes those who serve as subcontractors, consultants, or suppliers for contractors on government-directed projects.

Depending upon the nature of the contract and the goals of the engineers in undertaking the project, specification writers may use one or more of the government publications cited above as guides. Engineers can also make use of handbooks and standards published by professional societies, associations, and institutes. Several voluntary professional organizations publish standard specifications and codes that can help simplify and improve the writing process, as we'll see.

Construction Projects. Engineering specifications are often considered to be separate from construction specifications, although most construction specifications are written by engineers and architects. A basic distinction between the two types is that construction specifications guide work at the field site, whereas most engineering specifications cover items purchased and delivered to the site. Construction specifications normally originate with the engineers and architects designing a building, plant, or other facility.

Here, again, the writer may consult government publications such as Military Standard 490, if necessary. Not surprisingly, these documents can be used to greatest advantage on military and defense-related projects. However, such government publications were among the first formal efforts to promote standardization in construction specifications. Hence, they are used on a wide range of construction jobs. Today much of the standardization effort is governed by the Construction Specifications Institute.

Project Phases

Regardless of the type of project, the overall engagement normally has three major phases: design, manufacture, and testing. All engineering specifications play an important role in at least one of the phases. Thus, we can define a specification according to its main focus. Based on this scheme, four basic types of engineering specifications are commonly recognized: (1) design, (2) purchase, (3) component, and (4) system.

Additional specification types can be identified by further subdividing the project phases. For example, design and testing specifications can be divided into system, component, equipment, and material subtypes. Manufacturing specifications can be divided into process, material, and tooling subtypes. In some instances, an understanding of all these may be necessary for the general satisfaction of project requirements; but the steps for writing most of the subtypes are the same as for the basic specification types. In any instance, the writer must know the four basic types before he or she can proceed with the work of preparing a complete specification draft.

Design Specifications. This type of specification states what end results are to be achieved by a piece of equipment or other hardware. Design specifications do not normally tell how the item will be built, tested, or used. Instead, they define the item's functions—what the hardware will do, not how it will be done. A typical design specification might detail various design criteria, including operational, performance, and structural requirements. If necessary, a design specification may also include safety and maintenance requirements, and it may call for drawings or reports from the potential suppliers.

Purchase Specifications. These often form the primary basis of the bid price on a procurement. Usually they call for a much broader range of detail than design specifications. This is because purchase specifications typically represent the engineering staff's approach to a combination of requirements, including functionality, manufacturing (or construction) procedures, delivery, operation, and other factors. Most purchase specifications are used for standard components and "off-the-shelf" items that meet the requirements of the more complex system or project that the staff is designing.

A purchase specification may govern all major phases of hardware production and shipping—from design through manufacture and testing. There are some exceptions to this, however. If, for instance, the engineers know the prospective supplier as an excellent and reliable source, they may tailor the design section to give the supplier greater freedom in deciding how to meet functional requirements.

Since purchase specifications usually cover components, equipment, or materials, the titles *component specification, equipment specification,* or *material specification* are sometimes used instead of *purchase specification.*

Component Specification. A component is a constituent part of a system or subsystem that performs a specialized function within the larger entity. A component may be simple or relatively complex. Examples are the glass-wool filter elements in an air conditioner, an electromagnetic relay in a telephone switching system, the fuel injection pump in a diesel engine, and the central processing unit in a computer. Because distinctions between components and systems are not always clear-cut, the specification writer must find a practical way to divide the required hardware into components.

This task need not be difficult. Usually, the component specifications are developed from performance standards and design requirements for well-known items. Most engineers and specification writers are familiar with components normally supplied within a particular area of industry. Thus, they can easily determine whether they will be writing a specification for one or more components. For example, you might specify a predesigned thermostat for a heating control system. But you probably would not buy the heat-sensing unit separately from the electric contacts, or the adjustment dial separately from the thermostat casing. So you would write one component specification for the thermostat—there would be no need to document each part separately.

Among the topics the component specification normally covers are: structural, mechanical, and electrical details; operating conditions and parameters; and materials selection.

System Specifications. A system is a combination of several pieces of equipment integrated to perform one or more functions. The capabilities of the system are greater than those of any of its constituent parts or components. A series of burglar alarms and building monitors connected to a central power source is an example of a perimeter security system; a hand-held calculator is a small computing system; a network of ventilation ducts is a ventilation system.

System specifications define a variety of performance requirements, operational details, and documentation needs. Like component specifications, they may cover design and testing criteria separately or in one document.

Stand-Alone versus Guide Specifications

In covering the various phases of a project, specification writers may choose to write a separate specification for each, or a single document including all phases. The latter is commonly known as a "stand-alone" specification. This is the most common type written today. Stand-alone specifications can be extremely complex, however, and this sometimes leads to problems in writing or reading the specification.

One problem is time wasted through unnecessary duplication. This often occurs when an organization engages in a series of similar projects or buys many of the same items repeatedly. If the engineers write an entirely new set of specifications "from scratch" each time, much energy will be spent merely repeating information. The easiest solution to this problem is to take the data that are used most frequently and put them into guide specifications. The new document is a convenient reference source. Not only does it save time but it can also lead to better specification writing.

The titles *guide specification, primary specification,* and *master specification* are often used interchangeably. Basically, they refer to a document containing commonly used statements and technical requirements for a particular item or type of project. When a writer wishes to specify a certain product, he or she may simply incorporate the appropriate portion of the guide specification into the new document. Or the writer may use the guide specification as the basic document and make references, as necessary, to new requirements covered in an accompanying data sheet or "supporting" specification. Some guide specifications are computerized; the automated system makes it even easier for the writer to store, change, and produce new documents without a lot of repetitive paperwork.

When the guide specification provides blanks for the insertion of numerical data directly into the document itself, it is known as a *skeleton specification.* This type may also be referred to as a *generic, general, base,* or *sample specification.* Organizations that produce many documents containing similar data often write their own guide specifications, but some manufacturers and suppliers will furnish similar specification writing aids on request.

Choosing between Types

On many private and government-directed projects, the customer's requirements may govern your choice of specification types. Further, your selection may be determined by policy or tradition within your organization. Beyond this, logic and practicality must dictate. Most large projects begin with system specifications and proceed through component design and purchase specifications. A stand-alone document is usually best for unique types of projects, while guide-plus-supporting specifications are better for frequently encountered applications.

There is one area of specification selection, however, in which the writer often has a great deal of freedom. This is in the selection of an open, restricted, or closed approach to hardware requirements.

The Open Specification

Sometimes referred to as *performance specifications,* open specifications never cite an item by a supplier's name. The specifications are "open" for any manufacturer or supplier to tender a bid. This approach is most prevalent during the early design stages of large or complex projects, particularly in work for government agencies.

Generally, federal procurement regulations mandate the use of open specifications for government procurements. The main intent of these laws is to discourage favoritism and other forms of corruption in the awarding of government contracts. Most federal agencies require, therefore, that specifications written for them or their contractors be open whenever possible.

In practice, this goal is often difficult to attain. While respecting the intent of the procurement laws, many consulting engineers contend that it is impractical to write absolutely open specifications for any but the most general applications. As one engineer has said, the only time a specification is truly open is before it has been written—before the engineers have decided what they want.

Completely open specifications are not very common. When specifying many types of components and equipment, engineers must describe what they want in enough detail to permit the contractor or suppliers to bid, and this means that the type of equipment to be used must be basically established. Thus, some manufacturers and suppliers are eliminated automatically, without a name or brand being mentioned.

Materials are more adaptable to open specification than equipment—glass, concrete, aluminum, or paint can be described as to quality without regard to manufacturer. Similarly, the open approach may often be used in design specifications for large and complex systems, since they emphasize overall performance requirements rather than design features peculiar to any one manufacturer's or supplier's products. Many engineers do use open specifications for these items, even on work for private clients.

While engineers sometimes hesitate to use an open specification because it reduces their control over hardware selection, there are some advantages. When a specification is open, or as open as possible, there is broad competition among manufacturers and suppliers. This can lead to lower prices for good hardware, under certain conditions. Then too, open specifications allow new or relatively unknown firms to bid directly and competitively against old and well-advertised firms. This gives a new company, or a good new product, an equal chance.

As far as the engineer is concerned, the open specification is more expensive to prepare. The description must be detailed enough to eliminate all improper equipment—yet no names can be mentioned. Then when the contractors or suppliers propose a particular product, each item must be investigated in detail to determine that it meets all requirements. This increases the general burden of paperwork within the engineering organization and can lead to construction delays.

The Restricted Specification

Sometimes called a *bidder's-choice specification,* the restricted specification is one in which the material or equipment is described and then limited to one or more predesigned products. Most restricted specifications reference the desired quality or models by the names of qualified suppliers and a list of their hardware catalog numbers. With this approach, engineers sometimes add to the description the phrase "or approved equal." In theory this permits a contractor or supplier to replace the specified item with a comparable substitute, if the engineer approves the choice. In practice the engineer writing the specification usually has in mind those products he or she will or will not accept—far beyond the actual wording of the specification.

It was once customary to use the phrase "or equal" in these kinds of specifications instead of "or approved equal." Some engineers and technical writers have dispensed with the use of these phrases altogether. These practices reflect the evolution of two important views on specification writing.

On the one hand, the engineer is not expected to accept blindly the supplier's statement that a substitute is equal to the specified item. Thus, the engineer may use the phrase "or approved equal" to show that he or she reserves the right to examine, and accept or reject, any substitute before it is used.

On the other hand, some engineers have shunned both phrases—preferring, instead, to control the selection of equipment by citing specific alternatives themselves. For example, a specification might state, "Rivets shall be aluminum alloy 6061-T6 or aluminum alloy 5451-T6." This is a more definite approach than writing, "Rivets shall be aluminum alloy 6061-T6 or approved equal."

Many specification writers feel that giving one or more specific alternatives saves time and energy that might otherwise be required for evaluating unfamiliar substitutes. The restricted approach is thus considered best for most work done for private industry. It allows competition among contractor, supplier, and manufacturer, within a limited range, while permitting the engineer to design with good knowledge of what the final selection will be.

Restricted specifications are particularly adaptable to items such as plumbing, piping, wiring, and fixtures—that is, most components and materials that have predetermined performance features and that are standard in size and application. Most large and complex systems and subsystems cannot be handled so well with restricted specifications because a conceptual or preliminary design is often all that the engineering staff has to work with in the early stages of a complex project.

Contractor's Preference

Generally, contractors seem to prefer restricted specifications. They often know exactly what products will be accepted, yet the restricted specification leaves room for substitutes when needed. Where the open specification puts a great deal of responsibility on contractors, often making them decide what equipment will meet the performance requirements, restricted specifications shift much of this responsibility back to the engineer where it belongs.

The Closed Specification

The closed, or *base-bid specification,* is just what the name implies. The engineer writes one particular product, usually by catalog number, into the specifications, and that is the end of it. When a component purchase specification is written for predesigned equipment, it is usually a closed specification. While this sounds as though it eliminates competition and could cost the client extra money, that is by no means always so. A careless engineer who hurriedly picked some product from a convenient catalog and then closed the specifications could cost a client money, but that is not the way a closed specification is written—no matter what the eliminated supplier may think.

The difference between an open specification and a closed specification is much a matter of timing. When the design concept of the project is in its early stages, all specifications are open. The engineering staff may want to study more than one design approach for the entire system, or for one or more subsystems and components. At that point, a system specification and several component design specifications may be written and investigated. Once a suitable design approach has been found, the component specifications are restricted or closed, and the engineering staff may begin to write the purchase specifications.

Some engineering organizations make it a practice to get prices on all standard components and principal equipment before they select the combinations they feel are best. Then they decide, based on the design features and prices of the hardware, whether they will write a restricted or closed purchase specification. This method has a number of advantages. It allows for competition on price while reducing opportunities for deals or kickbacks between suppliers and contractors. It means that the selection of the equipment is entirely the responsibility of the consulting engineer—as it should be, for he or she acts as the owner's agent.

From the engineering staff's point of view, the closed specification has advantages. The staff knows exactly what materials and equipment are going into the project, so it can make detailed drawings and have them ready when the bids are called for. There will be no need for redesigning several sheets of plans because some items require different space or different fittings from those first specified. The client often gains because he or she can thus get good engineering at lower costs.

With closed specifications the contractor has the advantage of being relieved of all responsibility for guarantees on materials and equipment. The engineer and manufacturer or supplier are entirely responsible. The contractor need guarantee only workmanship—which, again, is as it should be.

Importance of the Writer's Approach

It may appear that we have given undue emphasis in these pages to the use of open, restricted, and closed specifications. This emphasis is deliberate. It is in choosing one type of approach over another in a given situation that specification writers sometimes make their biggest mistakes. At one extreme is the engineer who gives potential bidders so much leeway in their choice of materials or fabrication methods that the quality of the hardware is compromised as a result. At the other extreme is the engineer who closes the specification so firmly that it stifles competition and prevents costs from being lowered.

In order to write specifications successfully, the writer must carefully decide—based upon an analysis of the client's requirements and the adaptability of certain types of hardware to a given approach—whether a specification will be open, restricted, or closed. This will help to ensure that customers get what they need at reasonable cost.

How Firm Should a Specification Be?

A specification should be as firm as the engineering staff's opinion of its work. During the early design stages of a project, when much of the paperwork deals with conceptual or preliminary designs, revisions may be necessary to bring the specification in line with a selected approach. If the engineers are unsure of their decisions, they are open to suggestions from reviewers and others. Once the staff has selected the most promising approach, however, the final design or purchase specifications should be reviewed, sent to the purchasing department, and delivered to the prospective bidders. No changes should be permitted in the documents at this point, except in the most unusual circumstances.

If substantial weaknesses or errors are discovered in the specifications, the engineering organization must inform all bidders immediately. The Uniform Commercial Code requires that any explanations or changes sent to one prospective supplier must be sent to all. This helps ensure equal opportunities among bidders.

What Factors Influence Hardware Selection?

For the engineer, the choice of one supplier or one particular product over another generally depends on: (1) the suitability of the particular piece of equipment or brand of material for the specific installation being designed; (2) the manufacturer's or supplier's reputation; (3) the performance history of similar equipment; (4) the engineer's opinion of the design, material, and workmanship of the equipment; (5) the cost in relation to quality; (6) the promised delivery date and the supplier's reputation for meeting that promise; (7) the availability of engineering data on the product; (8) the reputation of the local supplier or representative; (9) the availability and quality of local service facilities; (10) the manufacturer's reputation for attempting sincerely to comply fully with the specifications; (11) the general acceptance by engineers of the brand name as a symbol of quality; and (12) the reputation for prompt attention to complaints.

These factors are not listed in the order of importance, and the most influential factor in selecting one type of hardware might fall far down the list when another type is being considered.

SOURCES OF DATA

Now that we have looked at ways in which an engineer might approach the general task of writing several types of specifications, we should consider briefly the sources of data the engineer may use in preparing these documents. The most widely used material comes from any of eight sources: (1) the client, who furnishes basic requirements in procurement documentation, written agreements, and discussions with the engineering organization; (2) the engineering staff, which conducts detailed analyses of systems and components for each project; (3) other sources in the engineering organization, such as the purchasing department; (4) architects and consultants working directly with the engineering organization; (5) all applicable laws, codes, and industrial standards; (6) voluntary professional groups, which publish standard specifications and writing guidelines; (7) other published reference works and handbooks; and (8) manufacturers and suppliers, who may provide catalogs, manuals, guide specifications, and advice on technical matters.

Careful selection and use of material from these eight sources will help the engineer to prepare a complete and detailed specification even when he or she does not have extensive experience with the specified items. Let's take a closer look at each source.

Basic Requirements

The essential requirements of any technical project originate with the customer. When the customer knows the subject in great detail, he or she may do much of the initial work of defining the problem and its technical requirements before selecting an engineering firm to complete the job. This often occurs in projects involving the military and government agencies.

As noted earlier, a customer may furnish certain specifications as part of a solicitation package or RFP. These requirements may then be modified or refined during contract negotiations with proposers. As a result, system design specifications and other technical requirements may flow directly from the discussions between the customer and the engineering organization.

In most cases, however, further investigation is needed. The customer may not have a completely clear or accurate idea of what the basic needs or problems are. Or the specifications submitted with the procurement materials may be too sketchy to provide a basis for any but the most general design. The engineering staff, therefore, must turn to other sources. When the engineers are working as consultants to an architectural organization or other commercial firm, they may often obtain important technical data from the higher-level source. Usually this information serves as a general guide for the engineers as they begin preparing the specification drafts for subsystems, systems, or components within the larger project.

If several parties are involved in the project, the relationships between the various entities and technical requirements may become quite complex. In this case, a flowchart or *specification tree* may be prepared. This chart enables the engineering staff to trace the flow of technical requirements from one organization or department to another, or from one level to another within the project hierarchy. The flowchart also helps to clarify the specification writing responsibilities of the various groups.

Together with data developed from systems analyses and applicable laws, codes, and standards, the client's specifications (or other high-level requirements) make up the materials that are absolutely essential for the preparation of a reasonably good specification draft.

System Analysis

Any technical project requires the engineering staff to analyze the potential subsystems or components of the system under consideration. On this basis, one or more models may be developed. These models usually consist of paperwork, or software, and do not call for the construction of physical prototypes. The models are useful tools for showing the interconnections between components within the system and often point to additional requirements that are not covered in the original system specification. Here, then, lies much of the engineering staff's responsibility for determining exact requirements relative to the new specification. In analyzing the model, the engineers may draw conclusions from a wide range of investigations, including mathematical calculations and studies of the interaction of physical laws with the properties of various materials.

Other Internal Sources

In addition to deriving important technical requirements from the system or model analysis, the specification writer may sometimes obtain key information from the purchasing department, quality-assurance department, and project managers in his or her organization. These sources will often furnish useful data regarding not only the current project requirements but also concerning previous projects with requirements similar to the new specification. Other related or "precursor" specifications can also be of assistance. Many organizations store such information on card files, though word processing systems and microfilm are also used. It is the specification writer's responsibility to seek these sources out and to stay in close contact with members of the purchasing and engineering departments throughout the specification writing process.

In order to promote greater consistency in writing practices and reduce the total engineering workload, some organizations publish specification writing guidelines for use by their staffs. These documents are sometimes referred to broadly as *engineering procedures* and may include: standard procedures for determining particular types of technical requirements, explanations of management policy as to design review and procurement, sample flowcharts and guide specifications, suggested formats, and listings of applicable codes and standards published by voluntary professional groups. Engineers have found these in-house guidelines very helpful in the total writing effort.

Consultants

On many types of projects, the engineering organization may utilize the services of a number of consultants or subconsultants, including architects and engineers who specialize in areas in which the larger organization may not normally practice. The consultants' role in determining technical requirements is similar to that of the basic engineering staff. Their main responsibilities involve system analysis and the development of related specifications.

Laws, Codes, and Standards

Specification writers must make certain that their documents comply with all applicable codes and laws. Not only is this important because of legal concerns but also because the overlooking of regulations gives contractors good and necessary reason for making changes. At best, such changes could lead to costly and time-consuming revision; at worst, contract renegotiations or lawsuits could result. Further, customers may insist that certain codes and standards be applied to selected pieces of equipment for insurance purposes. If the specification writer chooses a different set of requirements than the customer or insurance underwriter has dictated, the customer will be sure to complain. The best approach for the writer and his or her organization is to follow the customer's requirements exactly.

Professional Societies

Several voluntary professional organizations have prepared sets of standards to guide engineers in the determination of technical requirements. Among these organizations are the American Society of Mechancial Engineers, the American Society for Testing Materials, the American Water Works Association, and the American Petroleum Institute.

There are at least two reasons why engineers benefit by using the codes and standard specifications prepared by these groups: (1) Bidders prefer to work with codes and standards with which they are most familiar; and (2) standard specifications assure clear, concise wording, while they reduce the possibility of omissions and of misunderstanding by contractors. Also, they save time for the engineer.

Often, the engineer can furnish a full description of the desired equipment by merely referencing the standards it must meet. This does not mean that the engineer should simply reproduce an old set of specifications for a new project. Nor does it mean that standards are substituted in place of careful consideration of the special requirements of the project. It does mean that the adoption of the standard specification eliminates excess verbiage and makes possible the preparation of documents according to nationally recognized procedures. The standard specification should be thought of as a form of writing, never as a replacement for detailed study and investigation.

Other Reference Works

Engineering handbooks and related reference works are handy sources of ready-to-use data, practical examples, and background information. They permit quick access to existing analyses of selected equipment and materials. Commercial publishers are the main source of handbooks, though some voluntary professional groups publish their own reference books.

Manufacturers and Suppliers

Many engineers use the material published by manufacturers and suppliers while writing a specification. These include catalogs, design manuals, and guide specifications. Their treatment of a particular piece of equipment can often help the engineer to determine the necessary design parameters of a system and to select the correct dimensions and operating characteristics for components and parts. If necessary, the engineer may also seek technical advice directly from the prospective supplier or sales agent.

Relative Importance of Sources

Of the eight sources of technical data listed above, the first five—clients, engineers, purchasing specialists, consultants, and applicable codes and standards—are essential to the determination of technical requirements for most projects. No engineering specification should be written without checking with these sources. Industrial standards, reference books, and suppliers—the last three sources—are optional even though they furnish much of the data used in specification writing today.

SPECIFICATION FORMAT

Specifications may be sent out in a number of different forms, depending upon the customer's requirements, types of equipment covered, management policy, and other considerations. In any case the way an organization normally purchases goods and services will determine, in part, the format of the overall document.

If the specification writer and his or her organization do most of their work for the military and government agencies, they may adopt the format preferred by their customers—as defined in Military Standard 490, for example. If the organization performs a large number of contracts for commercial industry, it may use the standard military format with modifications. In any organization, the basic format is usually fixed, though other schemes may be developed in order to meet special needs.

In many instances the size and complexity of a project will influence the way in which the specification is sent out. Hence, there are many different approaches an organization may take. For a large project, the complete specification might comprise a package consisting of a general specification, supporting specifications, and several data sheets. For a simple component, a stand-alone specification might be used; this document can cover all aspects of design, manufacturing, and testing. When the engineering specification is sent to the contractor or bidder, it may be accompanied by additional materials, including control drawings and transmittal memoranda. If necessary, a detailed listing of administrative requirements may also be sent; this listing is sometimes called an *administrative specification*.

Using a General Format

Because specifications are subject to such a wide range of variation, it would be impractical to try to give actual examples of all types here. Such a treatment would tend to obscure the basic structural similarities found in most types of specifications. We can outline a general specification format in terms of these similarities, however, and we can offer some practical guidelines for its use. Most of the following discussion is based on specification writing practices that have become standardized in several segments of government and industry.

The general format given in this section can be used to prepare a reasonably good draft for most types of specifications. But keep in mind that it is only a guide. It is the specification writer's responsibility to arrange his or her material in the most practical manner possible, while following the customer's wishes and abiding by management policy. In any event the writer must know the items to be covered in the specification, must be familiar with any documents or standards applicable to the particular project application, and must use logic in organizing the data.

Organizing by Section

The decimal numbering system is generally accepted as the standard scheme for organizing specifications in all areas of industry and government. This method enables the writer to divide the material into any number of sections, subsections, and paragraphs. Technical requirements and other selected topics can be identified easily by referencing the appropriate numbers. Letters and Roman numerals, by contrast, are used rarely except in appendixes.

Sections are the major divisions of the specification. It is natural, therefore, to make the title of each section a first-order, or primary, heading and to designate it by an Arabic number. Sections may then be subdivided into an indefinite number of subheadings; but three is the standard limit. Thus, any section will be identified by a first-order heading and may contain second-order, third-order, and fourth-order headings as necessary.

Generally, specifications have a maximum of four levels. The first level of headings constitutes the basic specification format. As noted previously, many engineering organizations dictate the basic format for certain types of specifications, though the customer may define a preferred structure in the contractual requirements for the project. The contract may also call out various subheadings. Subsections and their contents, however, are more commonly determined by the specification writer since he or she will be most familiar with the special requirements and applications of the equipment to be specified. In any case the decimal numbering system helps to ensure logical organization of the material and, thus, a good specification format.

The Rule of 2. Whatever the format, the "rule of 2" applies: At any given level, there must be at least two headings. Here's a partial specification which serves as an example:

Specification for Prefabricated Insulated Pipe Units

3. Technical Requirements
   3.1. Underground insulated piping systems
      3.1.1. Conduits. All conduits shall meet the following requirements:
         3.1.1.1. The conduit shall be iron metal...
         3.1.1.2. Conduit ends shall consist of a...
      3.1.2. Pipe insulation.
         3.1.2.1. Insulation of pipe in conduit shall consist of...
         3.1.2.2. Insulation thickness for service requirements shall be...
   3.2. Overhead insulated piping systems
      3.2.1. Construction.
         3.2.1.1. Prefabricated insulated pipe conduit shall be furnished...
         3.2.1.2. Alternate: Prefabricated insulated pipe conduit...
      3.2.2. Hangers and supports.
         3.2.2.1. Supports shall consist of concrete piers...
         3.2.2.2. Hangers shall consist of steel posts...

Preferred Sections. A specification may contain as many sections (or first-order headings) as are needed to provide comprehensive, detailed, and complete coverage of all project requirements. In practice, a large segment of industry and government prefers working with a basic format consisting of six sections. These are typically identified as follows:

1. Scope
2. References
3. Technical requirements
4. Quality assurance
5. Other requirements
6. Ordering data

Depending upon the particular phase or phases of the project to be covered, the specification writer may modify the basic format to highlight certain requirements. For example, when necessary, the general title "Technical Requirements" may be replaced by "Design Requirements," or the whole section may be separated into three sections covering, respectively, design, manufacture, and testing. Other sections that may be added to the first level include: packing, shipping, and storage requirements and material fabrication requirements.

Format Considered in This Handbook

The general specification format given in this book is a modified version of the basic format shown above. The modified version has been chosen in order to delineate a wider range of possible elements and subheadings. Figure 6.1 shows the

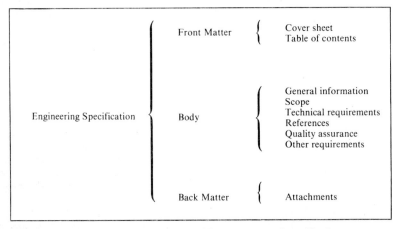

FIG. 6.1 A general format that can be used for most types of specifications.

basic elements of the general format, while the outline below defines each element and a number of possible subsections. Third-order and fourth-order headings are not shown since not all specifications require them; further, such subsections are usually too specific for inclusion in a general format. Engineers and technical writers are well advised to become familiar with any special customer requirements or organizational policies regarding particular specification formats before using this outline as a guide.

General Specification Format

Following is a description of the format elements listed in Figure 6.1.

Cover Sheet. Like technical reports, specifications may be bound in various types of covers. Whether or not a special binding material is used, the document must have a cover sheet for displaying transmittal data and other key information, as shown in Figure 6.2. In some organizations this sheet is referred to as the

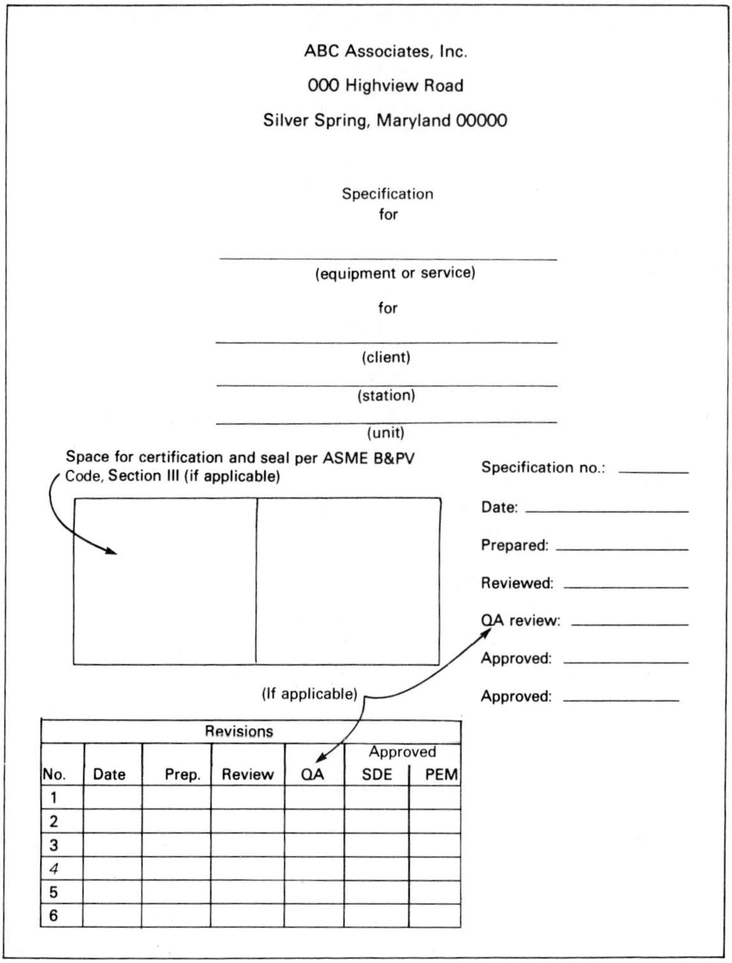

ABC Associates, Inc.
000 Highview Road
Silver Spring, Maryland 00000

Specification
for

(equipment or service)

for

(client)

(station)

(unit)

Space for certification and seal per ASME B&PV Code, Section III (if applicable)

Specification no.:
Date:
Prepared:
Reviewed:
QA review:
Approved:
Approved:

(If applicable)

Revisions						
					Approved	
No.	Date	Prep.	Review	QA	SDE	PEM
1						
2						
3						
4						
5						
6						

FIG. 6.2 Typical specification cover sheet.

"title page" of the specification. Among the typical items included on the cover sheet are: the document number used by the engineering organization to identify the complete specification, the general name of the goods or services specified, and blanks for insertion of signatures or comments by the main parties responsible for reviewing, editing, or using the specification. For more information on the selection of cover materials, see Section 2 on technical reports.

Table of Contents. Some organizations refer to this part of the specification as the *index*. Whatever name is used, the first-order, or first- and second-order headings must be listed. The table of contents should be detailed enough to enable the reader to locate particular requirements easily but general enough to permit a quick scan of the first-level, i.e., the major, sections of the document.

In most specifications the table of contents is followed by the first numbered section, as shown below.

1. *General Information.*
   1.1. *Owner:* Name of principal owner of plant or system for which equipment or services are to be furnished.
   1.2. *Engineering supervisor:* Name of supervisor or other parties with inspection access rights (e.g., customer or purchasing agent).
   1.3. *Engineering organization:* Name of firm or agency providing engineering services.
   1.4. *Construction manager:* Name of construction manager, if any.
   1.5. *Work location:* Geographic location of work site.
   1.6. *Plant description:* Brief description of plant, facility, offices, or site at which work is to be done.
2. *Scope:* This section usually offers a brief overview of the items covered by the specification (e.g., names of any standard equipment specified or a general description of component or system functions). The section can be subdivided as necessary. One way of doing this would be as follows.
   2.1. *Equipment covered.*
   2.2. *Work included:* General description of the tasks (such as manufacturing, testing, or construction) to be accomplished by the party working to the specification.
   2.3. *Related work not included:* Restrictions on the use of the specification; acceptable and unacceptable applications. This often includes a description of any work that is relevant to the specification but which is not included in the scope of work.
   2.4. *Codes and standards:* Any key industrial codes or standards used in writing the specification may be referenced. Alternatively, these may be given in a separate references section, as shown below.
3. *Technical Requirements.*
   3.1. *Design requirements:* All applicable design criteria, including specifications, codes, standards, desired features, and related conditions are listed in the most logical arrangement possible. Often the items are listed in the chronological order in which they will be used since this permits easier comprehension and planning. Otherwise, the requirements should be arranged in order of relative importance, with the more important items preceding those of lesser importance. The arrangement of data in all sections of the specification must follow a logical order, of course. Generally, the ordering scheme in subsequent sections will follow the logic used in the first major section of technical requirements. Other items that might be covered under "Design" are structural criteria, including various types of loadings and mechanical or electrical details.
   3.2. *Operating requirements:* Normal and abnormal operating conditions and parameters are defined.
   3.3. *Environmental conditions:* Relevant environmental conditions that may affect the application of the equipment are described. These may include such factors as atmospheric pressure, temperature, humidity, wind velocity, and seismic conditions.
   3.4. *Materials:* Materials are specified as to any of several designations, including type, durability, etc.
   3.5. *Manufacturing requirements:* Materials procurement procedures, fabrication processes, and applicable codes or standards are specified.

3.6. *Erection:* Any requirements for erection of structures are specified.

3.7. *Identification:* All labeling, safety markings, and other identification requirements are specified.

3.8. *Testing requirements:* These specifications may cover one or more of the testing phases, from prototypes to final production and operation. Required documentation, such as plans, drawings, and certification documents, should also be specified.

3.9. *Documentation requirements:* All reports, drawings, lists, manuals, and other documents necessary for specific project applications are specified.

3.10. *Finishes:* Painting, coating, or plating requirements are specified.

3.11. *Cleaning:* Required cleanliness and cleaning procedures for equipment or parts may be specified.

3.12. *Handling, Shipping, and Storage Requirements:* Packing, labeling, delivery, and other applicable specifications are listed.

3.13. *Samples:* Define any requirements for submittal of samples; e.g., weight, dimensions, etc.

4. *References:* This section lists any additional documents or references necessary for proper use of the complete specification. Thus, all codes, standards, regulations, drawings, and related specifications and other documents that are referred to, or included in, the text of the specification may be listed. Each item should be referenced by source, identification number, and title. All specifications contain a section on references; but some organizations prefer to place it directly after the scope section—as shown previously—especially when the references are crucial to a thorough understanding of the technical requirements. Here we have chosen to put the references section after the technical requirements. One way of arranging this section is as follows.

  4.1. *Organizational documents:* Applicable documents that originate in the engineering organization and which are used by engineers or writers in preparing the specification are listed. These may include guide specifications and supporting documents.

  4.2. *Drawings:* Any drawings not mentioned in subsection 4.1 are listed.

  4.3. *Regulations:* Titles and numbers of applicable regulations are given, along with statements regarding their applicability.

  4.4. *Codes:* Applicable codes and statements governing their use are listed.

  4.5. *Standards:* Applicable standards and statements governing their use are listed.

  4.6. *Related specifications:* All applicable specifications not cited in subsection 4.1 are listed, with statements governing their applicability.

  4.7. *Other documents.*

5. *Quality Assurance and Quality Control:* In some specifications the requirements for quality assurance and control can be written directly into the technical requirements section at the appropriate subsections. Many organizations prefer to describe these requirements in a separate section, however, particularly when the project involves matters of public safety—as in the nuclear power industry, for example. In this case, the applicable quality-assurance program is described in great detail as it relates to design, manufacturing, testing, or other phases of the project. The section may be subdivided as necessary for logical order and ease of application.

6. *Other Requirements:* This section is typically reserved for defining any requirements that are not within the scope of work but which are essential for completeness of the specification. Examples are: description of the procedures for filing data sheets and reports with government agencies, delivery requirements, and final documentation needed for installing or operating the equipment. In some specifications, however, the sixth section may contain actual data sheets to be used in the procurement; in this case, the section is titled "Ordering Data," as mentioned previously. As with most other sections, the sixth section is subdivided as necessary.

Additional Sections. The general format outlined above may be modified, of course, to suit particular projects and users. As noted, one of the most common modifications is to divide the technical requirements into separate sections for each phase of the project. Another is to add a section that covers exclusively the preparation of equipment for delivery. If the customer or the engineering organization require that several related documents be sent out with the specification, these are sometimes placed in a separate section titled "Attachments."

General Conditions. The preparation of separate administrative specifications by purchasing departments and project contract administrators has almost entirely supplanted the older practice of requiring engineers to begin each specification with a section titled "General Conditions."

Such contractual provisions as ownership rights, report scheduling, cost accounting, and management are now widely recognized as belonging to the administrative side of the procurement process, as opposed to the engineering side. Therefore, engineers and specification writers are rarely required to prepare the legal documents or administrative specifications for a project. In a small firm, however, the principals or top engineers may prepare these documents as part of their management responsibility.

As a matter of general interest, we have listed below some typical subheadings that might be included under "General Conditions" in one or more procurement documents. Remember that these subjects are typically the province of purchasing departments and contract administrators today. Therefore, we have not included them as part of the general specification format, though a few of the same items typically are covered in the general information section at the beginning of the engineering specification. Note that few actual projects include every item listed.

1. Owner
2. Engineers
3. Work location
4. Intent of contract and specifications
5. Definitions of terms
6. Contract guaranty bond
7. Contractor's insurance
8. Owner's insurance
9. Fire insurance
10. Unemployment compensation
11. Social security
12. Patents and royalties
13. Payment for permits
14. Damages
15. Liens
16. Planning and scheduling requirements

17. Construction reports
18. Document submittals
19. Cost accounting procedures
20. Partial payments
21. Application for payments
22. Certificates of payments due
23. Owner's right to withhold payments
24. Deductions for uncorrected work
25. Work correction after final payment
26. Definition of notice
27. Extra, additional, or omitted work payments
28. Claims for furnishing labor and material
29. Final payment
30. Owner's right to terminate contract
31. Contractor's right to terminate contract
32. Violation of contract provisions
33. Assignment of contract
34. Subcontracting
35. Owner's right to do work
36. Other contracts
37. Work under protest
38. Arbitration
39. Drawings
40. Shop drawings
41. Ownership of drawings and models
42. Operating instructions
43. Samples
44. Materials and workmanship
45. "Or approved equal" clauses
46. Surveys, lines, and grades
47. Mutual responsibilities of contractors
48. Relations of contractors and subcontractors
49. Real estate and rights-of-way
50. Inspection of the work
51. Inspection and testing of materials
52. Authority and duty of the inspector
53. Removal of condemned material and work
54. Superintendence
55. Engineer's status
56. Architect's status
57. Company coordinator

58. Decisions of the engineer
59. Temporary offices
60. Job accounting
61. Control of methods and procedures
62. Order of doing work
63. Accident prevention
64. Protection of work and property
65. Emergency work
66. Temporary heating and cooling
67. Use of job site
68. Sanitary conditions
69. Drinking water
70. Sewage disposal
71. Delays
72. Extension of completion time
73. Testing for operation
74. Cleaning up
75. Final inspection and acceptance

Automated Formatting. The process of preparing a good specification draft can be facilitated greatly through the use of a word processor or other computerized system. A key benefit for many specification writers and their organizations involves automated formatting. By programming the system to process data according to the requirements of a selected format, the writer can concentrate on other details instead of worrying about the arrangement of sections, subsections, and paragraphs. Optional formats can be stored in the computer's memory and retrieved quickly and easily. In addition, standard statements, codes, boilerplate paragraphs, and other frequently used data can be stored on a disk and added to any specification as needed. When the writer wishes to insert a particular industrial standard or product in the specification, the text does not need to be rewritten—the computer does the work. Modifications and revisions proceed much more quickly on computer than in a document that has been written exclusively by hand.

WHAT ARE THE SPECIFICATION WRITING STEPS?

There are many ways to write a specification, but most methods use the same general pattern. In today's busy conditions you will seldom have to prepare a specification without some kind of guiding, or "master," document. Earlier in this section we discussed typical guide specifications and other sources of data available to the specification writer. It is with these sources that the process of preparing the new document usually begins.

To prepare a specification, follow these steps: (1) Obtain all applicable data available from the sources listed earlier in this section under "Sources of Data." (2) Get a set of the latest drawings covering the project or product. (3) Study the guide specification and drawings. (4) Prepare an outline for the specification you

propose to write. (5) Confer with the engineers with whom you are working to learn their views of how the specification is to be written. (6) Write the specification draft. (7) Have the specification reviewed, edited, and approved. (8) Submit the approved specification to the document production department for final typing and reproduction.

To make the most of the preparation process, let's take a closer look at these eight steps.

1. Obtain Guide Specifications and Other Applicable Data

Be certain to make full use of all sources of material, including: basic requirements furnished by the client; all applicable laws, codes, and standards; engineering handbooks; and manufacturers' catalogs.

Despite the applicability of these materials, new specification writers sometimes are reluctant to use standard specifications or skeleton paragraphs. These writers believe that such master specifications inhibit creativity. While this is probably true, the time and labor that these sources save make any resulting reduction in creativity negligible. Also, the specification writer is not expected to be a creative writer; he or she is part of a team working to produce the best specifications in the least time with the least expense. By exercising good judgment in the use of various sources, you will have greater opportunity to be creative in those parts of the project not covered by industrial standards.

2. Get a Set of Drawings

It is nearly impossible to write a correct specification without a complete set of working drawings of the project or product to be covered. A specification amplifies the drawings, explaining many items that are impossible to represent graphically. There are many types of drawings, of course, and it is the writer's responsibility to obtain the proper drawings for any given application. Some of the most common types used by specification writers are: installation drawings, process flow diagrams, equipment requirement drawings, fluid power diagrams, and interface control drawings.

Make certain you get the latest issue of the drawings; changes are often made from one set to another. If you work with an outdated set of plans, the specifications will almost certainly require revision. Revisions or amendments to an approved specification often necessitate, in turn, the issuance of a *drawing-change notice* (DCN)—which means an even greater expenditure of time and energy by everyone involved. Having a specification sent back under these circumstances is one of the least enjoyable experiences a writer can have.

3. Study All Applicable Documents and Plans

Examine your materials. See how the guide specification, or a similar specification from a previous project, is written. Note the sequence of data used. Visualize how you can present your material in the same or a similar order. Look carefully at the language and typical phrases used.

Some general phrases used in many types of specifications are "shall be," "shall provide," "shall furnish and install," "will review," and "will supply."

Note that the word *shall* designates a requirement that must be met by the contractor, supplier, or manufacturer, whereas *will* is used to indicate requirements that the buyer himself or herself proposes to meet. Here are some examples which illustrate the difference:

> The contractor shall furnish two ABC model 30D reciprocating water cooling machines. Unit capacity shall be 12 tons.

> The manufacturer shall supply all assembly drawings necessary for assembling the specified basic components into a complete unit.

> The client will review all unit assembly drawings prior to final assembly of the complete unit in the field.

Some organizations use abbreviated or streamlined forms of specifications. These forms eliminate the need for constant repetition of expressions like *shall be* and *shall provide*. Thus, you might write, "Insulate steam pipes with 85 percent magnesia," instead of, "The contractor shall furnish and install 85 percent magnesia pipe insulation on all steam pipes." While the abbreviated form is more concise, it must be used only where it does not reduce clarity. In any case the language, terminology, and organization of material must be consistent throughout the specification.

In addition to scrutinizing language, make certain to study every applicable drawing. Try to become so familiar with the drawings of the system or components that when you close your eyes, you can visualize the entire project. If you are writing the specifications for an industrial plant, start the study of the plans at the plant entrance. Then "walk through" the plant, using the plans as your guide. Try to relate what you see on the drawings with other plants with which you are familiar.

Use the same technique for equipment. If the drawings show a series of ventilation ducts, use your imagination to "enter" the ducts at the point where air will enter the ducts. Visualize the flow of air through the system. Imagine that you are working the various controls for the exhaust fans and intake mechanisms. Only by careful study like this can the writer become thoroughly familiar with a set of drawings.

4. Prepare an Outline of the Specification

Review the client's requirements to see if a particular format has been prescribed. If not, use an outline similar to that in your guide material, if the requirements are nearly identical to those for the current project. Or use the general format given earlier in this section as a guide, then modify it as necessary to suit your particular requirements.

Outline every major and minor section you expect to include in the specification. Don't worry if you seem to have too many items in the outline. It is usually much easier to delete excess material later than it is to find subsections or paragraphs you have left out. For more detailed guidelines on outlining, see Section 2 on technical reports.

5. Confer with the Engineering Staff

Show the engineers your proposed outline. The staff will probably be familiar enough with the project to be able to spot any missing items.

Discuss your writing plans with one or two key people on this staff. Be certain to obtain the results of their model analyses, and study these for complete understanding. Show the engineer any guides you expect to use. Ask the top person on the staff for his or her general approval of your writing aims. If you feel unsure of your plan, or if there is some doubt as to the best writing approach, arrange to have the engineer check the first few pages you write. Then there is little likelihood that the entire specification will later be declared unsuitable.

6. Write the Specification Draft

Start with the part of the project or product that you know best. Keep your standard or guide specifications and the plans nearby while writing. Prepare the draft in whatever manner you find most comfortable. Many writers prefer to write the first draft in longhand; only an accomplished typist will feel fully at home typing a long and complex specification.

As you write, there are many rules that you must keep in mind. These include the proper language to use in identifying specified requirements, the correct format to follow throughout the document, and the customer's requirements, if any, regarding the organization and structure of data. In addition, there are a number of helpful tips that you can follow to further ensure that your specification draft is detailed and complete. For easy reference we've listed these tips separately, following the remainder of the discussion of specification writing steps.

7. Have the Specification Reviewed, Edited, and Approved

After you write the complete specification draft, submit it to the engineers or other experts responsible for reviewing specifications within your organization. Also, make sure you furnish any related documents or drawings required to make the specification complete.

Most companies have some type of management policy that dictates who will review particular kinds of specifications. For example, reviewers may be assigned according to their major areas of concentration; there may be one reviewer for each project phase, including design, manufacturing, testing, safety, etc. These individuals may include the project manager, quality-assurance engineer, and various specialists.

In most firms the process of reviewing the document consists of the following phases. The writer drafts the specification and has it typed. The reviewers read each section of the specification, concentrating on their assigned areas, and check the document against the project plans. They write down their comments, then send the copies of the draft and the comments back to the writer. The writer studies the comments and makes any necessary changes in the document before sending it to the editor. Then the editor performs a thorough review of the entire specification and edits the document, or has the writer revise it, to resolve any remaining problems. Finally, the writer determines who must furnish the approval signatures, according to the organization's rules, and submits the specification for approval. If all the required approval signatures are obtained, the draft may then be submitted for final typing or printing.

It takes time to check a specification, but the actual number of hours spent is often far less than if the engineer were to prepare the specification alone. Some organizations allow the writer to exercise some control over the total time spent on reviews, while others permit no variation from a standard procedure or schedule. Regardless of the procedure followed in your organization, remember that you are not imposing on the reviewer's time, for this is one of the key steps in verifying the overall quality of the specification.

8. Submit the Document for Production

The approved specification is not ready for issue until it has been typed, reproduced, and rechecked by the writer. Before you submit the document for production, make a final check. Then take the manuscript to the typist or supervisor of the publications department. Give specific instructions for typing. Instructions may include specifying paragraph indentions, marking particular words or headings for italic type or other special typefaces, and selecting the quality of paper to be used.

As the pages are completed, read them carefully, comparing each page against the corresponding manuscript. By reading a few pages each day, you can keep a running check on the typist's progress. Also, you will be able to catch any errors early enough to prevent a costly retyping of the entire document.

HELPFUL TIPS FOR GOOD SPECIFICATION WRITING

You can further ensure a high level of quality in your specification writing by keeping the following tips in mind:

1. Be thorough—incomplete specifications lead to errors and a waste of time.
2. Avoid nonessential details and ambiguous statements. Also, try to avoid using compound sentences. Using the simplest words and the shortest sentences will best convey your intended message.
3. Be realistic in the design conditions you specify; if you doubt any of the requirements, check with the engineer.
4. Balance usefulness, cost, and performance; if you do not, the installation may be extremely costly.
5. Where possible, allow manufacturers to quote on standard equipment; usually this saves money.
6. If possible, use a standard specification form and supplier's data sheet. These are available from most equipment builders. The forms help to eliminate errors, improve bid evaluation, and cut specification writing time.
7. Use code-approved materials wherever possible.
8. Provide enough detail to describe comprehensively the characteristics of the equipment. But try not to go overboard in specifying small details that are not likely to affect the desired functions or applications. For example, avoid explaining or justifying the specified requirements; the supplier needs to know only what the requirements are, not how they originated.
9. Try tabulating your data where it would otherwise be too unwieldy or where

TABULAR DATA FOR LOAD ANALYSIS OF AN AIR-CONDITIONING UNIT

Cooling
(Supply air temp. = 62° F. Density CFM Fac. = 1.02)

	Peak		Average
Total CFM	25,200	—	—
Supply area, Btu/H	593,000	—	—
Supply area, SHF	0.91		
Total tons	79.80	—	—
Fan, BHP	19.6	—	—
Fan, kW	17.2	—	14.5
Refrigeration, kW	89.1	—	37.4
Total demand, kW	106.2	—	—
Average free cooling temp.	70.0	—	—
Total refrigeration, h/yr	—	1,595	—
Approx. refrigeration, kWh/yr	—	60,000	—

FIG. 6.3 A table like this one permits easy reference to numerical data for certain specified requirements. This table shows an air-conditioning load analysis.

it might get "buried" under other material in the text of the specification. Figure 6.3 is a typical table used in a specification; this one shows an air-conditioning load analysis.

10. Complete all the appropriate blanks in any skeleton specifications and data sheets.
11. Permit the supplier to make alternate offers for consideration, but point out unacceptable features for a particular application.
12. Check and double-check the operating conditions.
13. Give manufacturers enough process data so they don't have to guess; but try not to restrict them unnecessarily.
14. Ask for enough literature from manufacturers to enable you (or your supervisors) to fully analyze the bid.
15. Allow contractors to use their standards or suggest improvements in your design, pending your approval.
16. Make a final check before submitting the specification draft to the reviewers or editors. Ask yourself, "Does the specification tell the contractor or manufacturer exactly what my organization wants?"

AFTER THE SPECIFICATION IS PRODUCED

The writer's responsibilities do not end with the writing and production of the specification. To some extent, depending on the organization's rules, the writer may be involved in subsequent phases of the procurement process. Generally, he or she will be required to prepare a materials requisition and send it to the purchasing department. Purchasing may then prepare a purchase order, an administrative specification, or a proposed contract. The engineering specification and related documents are sent to prospective bidders with a request for quotations, and the purchasing department oversees all contacts with the bidders.

If the bidders observe any flaws in the engineering specification or if they want to suggest modifications, the writer must evaluate their comments. As a result, the writer may need to confer with the engineering staff to propose changes. This may make it necessary for the writer to prepare a formal document, such as a *change proposal,* and send it to reviewers. When revisions are to be made, some organizations require the writer to send editors and reviewers a *review request form,* as shown in Figure 6.4. Using this form helps to coordinate comments and speeds the revision process.

Document review request for comment

Client: __________ Project: __________ Job no.: __________

Date: __________ Please review: __________
(Title)

To: __________ __________ (Document no.) __________ (Rev. no.) and

__________ return to __________ (name)

__________ at __________ (Location) on or before __________ (Date)

__________ Explanatory remarks attached: Yes ____ No ____

__________ Safety related: Yes ____ No ____

Comments	Resolution

____
(Reviewer) (Date)

Page ____ of ____

Comments resolved: __________
(Responsible engineer)

Discussed with commenter? ____ Yes ____ No

Copy of this form sent to commenter? ____ Yes ____ No

FIG. 6.4 A typical review request form.

Occasionally, additional revisions may be mandated as work proceeds at the manufacturing or construction site. While these are a normal part of the procurement process, no writer looks forward to the possibility of having to rewrite and reorganize a specification. To reduce the likelihood of revisions becoming necessary, the writer must make every effort to know exactly what the reviewers and editors expect from the initial specification. Using the checklist below and the guidelines listed in this section should help you adhere to the most widely used general rules for writing many types of specifications. Beyond that, it is your responsibility to learn what your organization requires and to use good judgment in your writing at all times.

OTHER CONSIDERATIONS

To become a competent specification writer, you must know much more than just how to use correct English. You must be able to read and comprehend a variety of drawings—mechanical, electrical, structural, architectural, plumbing, etc. Skill in reading and understanding drawings comes only after much practice and training.

A wide knowledge of the project or product about which you are writing is also important, if not essential. This, too, comes only from experience and study. You should also be familiar with laws and code requirements that may apply to the project. And do not overlook a knowledge of the work performed by the various skilled workers who are employed at the work site. In sum, to continue to improve your specification writing skills, you must study widely.

CHECKLIST FOR SPECIFICATION WRITING

Before You Write

1. Have you studied thoroughly the component, system, material, services, or other items that the specification will cover?
2. Have you determined exactly which project phases the document must cover (e.g., design, manufacturing, testing, or subdivisions of the major phases)?
3. Have you established whether the project or product will be covered best in a stand-alone specification, or in a primary specification with supporting documents for separate applications or phases?
4. Based on your examination of the items to be covered, have you classified the type of specification you will write?
5. Have you decided on the proper approach to use in specifying the technical requirements? That is, will you use an open, restricted, or closed specification?
6. Have you reviewed all appropriate engineering procedures or standard rules used for writing specifications in your organization?
7. Have you determined what documents, codes, industrial standards, or laws apply to the goods or services you will be covering? Have you obtained copies of all these applicable documents?

8. Have you obtained a complete set of the latest drawings of the product or project?
9. Have you obtained copies of any master or guide specifications that can be used?
10. Have you conferred with the engineering staff, purchasing department, customer, and other sources of data to make certain you have all the information you need to put the technical requirements into writing?
11. If the project is large and complex, have you or other members of your organization considered preparing a flowchart (or specification tree) to show the arrangement of and relationships between all the project specifications? Such a flowchart may simplify the tasks of scheduling, determining the specification writing workload, and overseeing the whole writing process.

Preparing the Specification Draft

1. Determine the format and structure that you will use in your document.
2. Prepare an outline of the specification. Use first-order and second-order headings.
3. Study all applicable drawings for the items to be covered. Get the "feel" of size, location, major and minor parts or components, etc.
4. Write the first few pages of the specification. Have the text and structure checked.
5. Complete the text and check it against the drawings.
6. Review the list of tips for good specification writing given earlier in this section. Have you followed all the guidelines?
7. Prepare the front matter—cover sheet, table of contents, general information, etc.
8. Have the specification typed. Read the copy, checking it against the original.

Reviewing the Draft Prior to Submittal

1. Are the data organized and subdivided logically? Have you used tables and lists, when possible, for greater clarity and ease of use?
2. If the product or project is for the federal or state government, have you followed all applicable procedures, standardized practices, and standard specifications?
3. Are all specified requirements in the correct places in the document?
4. Are there any company documents, master specifications, or other reference sources that you should include but have left out?
5. Have all unnecessary words, descriptions, or explanations been eliminated from the text?
6. Have you followed the rule of 2 throughout?
7. Have all errors of grammar and spelling been corrected?
8. Are all the specified requirements clear and unambiguous?

SECTION 7
PREPARING WRITTEN DIRECTIVES AND PROCEDURES

Technical specialists and administrators in most organizations have to spend some part of their day writing. But the role of writer is only one of many these people fill each day. Among their most important duties are the coordination of technical tasks and the control of procedures necessary for successful operation. Not writers by profession, many technical managers see writing as a task divorced from their "real" work and as a distracting chore that removes them from their normal endeavors at the field site, lab bench, or planning table.

Nearly everyone scoffs at paperwork now and then, and the engineer, scientist, or manager has a legitimate concern about being swamped with paper in order to perform a simple task. But there is also a real need for certain written communications in technical management. Experienced executives recognize this fact. They know that good technical writing will not interfere with their most important duties. Instead, it helps them manage people and make decisions more effectively.

Two kinds of documents are considered especially valuable in successful technical management and decision making: directives and procedures.

Although the preparation of written directives and procedures is primarily a management concern, it is recommended that technical specialists and businesspeople be familiar with the basic requirements of these forms of writing. Lacking this familiarity, you may find it hard to provide the coordination needed to fulfill your organization's technical and business objectives.

If you have ever had to issue instructions orally, you have probably seen how easily they can be confused or forgotten. By contrast, a well-written directive or procedure will ensure that the required data is at hand when people proceed to do their jobs. Once you know the requirements for these kinds of documents, you should be able to get your message across not only with less confusion but also faster and at reduced cost to your organization. This is true regardless of whether you are in engineering, science, management, or technical support.

DEFINITIONS OF DIRECTIVES, PROCEDURES, AND RELATED DOCUMENTS

Directives and procedures share certain features with other technical writing tasks—yet they are unique in several ways. That's why we've devoted an entire section of this handbook to these two forms of technical writing.

What Is a Directive?

In general, the term *directive* refers to written orders and instructions prepared and issued by management for internal distribution within an organization (or one of its divisions, departments, or other units). In its simplest form the directive might be only a few words, as a brief memorandum: "All personnel must report to the conference room at 3 p.m." More complex directives may contain multiple sections and run to many pages, as for a discussion of engineering responsibilities or techniques.

Purpose of Written Directives. Each directive has a specific function. Typically, the written directive conveys a special message not meant for out-of-company readership. Occasionally, however, a directive may be exhibited to a client to acquaint the client with the manner in which the technical organization manages or performs certain operations. A directive may also define a responsibility, establish or confirm policy, dictate commands, or otherwise serve to guide, control, or point out details of the organization's work.

Flexibility of Directives. Directives are a flexible type of document and may be referred to by a number of titles, depending on such factors as length and the intended audience and application. If the directive is brief—one page, say—it may be referred to as a *directive memo*. When it establishes a standard technique for a particular task, the directive may be called a *standard practice instruction*, as the printed form in Figure 7.1 shows. Other commonly used designations include: *engineering directive, production directive, research directive,* and *purchasing directive*.

What Is a Procedure Document?

Procedure documents are the written means whereby management seeks to control basic functions within an organization or unit in the most cost-effective and qualitative manner possible.

Purpose of Written Procedures. The aim of most procedure documents is (1) to furnish a record of the steps required for satisfactory performance of a certain task (or for satisfaction of all requirements of a project) and (2) to promote a clear understanding of the methodology of the work processes performed. In general, management considers the procedure document key to obtaining full cooperation of all personnel responsible for successful operation of the organization, unit, or project.

Following the formulation of a procedure, management may submit the document for review and approval by supervisors and others whom the procedure will

ABC Engineering Corporation, Inc.
Field Division

Division Standard Practice

Directive No.: 12
Page 1 of 3 pages
Date Issue: 3/3/XX

Subject: Dissemination of technical data

Please destroy all copies of directive 11, dated 11/3/XX.
This directive supersedes no. 11.

1.0 Purpose and scope

1.1 To establish methods for controlling the dissemination of technical data incident to field operations.

1.2 The methods defined apply to all technical evaluations and scientific studies, environmental assessments, environmental impact studies, topical reports, and cost summarizations generated in engineering. All departments of the field division are affected.

2.0 Procedure and responsibilities

2.1 The study manager shall be responsible for compiling the list of applicable documents — including all references, bibliographies, transcripts of conferences with consultants, calculation sheets, etc. — needed to substantiate the validity of field evaluations.

2.2 The study manager shall prepare a list of documents in accordance with the required format (see directive 4). This list will be used during formulation of the study.

2.3 The study manager shall follow the procedural flowchart provided herewith (see attachment). This identifies the proper flow of data.

Prepared by: J.O. Smith Date 3/1/xx ______ Date ______

Approved by: J. A. Doe ______ ______
J. A. Doe

Authorized by: W.A. Crenshaw 3/1/xx
Division Manager

FIG. 7.1 Directive designed for use as a standard practice instruction.

affect. The procedure is then issued to all personnel affected directly by its provisions. When a question arises regarding the mechanics of a certain operation, the persons responsible for a decision need only to refer to the written procedure.

Variability of Procedures. Like directives, procedures come in various forms. The procedure might be short and simple, as: "Dissolve residue in 25 cc distilled water. Then add benzene." Or it may be long and complicated and consist of several volumes, as for the operation of a cooling system or aircraft. Various titles may therefore be used, including: *operating procedures, maintenance procedures, design procedures, test procedures, manufacturing procedures, installation procedures,* etc.

A large document intended for general use throughout an engineering department will usually be called an *engineering procedures manual* or *standard procedures manual,* while one developed mainly for use on a certain engagement will be called a *project control procedure, project procedure,* or, simply, *control document.* Figure 7.2 shows a page from a typical engineering procedures manual.

In many instances, procedures manuals will be compiled from various industrial codes, standards, and project procedures. Since the requirements for writing these manuals are practically the same as for writing a single project procedure, we emphasize the simpler form—the project procedure—in this section. For a detailed discussion of instruction manuals, see Section 8.

ABC Engineering Corporation, Inc. **Procedures**	No. Page Issue
Engineering Department	Supersedes
Subject: Drawing Release (Accessories)	Effective Date:

1. General
    1.1 Drawings completed by the accessories engineering department will be forwarded to the release group with a "release request" marked either "experimental" or "production" in the space provided for indicating the release desired. This release request will also serve as a "job ticket," by placing the job number in the upper right-hand corner and obtaining the project engineer's signature in the lower right margin as authority for release.

    1.2 Accessories engineering department releases will be handled on the basis that all necessary checking is accomplished prior to release, and they will be printed and distributed without passing through the regular engineering release procedure.

2. Experimental Release
    2.1 All prints on an experimental release will be stamped both "accessories" and "experimental," in addition to other stamping, and will be red-line prints.

    2.2 The experimental department will notify the engineering department by "change requests" of all changes necessary to correct the experimental drawings for production. Upon these changes being incorporated, tracings shall be forwarded to the release group for a "production" release.

3. Production Release
    3.1 All production prints shall be blue-line and shall be stamped both "accessories" and "OK for production," in addition to other stamping.

    3.2 Experimental drawings rereleased without design change for production shall be assigned the next change letter for record purposes. A notice of change is not necessary, and "released for production — no change" shall be entered in the alternate black.

4. Release Quantities and Destinations
    The quantities of prints listed on page 2 of this procedure shall be forwarded to the listed destinations on each release of a drawing.

                                                 John Doe, Administrative Engineer

FIG. 7.2 A page from a typical engineering procedures manual.

Differences between Directives and Procedures

In one sense, the written project procedure is a form of directive: It tells the people concerned with a particular operation what steps management deems necessary for successful completion of the operation. Thus, the procedure shares some basic functions with the written directive: to instruct, to transmit an order, to establish policy, and to provide a handy and authoritative reference. Directives and procedures are virtually identical in regard to origin, readership, and aim. The same guidelines for good directive writing apply to most procedures.

The main difference between directives and procedures is that the directive typically requests or dictates an action, while the procedure outlines a series of steps. In effect, the procedure defines the ways and means of performing several actions, all of which may be integral to a particular task that has been (or will be) assigned. Often the written procedure is intended primarily as a reference tool; therefore, it may not require immediate application by the reader. By contrast, the typical directive treats a specific problem or question regarding some phase of a task or operation and mandates immediate action.

For example, a design supervisor may issue a directive that gives the design staff the go-ahead to start work on a new engagement. But if the message details the various design steps the group must follow, the document usually is said to be a *project procedure* or *control document*. Some technical directives are more complicated, however, and define a number of actions that the writer expects the reader to take; in this case, the actions are presented in list form, similar to the manner in which the steps of a procedure would be arranged.

What's the Message?

In preparing a directive or procedure, you must ask yourself, "What message do I wish to convey?" If you are writing a directive, the answer is simple: The message is the action you expect—the response you desire from the reader. If you are writing a procedure, the answer is more complex, for the message explains the ways and means of getting things done. Because you must explain all the various steps involved and arrange them sequentially, the procedure typically takes longer to write than the directive.

Memos, Bulletins, and Policy Statements

Directives and project procedures both belong to the same broad class of technical documents which includes memos, bulletins, and policy statements. The main similarity among all these technical publications is their origin and readership: In general, they are in-house documents—developed by management for distribution within the organization, department, or unit. The various in-house publications also share the same basic rules of good technical writing.

Applying Distinctions. Together, directives and procedures are distinct from other in-house publications in several ways. For example, although a brief directive may at first glance appear no different from a traditional office memo, careful examination shows the directive to be more strictly functional. As noted, the directive normally requests or dictates some kind of action. This message will be contained in the first few sentences of the body of the directive. The traditional memo, on the other hand, may give several paragraphs of preliminary details before suggesting an action; often it will merely report information and recommend no action at all. Further, the directive will usually have a higher degree of technical content than the memo.

Another difference is length. Technical directives will often run to several pages in length, while the typical memo is confined to a single page. However, a lengthy technical directive or procedure may be accompanied by a memo which amplifies or explains the technical document itself.

A second type of in-house publication that is closely allied with directives is the bulletin. Like directives, bulletins are flexible and often quickly produced and distributed. The writer may choose an informal approach ("From the desk of So and So") or a more structured format. Unlike directives, however, bulletins often are concerned with covering "news" rather than promulgating rules and instructions. Occasionally a writer may use the terms "bulletin" and "directive" interchangeably. Bulletins and directives are usually shorter than project procedures.

Divorcing Policy from Other Documents. Finally, we ought to note the relationship that directives and project procedures have with another in-house publication: the policy statement. Directives and procedures are control-oriented documents that frequently convey policy-related information. To the reader of the technical document, however, this may not always be clear; therefore, management must take special care to highlight and explain any policy-related information contained in the directive or procedure. Winning staff members' cooperation is often much easier if you label such information "Policy" and say why the policy is being established or changed.

Despite these efforts, the policy may still be obscured by the control-oriented emphasis of the procedure or directive. Thus, many organizations prefer to issue policy statements separately. Writers then update the policy statements regularly, incorporating relevant new data from written directives and procedures as required.

Using Directives and Procedures in Client Relations

Although directives and written procedures are intended primarily for in-house use, they also play an important role in client relations. Further, such documents are sometimes used to enhance the process of contract negotiation.

A well-written directive or project procedure can foster better understanding and appreciation of your organization's methods, leading to more complete satisfaction of contract requirements. Provided that you take adequate precautions against unauthorized use of the data, the documents can be used to advantage before, during, and after negotiation with clients.

Prior to Contract Negotiation. Along with contracts and specifications, directives and project procedures form the backbone of documentation in a technical or scientific organization. These documents will illustrate not only what methods have been applied to a single operation, they will also reflect the general quality of the work performed and the competency of the group as a whole. So don't be afraid to show your prospective client a copy of a directive or procedure if you feel it will help your organization win the contract. But first be sure that the items you choose to show (1) can be applied—directly or indirectly—to the proposed engagement and (2) represent your organization's best efforts. Otherwise they'll be of little value to you or to the client. Useful documents may come from past, existing, or proposed work.

While the Contract Is Being Drawn Up. Just as clients may furnish some of the basic technical requirements for an engagement, so may they specify certain methods and procedures which they want the scientific or technical group to follow. Such requests and requirements constitute a portion of the issues management normally expects to address while negotiating a proposed contract.

If management accepts what the client has requested, this agreement normally will be written into the contract provisions. Such a provision helps ensure that the contract will be performed to the client's ultimate satisfaction. Occasionally, however, one of the parties may not deem the contract provision sufficient per se. In this case, the simplest solution is to write one or more separate control documents which detail the specific procedures agreed to. These documents—the project control procedures—are then issued to the personnel directly responsible for conducting the engagement. If desired, management forwards copies of the procedures to the client. The advantages of this approach are: (1) the directives provide authoritative guidance to technical personnel in satisfying the client's requirements and (2) the client knows that management is making every effort to ensure the organization performs as expected.

Occasionally, other kinds of documents may be written for the same purposes. As noted in Section 6 on specification writing, the nontechnical requirements for some projects may be defined in a separate set of documents known as *administrative specifications*. These include requirements for organization and assignment of personnel, planning, scheduling, licensing, cost accounting, and documentation.

In some instances, the administrative specifications will not be amplified by separate project control procedures. In any instance, written procedures will furnish additional protection—for both the client and the technical organization—against potential complications. Further, the preparation and distribution of control documents may be required as part of the contractual provisions. Management and authorized technical personnel are advised to use project control procedures whenever feasible.

After the Contract Is Signed. As a project progresses, management may have to issue a number of new directives and procedures. These may be necessitated by changes in conditions at the work site or many other kinds of developments. When necessary, copies of new directives and procedures may be sent to the client to keep him or her abreast of the changing situation. In any event, the technical organization must record the changes and store copies of the documents in an appropriate file. Applied conscientiously, this system of documentation protects both the technical organization and the client against potential disputes and liabilities. If questions, disagreements, or other problems arise concerning the satisfaction of project requirements, the documents may serve as valuable evidence of events leading up to the dispute.

Why Write a Directive or Procedure?

Aside from the situations discussed above, there are a number of other times when writing a directive or procedure is suggested.

Usually, these documents emerge in an effort by top or middle management to resolve a specific problem or satisfy a certain requirement. Determining that there is a need to write a directive or procedure on a particular subject is often accomplished by conferring with supervisors, clients, or staff and listing subjects that in their judgment need to be put in writing. Discussion with the structures supervisor, for example, may result in a decision to write procedures on structural analysis and testing. Similarly, the need to write a directive might become apparent during management review of an engagement or while instituting changes in certain technical programs. Other topics that management may be able to address effectively in a directive or written procedure include changes due to new regulations and industrial standards, loss or departure of key personnel, organizational restructuring, market variation, and financial pressures.

There are many other reasons you might want to write a directive or procedure. Some of these are: to establish a project network, to define schedules and milestones, to clarify authority, to present a scenario or contingency plan, to furnish references, or to highlight the most efficient techniques for accomplishing a particular task. Whatever the subject, a well-written explanation will improve the flow of data from management to staff.

Experienced managers view the periodic issuance of directives and project procedures as necessary for maintaining productive and harmonious personnel relations. In many organizations, it is not unusual for management to issue several control documents each week.

The documents need not be fancy or slickly produced. Even a brief photocopied page defining a basic rule or procedure is preferable to the daily confusion and interpretation of a vague policy by both supervisors and staff. Printed on looseleaf paper, the messages can be filed conveniently in a notebook. Employees may then consult the book whenever an operating problem occurs.

Sources of Information

It doesn't matter who provides the ideas for the document—the company president, the technical specialist, and the think-tank team are a few of the sources who might point out the need for preparing a directive or written procedure. Most directives and procedures seem to work best, however, when one person in management is designated to control the preparation, editing, and revision of these kinds of documents. If staff members are familiar with the general requirements of these publications, they may be able to offer constructive suggestions to management during preparation and revision.

Value to Management

In sum, directives and procedures have a wide range of uses in a technical or scientific organization. They help management to: (1) free itself of detail, (2) define objectives clearly, (3) show clients how a certain job will be done (or was done), (4) clarify staff responsibilities and coordinate activities, (5) provide on-the-job reference to standard practices and procedures, (6) orient new employees to basic job requirements without wasting other people's time, (7) obtain better organizational and personal commitments, (8) control the effects of external and internal changes on a project (or on the organization as a whole), (9) eliminate unnecessary and costly duplications, and (10) utilize constructive suggestions from personnel.

Document Distribution

Generally, directives and written procedures are distributed only to the personnel who will be affected directly by the provisions they contain. The primary recipients, therefore, are department managers and supervisors. A typical engineering directive will be distributed, as applicable, to the chief engineer, chief draftsperson, project engineer, and relevant supervisors. These are the main people responsible for carrying out the provisions.

If the directive will affect other personnel immediately—such as structural and materials engineers—then these people become the secondary recipients. While the typical pattern of distribution is from top management to middle management and supervisory personnel, an appropriate directive or procedure can, of course, be sent to any part of the organization: technical departments, pricing and estimating, procurement, quality assurance, manufacturing experts, etc.

Know Your Audience

An important point to remember here is that the various levels of the organization represent separate audiences, each with its own interests and concerns. So you must tailor your message to the mind of the reader you wish to reach. In the following discussion of document types, you'll find some suggestions for customizing your message.

TYPES OF DIRECTIVES AND PROCEDURES

Several types of directives and procedures may be defined. In the end, all must contribute to the same general goal and must convey the "who," "what," "why," "where," and "when" details needed to put a given message into action. The distinctions among the types are not always clear-cut, and this can be a problem in a large organization that uses a small number of printed forms for its internal communications. For example, a directive or procedure that is not of a clear-cut type may defy limits imposed by the printed data; in this case, the writer must modify the existing form or prepare a new document from scratch.

Certain features are common to all well-written directives and procedures, despite the differences among the various types. This handbook treats directives and procedures as a single category of technical writing and covers the most widely used general types. Once a writer becomes familiar with these, he or she may apply this knowledge to any type used.

Single-Use and Standing Documents

A *single-use directive* or *procedure* is a document intended for one-time-only application to a specific project or phase of work. As part of the documentation process, such a document will normally be filed away after it has been used, even though the material as a whole will not be applied to subsequent work. Documents from past efforts may be used, however, as reference sources and as models when preparing a new release.

A *standing directive* or *procedure* is a document designed for general use—on more than one engagement, for example, or whenever a certain operation is performed. The standing directive or procedure remains in effect until such time as management issues a new document which supersedes the original. Figures 7.1 and 7.2 are examples of standing documents.

In practice, the standing document is most useful for identifying standard procedures and policy, while the single-use document works best for urgent instructions and temporary procedures. From such practical considerations, then, the writer must decide—based upon what phases of operations the message will affect—whether he or she will be writing a document for a unique, short-term use or for more general, long-term application. A single-use directive or procedure will emphasize matters requiring immediate action by an individual or group as opposed to permanent, organization-wide rules or guidelines. A standing directive or procedure will set out the major actions that management requires of a department or unit (or of the whole organization) and will place more emphasis on policy issues.

Note that care must be taken to ensure proper application of the material in either type of document. Each release must be labeled precisely as to whom it affects, what actions are required, where (on what projects or phases of work) the actions are to be taken, and when the order or procedure is to be put into effect. These items may be summarized near the beginning of the release in standard memo format. In a long document, the information may be defined further in a section titled "Scope." These details alone will often be sufficient to distinguish the message as being of a single-use or standing type.

In some organizations, managers depend on the use of printed forms. A well-designed form is effective in getting the message across clearly and quickly. Not only does it assure that a writer supplies the required information, but it puts each piece of data—source of the document, parties affected, subject, applicable dates, etc.—in a specific spot so the reader's eye can quickly locate the needed details. Further, a good form attracts attention and leaves no doubt as to what type of message is intended.

If your organization has its own standard forms for certain kinds of directives or procedures, be sure to abide by the local rules. This handbook offers some suggestions for the makeup of various in-house documents and shows how certain forms have been used by successful organizations; but we wish to emphasize that these are simply examples from a wide choice of acceptable forms and formats. Unless you can convince your superiors that the standard forms should be changed, you will probably get good results by using the standard forms, even though they differ from the suggestions in this book.

Technical and Nontechnical Communications

As noted earlier, a directive or procedure may be primarily technical or largely nontechnical in content. The principal difference is that the operations described in technical directives and procedures are usually highly complex. When complicated and numerous technical details are involved, selected information may be presented in the form of a list or chart. Still, the writer must strive to make the technical document clear, direct, and as easy to follow as, say, this brief directive governing computer use by employees:

To: All Engineering Department Employees

From: J. Doe, Company President

Subject: Personal Use of Computer Terminals

Date: 1/1/XX

To guard our system against accidental misuse and unauthorized release of engineering information, I am asking that all engineering department employees limit nonbusiness use of terminals to weekend hours when the general manager of data processing is here. If you are in doubt about the manager's schedule for an upcoming weekend, please check the list of dates posted on the bulletin board, Room 114.

Typically, the degree of technical content will be determined by the audience for whom the document is being written. Most directives and procedures intended for the engineering staff will be primarily technical, while most of those for administrative support personnel will be nontechnical. As a general rule of thumb, this sort of distinction will help a writer tailor his or her message to the interests of the reader. But what should the writer do when the audience is technical and the topic is nontechnical, or vice versa? What happens, for example, when you must write a directive telling staff scientists to use a new type of filing system? Or a procedure showing typists how to store computer data on magnetic tape? Here the distinctions become blurred. Will you be writing a technical document or a nontechnical one?

Ideally you should construct your message a little differently for each segment of the organization you wish to address. A research scientist may respond more favorably to a directive that emphasizes how the new filing system will enhance basic research as opposed to record keeping. A typist may be more likely to remember the computer procedure if it explains how tape storage will eliminate the need for carbon copies or retyping. Tailoring your message to the interests of your audience is an important part of any writing task—technical or nontechnical.

In the final analysis, the requirements for writing a technical directive or procedure are the same as for a nontechnical one. The documents written for the typist and the scientist must convey their messages in the same simple style as the memo governing computer use by engineers. The same is true for specific subtypes of technical directives and procedures, such as electrical engineering directives, mechanical engineering procedures, computer engineering directives, and chemical research procedures.

The writer of the directive above should not make the instructions needlessly complex or full of jargon simply because the audience (and the subject) are technical. On the contrary, instructions are always effective when they tell the reader what to do in the simplest, most direct language possible. Careful organization of information is another important aspect of effective procedures and directives, as we'll see.

Urgent and Nonurgent Messages

Directives are an excellent medium for conveying urgent messages. They are flexible, quickly produced, and can be distributed directly to their intended readers. The same is true of procedures—though to a lesser extent since procedures often take longer to formulate. If you need to issue instructions rapidly, directives are among the most practical means around.

There are many levels of urgency, however, and some organizations have developed fairly elaborate schemes for designating a document as extremely urgent, somewhat urgent, not urgent, or something in between. For example, an urgent directive might be marked "Critical" and thus be said to constitute a different type of message than a nonurgent directive with no special notation. A directive that demands immediate action to remedy a safety flaw in a piece of equipment will be more urgent than, say, a directive reminding the design staff it has a month to complete a project. And a directive that commands a proposal writer to submit a 60-page draft by noon may be more urgent still. In any event, the documents will vary somewhat—not only in content but also in the time factors involved in writing them and getting them read, and possibly in their appearance too.

Determining Urgency

The writer must decide—before writing—exactly how critical the message is. Once this is done, he or she must prepare the document in a way that clearly conveys the proper level of urgency. In short, the writer must make certain the document gets read in a timely manner without interfering unduly with a reader's more pressing concerns.

There are several ways to accomplish this. If you have a matter of critical importance that must be addressed in writing, ask yourself these questions:

1. What Time Factors Are Involved? It takes time to write, produce, and distribute any directive. Once distributed, the document must be read and understood—then readers must apply the data. If the message is long and complex or must reach a large number of people, it will take even longer to communicate. By taking such factors into account, however, you can still do a first-rate job in preparing a timely directive. This requires both a knowledge of directive writing practices and good judgment. This handbook can supply only the first. Good judgment must be supplied by the writer.

2. What Is the Time Differential? In other words, how much time will it take to get your message across, and how long do you have before the information must be put into action? You'll be in good shape if you can dash off a directive in a hurry, giving your readers ample time to act. On the other hand, if you have an hour to tell your department's cost estimators to revise a long list of costs but the directive will take you 45 minutes to write, you'll have to take special steps to get the job done on time.

3. What Are the Risks of Delay? In a technical or scientific organization, the costs of delay can be high—especially in engineering, where matters of public safety are often involved. A directive applied properly in an emergency may make the difference between finishing a project on time, within budget, and in a manner satisfactory to the client and finishing it late, over budget, and unsatisfactorily. Generally, the larger the risk of delay, the more urgent a message will be.

Once you've answered these three questions, you should have a reasonably good idea how critical your message is. The next step is to start writing. Waste no time going from preparation to review, editing, and distribution. Allow the urgency of your task to carry you through these stages quickly. But be careful. Any urgent technical document will have strict requirements for clarity, accuracy, and proper distribution. And remember: In a busy organization, people may fail to read material in a timely manner unless you make it mandatory. If you can't make it mandatory, at least try to make it interesting.

Getting Attention

To help ensure that your document will be read quickly and thoroughly, let's examine some attention-getting techniques you can use.

1. Choose an Appropriate Tone before You Write. Tone—the voice you use to present your subject—is very important in an urgent communication. To convince your audience that the message is crucial, your tone normally should be both serious and direct: "All design personnel must report to the supervisor's office immediately upon arrival." This does not mean that your writing must be extremely formal, for the tone should fit the audience. Thus, it is not unusual for a supervisor to use first names in an urgent directive (provided that the people who are referred to know each other): "Due to critical changes in management authority, effective immediately, all further design activities will be controlled through Phil."

A certain amount of informality may sometimes aid a communication, but a serious tone must not be sacrificed for the sake of wit or familiarity. If you start your directive with a joke, your readers may fail to attach much importance to the message. Worse still is the possibility that they will shelve the directive and forget it.

2. Keep the Message Brief, When Possible. It is a well-known principle of readability that people will sooner read and remember a brief message than a long and complex one. By holding your instructions to one page, you can greatly increase the chances that the whole document will be read carefully. Unfortunately, this is not always feasible with technical material. But you can still apply the same principal to a longer document. How? By keeping your sentences and paragraphs short.

3. Use Special Forms for Visual Impact. Aside from the message itself, the most important tool you can use to generate interest is visual impact. In some organizations, a number of printed forms and specially designed letterheads are available for this purpose. Typically, these forms achieve their visual impact through the use of large type, similar to a newspaper headline. Colored paper is seldom used; instead, all necessary details are set out clearly in black and white. For example, a form used for an urgent directive might have the word "PRIORITY," "CRITICAL," or "URGENT" printed at the top of the document in half-inch-high letters. Note that each of these words may be used, in some organizations, to designate a different level of urgency.

Special headings like these do not necessarily have to be produced by an expensive, professional printing service; with a word processing system or a small computer and printer, you may be able to print these forms when you need them—cheaply and quickly. *One caution:* The types of forms discussed here should not be confused with the forms used by many technical organizations to transmit advance engineering data; stop orders, drawing changes, engineering orders, etc., are highly specialized documents used to convey urgent technical information, but they generally require rather complex, printed forms. Further, this type of documentation is primarily a function of engineering as opposed to technical writing. Stop orders, drawing changes, engineering orders, and the like, therefore are not considered germane to the subject matter of this section.

4. Use Special Distribution Channels. If you want your message read quickly, don't depend on the company mailing system. That may take days, and you don't have this kind of time to waste. Designate a place for people to pick up urgent material. You may also find it useful to assign someone to control the access to that material. If an intended reader doesn't pick up his or her copy by a predetermined time, the reader may be contacted directly. This approach will help to impress upon readers the true importance of the material. Other distribution channels, such as bulletin boards and display tables, may be used when the material is not of a sensitive nature or does not required restricted distribution.

For any given document, you can use one or more of the four attention-getting techniques listed. Figure 7.3 shows how a few of these techniques may be combined for maximum effect in a directive memo. The writer has managed to pack a substantial amount of information into a brief, one-page document. Note how easily it reads. The sentences are short and simple. The tone is serious and direct, though not too formal. The information is arranged for clear and immediate visual impact. Capitalization and underlining are used, where appropriate, for emphasis. And the words in large type at the top and bottom leave no doubt as to the importance of this communication. Note that the directive is a single-use document designed for application to one project, rather than a general, standing type of communication.

Certainly not every directive or procedure you write will be an urgent, "priority" matter. Many in-house documents are confined to routine subjects of office policy and everyday administration. To help manage people effectively, however, a writer must know how to get a reader's attention—regardless of whether the message is urgent or not. Once the writer knows the techniques for generating interest, he or she can apply these to any document required.

STRUCTURE OF DIRECTIVES AND PROCEDURES

Every in-house communication should be carefully planned and constructed to get results. Remember, a directive and a procedure are a means not only of disseminating information but also of managing people. They are the answer to the question, "How do we get things done around here?" Thus, they should be written in a form that tells any reader what that reader needs to do to help the organization achieve its technical and business objectives.

No matter how management delegates authority, the directive or procedure must enable technical professionals to do their jobs in a qualitative and cost-effective manner. If this is not done, trouble will follow. At best, an unclear or poorly organized document will lead to a waste of the technical staff's time. At worst, serious errors, omissions, or lawsuits could result. One way to avoid such trouble—and to make your directive or procedure easy to read and understand—is to follow the general structural guidelines given below.

Remember that the structure, or makeup, of a directive or procedure will have

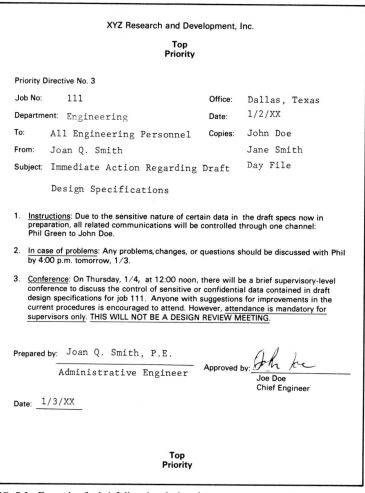

FIG. 7.3 Example of a brief directive designed to convey an urgent message.

a lot to do with the way a reader responds to your message. If the document is well-organized and contains all the necessary elements, the reader should have no trouble putting the information into action. On the other hand, if you omit certain data or fail to organize your writing properly, you will find yourself doing several follow-ups to get your point across. Since this wastes valuable time and energy, make every effort to structure your document as effectively as possible. To get the results you want from every directive or procedure, follow these structural guidelines:

1. Choose a basic functional format.
2. Arrange action details in list form.

3. Expand the format for lengthy or complex material.
4. Draw flowcharts, if needed, to simplify procedures.

Choose a Basic Format

Once you've determined the main message you want to give your readers, you will have to decide how to organize your material. There are a number of organizing schemes to choose from for a directive or procedure. But organize, you must—for the information will be of little value unless it can be easily read and understood.

Two basic formats are widely used, varying slightly from writer to writer: (a) a traditional structure based on the format for memos and short reports and (b) a functional structure developed especially for directives and procedures. For most simple directives and procedures (one or two pages in length), either basic format will do; but the functional format has several advantages. A long or complicated message, on the other hand, will usually require an expanded version of the functional format.

Traditional Format. The traditional format has four parts:

1. Transmittal data
2. Introduction
3. Discussion
4. Conclusions and recommendations

The trouble with the traditional format, as shown in Figure 7-4, is that it places the items of major concern near the end of the communication. It is a roundabout way of telling people what to do. And as we've noted, most technical specialists and managers are too busy to waste time wading through introductory information before getting to the useful facts.

To be effective, a directive or procedure must state the bottom line—the message—up front. Because a directive or procedure brings readers face-to-face with important, sometimes critical, and occasionally controversial information, it immediately triggers questions in their minds. The document therefore must answer these questions as quickly and directly as possible: *What* is the message? *Why* is the message necessary? *What* action or steps are recommended? *Who* should perform them? *When, where,* and *how?*

In addition, the document should be easily identifiable as to form or type, and

Transmittal Data	(Routine information)
Introduction	(Background, history)
Discussion	(Details, investigation)
Conclusions and Recommendations	(Reasons for the message, recommended actions or steps)

FIG. 7.4 In traditional memo format, the actions you wish your reader to take are buried.

the format should be able to accommodate optional material to support the message. The basic functional format meets these requirements.

Functional Format. The following six-part functional format is suggested for most simple directives and procedures:

1. Form identification
2. Transmittal data
3. Conclusions
4. Recommendations
5. Discussion
6. Optional Material

Figure 7.5 shows how this basic functional format places the items of most interest to your reader at the top of the document, while answering the reader's important questions.

Because this format avoids the background discussion that precedes a message written in the traditional format, you can keep most simple directives and procedures to one or two pages. To see how this format works in an actual practice, consider the simple procedure in Figure 7.6.

Notice that the titles of the various parts, or compartments, of the format do not appear in the document itself. Also, notice that in the functional format there is no introduction. Instead, the writer begins the procedure document with the answers to the most important questions, "Why?" and "What?": "XYZ is revising...procedures to improve energy conservation.... I suggest you implement the following...."

The author tells the reader "how to do it" in five short steps. Then he gives all the other details—"Who?" "When?" "Where?"—that the reader needs to take the recommended action. Finally, the writer weaves some background information into the discussion, giving further justification to the message.

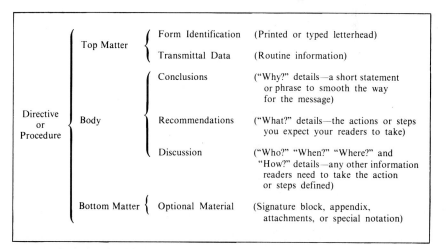

FIG. 7.5 Basic functional format for directives and procedures.

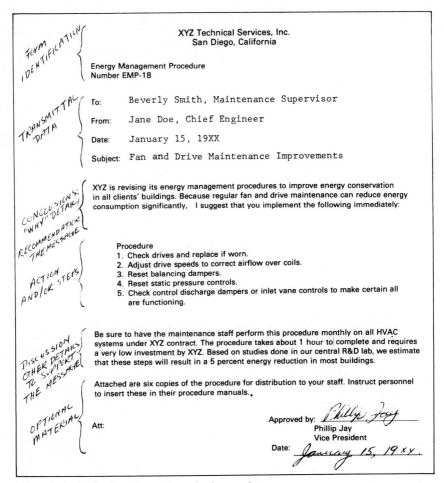

FIG. 7.6 Basic functional format for a simple procedure.

This basic functional format is well-suited to directives and procedures because it meets a fundamental management objective for which these documents are written: It tells the reader, "This is how we get things done around here." And unlike most documents written in the traditional format, it states the message at the very start.

Although the writer could have used the traditional format to convey his message, he would have run a greater risk of losing his reader's attention. Why? Because the traditional format is less direct. Further, readers often are already familiar with the material in an introduction. And much of the material in introductions is boring. Frequently, readers look over the first few sentences of a document written in the traditional format and ask, "So what? What does this have to do with me?" If they bother to go past the introduction, some readers may find the important details unfolding gradually. But they still have to dig out the message from the end of the communication. By the time the readers finally get the point, they might be so bored that they ignore or forget the directive. Or they may feel so frustrated that they actually resent management's suggestion that they follow a new procedure.

The functional format helps you to avoid these problems by organizing your material more effectively. To see how various pieces of information are arranged within this scheme, let's take a closer look at the typical elements of the basic functional format.

Basic Functional Format for Directives and Procedures

Top Matter. This compartment contains the form identification and transmittal data—that is, all the information needed to identify the document as a specific in-house communication. Some organizations provide a printed form for this purpose, while others use their regular printed letterhead and add other identifying data as needed. The form identification and transmittal data consist of several pieces of information, as follows:

1. Form identification

    Name and address of issuing organization, division, department, or unit
    Type of document (a general title, e.g., "Standard Procedure," or "Control Document")
    Document number
    Special notation (e.g., "Priority" or "Critical" for an urgent message)
2. Transmittal data

    To line (name of person to whom message is addressed)
    From line (name of author)
    Date
    Number of pages
    Subject

Although sometimes referred to as "routine information," these transmittal data are an important part of the record of the communication. Thus, they are standard in any good directive or procedure. Again, if the issuing organization does not provide a printed form, the writer or typist must supply these items.

Some general guidelines for transmittal data are: (1) Type the lines double-spaced for easy reading; (2) indicate the recipient's job title to direct the message to the right place quickly; (3) in general, omit courtesy titles such as Mr., Ms., Mrs., and Miss—they are usually unnecessary; (4) sign your initials in ink by the typed name to show the release is authorized; (5) be sure to include a "Subject" line—it lets you summarize the contents of the document and focus the reader's attention in a few words; (6) include additional identifying data as needed—for example, if the document is to be applied to a single project, specify that project by name or number on a separate line.

Body. This compartment is the heart of the document and the place where organization is most vital. By following the basic functional structure (conclusions, recommendations, discussion), you can present your central message directly instead of gradually working your way into the subject. Further, this format enables you to organize most simple directives and procedures in your head. Here's how you can make the basic functional format work for you in the document body.

1. Conclusions: Begin your directive or procedure with the answer to "why." *Remember:* Readers may not recognize the implications of the actions you're requesting or they may be skeptical that such actions are necessary. By including a short statement or phrase to smooth the way for the message, you can win your readers' cooperation much more quickly.

For example, if you intend to ask employees to go through the trouble of changing over to a new procedure for the release of engineering drawings, explain that you are introducing the procedure because past drawings have been lost or misused and that you want your staff to control the drawings as easily and effectively as possible. A logical reason makes almost any directive more persuasive.

2. Recommendations: After you give a plausible reason for the message, go directly to the "what" details: "We are changing the drawing-release procedure as follows...." "All employees must delete old data files to free up computer memory...." "These are the steps for calculating the amount of boiler blowdown needed to maintain the desired level of solids concentration...."

When your readers know what action(s) you want them to take, they can pay attention to your "how to do it" details with greater understanding and purpose.

3. Discussion: Based on that narrowly focused message, give any other details that your readers need to take the action or steps you've defined. But stick to the essentials. Usually the "why" and "how" will be the most vital details of your message and will already be contained in the message or action sentences earlier in the document. The "who," "when," and "where" details also may be stated near the beginning of your communication. If you have not mentioned them, however, elaborate on them here.

Provided that you omit any details that are obvious to your readers, the discussion compartment may also contain some data to further justify your message, such as a brief example of the results you expect: "This procedure will reduce peak power consumption by 10 percent...." "If each employee deletes all obsolete files by 4 p.m., we should have enough disk storage space to complete the estimating database on schedule...." "Results of past projects indicate that these steps will protect our organization from contract disputes...."

In most simple directives and procedures the discussion should be as brief as possible—two or three sentences usually are sufficient. In a lengthy procedure with an expanded format, however, the discussion may be extended to include as many sections or subsections as necessary.

Bottom Matter. In general, *bottom matter* refers to any special notation appearing at the bottom of the first page of the document. For example, the abbreviation *Att.* might be placed in the lower right-hand corner of the page to inform the reader that an attachment of some kind, such as a chart, table, or calculation, accompanies the document. Bottom matter may also refer to the attachment itself, to an appendix, or to other optional material. When this material is stapled, clipped, or bound to the back of the document, however, the term *back matter* may be used.

Another element common to the bottom matter in many directives and procedures is the signature block. Because these documents usually are subject to top management review prior to release, many organizations require that the bottom of the first page carry this item. The signature block makes it easy for top executives to indicate their approval of the writer's work. All they need do is sign their names in ink by the typed names in the signature block to signal that the document is ready for release.

Arrange Action Details in List Form

Whenever you plan to ask your readers to take multiple actions, try to put your details in a list. If the actions you wish to describe are clearly part of a step-by-step procedure, you will simplify the reader's job by arranging the actions as a series of numbered steps. Thus, if an employee needs to discuss the procedure, he or she can ask a question by referring to step 4 instead of "the tenth line down in the second paragraph on the first page, where it says to adjust the whatchamacallit meter."

The same principle applies to other possible details of your message, such as equipment, materials, forms, and references sources. When the items are numerous, arrange them in list form. Organization always aids clarity. Not only will a list help your readers to visualize your details, it will also help you to arrange your thoughts as you write.

Although you can list your details at any point in the process of preparing a document for issue, we suggest you begin as soon as possible once you know what your central message will be. Make a rough outline of the document as a whole. This might be a random list of all the details you feel your readers will need to know in order to put your message into action.

Before writing the directive or procedure, be sure to add all the explanations, precautions, illustrations, and other data that come to mind. Everything you put down in the outline will save you time later. Even for a one-page directive, an outline is a good idea if you have several details to include.

After completing the rough outline, you can see relationships in your details. Now is the time to reshuffle the items in your random list and put them in logical order. This is also a good time to weed out any unnecessary details and set them aside temporarily or drop them completely.

Put the action details (the steps of the procedure) in one list and make separate lists for equipment, materials, etc. Some details may fit into more than one logical arrangement, so experiment with a few tentative arrangements for each list. This will help you find the most appropriate arrangement for your purposes.

Typically, the steps of a procedure are listed chronologically, with each step indicated in a separate sentence or paragraph. Depending on the purpose of your document, however, you may prefer to arrange the details in some other manner—by order of the importance of each step, for example, or by order of the responsibilities of the various parties involved. Above all, arrange your details in a manner that will be comprehensive yet easy to read and use. You will only confuse your readers and make the result unpredictable if you forget to mention an important item at the right time. For a more detailed discussion of outlining, refer to Section 2 on reports.

Use an Expanded Format for Complex Messages

When the action or steps you wish to describe are lengthy or complex, you will need to expand the basic functional format to accommodate the numerous details of your message.

Most engineering and scientific procedures require an expanded format. In directives, the expanded version is used less frequently since the message has a very narrow focus. In any event, before you write, be sure to analyze your subject and decide to what extent your readers must be instructed. If a page will do, don't waste your energy writing a half-dozen pages of elaborate directions. On

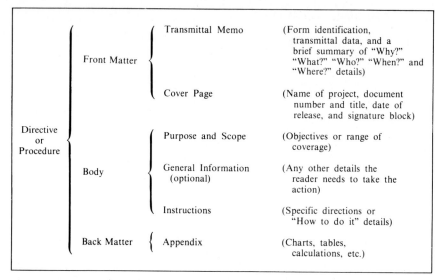

FIG. 7.7 Expanded format for multiple-page directives and procedures.

the other hand, if your details won't fit on one or two pages, the expanded format is recommended.

The typical multiple-page directive or procedure follows the expanded format in Figure 7.7.

In practice, the expanded formats use the same sequence of information found in the basic functional format: *Why* the message? *What's* the message? *What* action(s) should be taken? *Who, when, where,* and *how* details? Optional material?

The chief difference is that the expanded format spreads these items over a larger amount of material. Therefore, the elements suggested here should be useful to the manager and to the technical writer alike. Examples of selected elements are reproduced in the pages that follow.

Expanded Format for Directives and Procedures

A. Front Matter. The front matter includes the same elements (form identification, transmittal data) as the top matter of the basic format, but it also provides the reader with general orientation.

   1. *Transmittal memo:* A transmittal memo gives all information needed to identify the directive or procedure and to direct the document to the intended reader. It says, "Here is a listing of the actions you should take," or words to that effect.

   The transmittal memo must do more, however, than transmit a copy of the document. It should also summarize the salient points (conclusions, recommendations, discussion) typically found in the body of a document written in the basic functional format. For example, the transmittal memo may: (a) contain a brief statement of the purpose of the directive or procedure; (b) tell who authorized or

PREPARING WRITTEN DIRECTIVES AND PROCEDURES 7.23

ABC Engineering Associates
Memorandum

Job No.:	30	Office:	Dallas, Texas
Department:	Engineering	Date:	January 3, 19XX
To:	All Engineering Personnel		
From:	Jane P. Smith		
Subject:	Directive 30.1, Distribution and Control of Engineering Information		

Because this project involves classified information, we must employ strict control over the distribution and use of all data and publications that we produce. Therefore, I am asking all engineering personnel to follow the instructions in directive 38.1, attached.

Directive 30.1 is a single-use directive, applicable to job 30 only. Effective immediately, the directive applies to all evaluations, scientific studies, environmental impact studies, topical reports, cost summarizations, and other publications produced at the project site and elsewhere.

These instructions are designed to simplify the distribution and control effort while avoiding the potentially disastrous information leaks that have plagued other firms working in this area. If there are any questions or problems concerning this directive, please contact me as soon as possible.

Prepared by: Jane P. Smith, P.E.
Supervising Engineer

Att.

Approved by: _John Doe_
John Doe, Manager
Engineering Department

Date: January 3, 19XX

FIG. 7.8 Example of a transmittal memo accompanying a multiple-page directive.

requested the document to be written, and why; (c) make any pertinent comment about the document (for example, the transmittal memo might say whether the document contains single-use or standing instructions); (d) state how many copies of the document are being distributed, as well as to whom they are being sent, and tell where additional copies will be filed.

A transmittal memo is normally stapled or clipped to the top of a multiple-page directive or procedure. Figure 7.8 shows a typical transmittal memo.

2. *Cover sheet:* A cover sheet is used occasionally with technical directives and is considered essential in most project procedures and control documents. A typical cover sheet contains the name of the project, the procedure number and release date, and a signature block, as in Figure 7.9. The signatures of the writer, the project manager, and the quality-assurance engineer indicate that these people have reviewed the final document and given their approval prior to its release.

```
┌─────────────────────────────────────────────────────────────────┐
│                                                                 │
│                     ABC Engineering Associates                  │
│                          Dallas, Texas                          │
│                                                                 │
│                     Project Procedure No. ____                  │
│                         January 6, 19XX                         │
│                                                                 │
│                  Satellite Power System Project (SPSP)          │
│                         Palo Alto, California                   │
│                                                                 │
│                                                                 │
│    Prepared by: _____                 │
│                 Jane P. Smith                                   │
│                 Supervising Engineer                            │
│                                                                 │
│    Approved by: _____                 │
│                 Donald E. McGonagle                             │
│                 Quality Assurance Engineer                      │
│                                                                 │
│    Approved by: _____                 │
│                 Evelyn E. Kolman                                │
│                 Project Manager                                 │
└─────────────────────────────────────────────────────────────────┘
```

Status	Date	Approved	Page No.
Rev. No. 1	3/12/XX	REK	2
Rev. No.			
Rev. No.			

FIG. 7.9 Example of a project procedure cover sheet.

B. Body. In the expanded format for directives and procedures, the body contains a brief introduction, followed by a listing of the specific actions that readers are expected to take. If these instructions run to a large number of pages or if they are too complicated to allow quick reference, a table of contents may be included. In this case, the table of contents precedes the first section of the document body. See Section 2 on reports for help in writing a table of contents.

In most multiple-page directives and procedures, the decimal numbering system is used to designate the various sections and subsections of the body. For more details on the decimal numbering system, see Section 6 on specification writing.

1. *Purpose and scope:* Here it is advisable to include a brief summary of the objectives of your message. Also, tell how much ground the document covers and whom it affects. Give other "why," "when," "where," and "how" details, if these are relevant to the directive or procedure as a whole. Explanations or precautions that apply only to specific instructions may be given in the listing for those instructions.

2. *Optional sections:* It may sometimes be necessary to follow the first section with some general information that will aid your reader in following your specific instructions. For example, you might draw attention to the manner in which you've arranged the steps of the procedure. Or you might alert the reader to situations in which your directive does not apply. Also, at this point you may decide to cover certain critical features of the project or task, such as responsibilities of key personnel, equipment and materials, or policy. Such details should usually be placed in separate sections, particularly if the items must be devised or changed especially for the work you are covering.

A directive or procedure therefore may contain one or more of the following optional sections: (a) general information, (b) responsibilities, (c) equipment and materials, or (d) policy.

3. *Instructions:* The title of the main section of the body will vary from document to document, depending on the content of your message. Some writers prefer the more general "Instructions" both in directives and in procedures, but the titles "Procedures" and "Directives" are widely used.

Whatever the title of this section, make certain to arrange the desired actions and "how to do it" details in logical order. Steps are most often given chronologically, that is, in the exact sequence you want them to be followed or by the scheduled dates of project operations. However, as we've noted, it sometimes helps to experiment with a few different logical orders—especially for a complex, multiple-page document. Among the logical orders frequently used are: arrangement by project responsibilities, arrangement by order of the importance of various tasks, or arrangement by subdivisions of your main subject.

Specific instructions are usually numbered, with each step placed in a separate paragraph. Related items—explanations, safety precautions, examples, or supporting graphics—should be given as close to each step as possible, as long as they will improve the reader's comprehension.

Other devices used to aid clarity include indention, capitalization, underlining, and special typefaces. These enable you to highlight or otherwise set off specific, important details, thereby distinguishing them from the supporting items mentioned above. Examples, for instance, are often indented beyond the normal margin or put in separate paragraphs or placed in parentheses. If extremely lengthy, they can be put in an appendix.

C. Back Matter. The appendix may contain any of the supporting items mentioned above, as well as any calculation summaries, statistics, notes, tables, or specifications that are relevant to your instructions. Often the appendix will contain optional evidence to support your message. In short, anything that would interrupt continuity of the instructions should go in the appendix.

Occasionally, it may be necessary to include an additional section of back matter for applicable documents and references. This section may be titled "References," "Sources," "Bibliography," or "Applicable Documents," depending on content.

Use Flowcharts to Aid Comprehension

Whenever the operating steps of the directive or procedure are numerous, a procedure chart is suggested. In general, make it a point to draw up a procedural flow diagram for each multiple-page procedure and for most complex directives. If you develop the flowchart before you write, it will simplify your writing job. Why? Because a good procedure chart does these things: (a) It summarizes the most important details in quick reference form; (b) shows the sequence of operations to be performed; (c) fixes the work responsibilities of the various parties involved; (d) encourages you to choose concise descriptions to conserve space.

The flowchart will also aid reader comprehension. Because it is so compact, a chart often can be followed more easily than a lengthy narrative. Further, the procedure chart reduces the need for repetitious reading of routine procedure phrases once the reader has performed the procedure a few times. After the reader has become familiar with the basic routine, he or she can refer to the flowchart for a quick reminder, rather than reread the entire narrative each time.

The flowchart format illustrated in Figure 7.10 is often used for procedures that involve a significant amount of client input. This example covers the procedure for report preparation and distribution within a medium-size engineering firm.

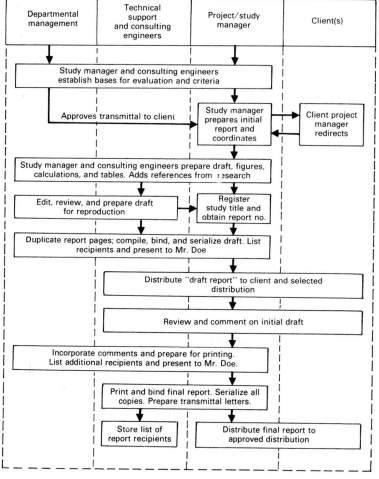

FIG. 7.10 The format used in this flowchart is often used for procedures that involve a significant amount of client input.

This book does not delve into the mechanics of flowcharting; however, several good flowcharts are reproduced throughout.

LANGUAGE AND STYLE GUIDELINES

Many experienced technical managers agree that to manage effectively, you must write effectively. To write a good directive or procedure, however, you'll need to know more than what format to choose or how to write clearly. You'll also need to know the particular language and style requirements that apply to these forms of technical writing.

In general, the language requirements for directives and procedures are more restrictive than for many other kinds of writing. Unlike a letter or memo, for example, a directive or procedure usually gives instructions, which should be stated in the imperative mood—the language of command. Furthermore, each directive or procedure has a highly specialized function: It aims to get profitable results through the efforts of others.

To get the results you want from a directive or procedure, you must:

1. Choose words that serve your readers' interests.
2. Write in an impersonal style.
3. State instructions or requests in the imperative mood.
4. Give explanations or precautions in the indicative mood.
5. Write short declarative sentences whenever possible.

Let's look at each of these suggestions in more depth.

1. Choose Your Words Carefully

Especially when you write your final draft, and also when you start planning your document, you must use language that serves your readers' interests. For example, in giving a nontechnical employee the procedure for operating an office computer system, you would not usually tell the employee, "Boot drive 1." Instead, you would use terms from the general vocabulary, possibly instructing the employee as follows: "To start the first disk drive, move switch 1 to LOAD. Wait for the READY light to come on. Then press RUN." Or you might define the term *boot* the first time it appears, and then give the procedure for booting the disk.

The language of the directive or procedure should be aimed directly at, and understandable by, the reader. As a general rule, don't choose words that reflect your advanced technical knowledge. Unless your readers know your shoptalk, such terms will lead to confusion and frustration.

On the other hand, you should avoid oversimplifying the communication and "talking down" to your readers. Shoptalk—the specialized technical vocabulary of a particular profession or discipline—can sometimes serve your readers' interests, if properly applied. Why? Because shoptalk often expresses in one word an idea that would require several words from the general vocabulary. And technical specialists are usually more interested in precise, economical coverage of technical details than in reams of elementary description.

Further, technical terms used accurately tend to give an air of authority to your message. This serves your readers' interests (and your interests) by helping them to understand fully the importance of the document.

This discussion of the readers' interests might be generalized into four rules:

1. When a general, nontechnical term conveys your message as briefly and as clearly as a technical term, choose the general term.

2. Even when a technical term seems more economical or more exact than a general term, don't use it unless you're certain your readers will know it. (If you are going to express the concept repeatedly, you can use the technical term if you define it where it first appears in the document).

3. If you want to say the same thing to several readers who have widely divergent backgrounds, it may sometimes be necessary to write a different version for each reader. This helps you to visualize each reader and choose words that are especially well-suited to his or her area of specialization.

4. When you write, try to imagine the reader's point of view. Make an effort to get some firsthand experience with the action or steps you will recommend. If you can't perform the procedure yourself (or watch it being performed), at least try to visualize it.

Only by addressing the interests of your audience can you achieve profitable results through the efforts of your readers. By following these four rules and choosing your words with care, you should be able to address those interests more effectively.

2. Use an Impersonal Style

Most directives and procedures require a certain amount of impersonality in order to be fully effective. Because your aim in issuing a document is to get results, you are interested in action, not in talk. Your readers are interested in your suggestions or commands, not in you. Thus some writers prefer to omit all personal names and put only job titles in the written directive or procedure. This is often a wise approach, especially in a lengthy standing document, since the occupants of various positions do change. Likewise, the first-person pronouns "I" or "we" and the second-person pronoun "you" are used sparingly in most directives and procedures.

Often, however, you need to introduce the concept of yourself to describe the reasons behind your message or to inform your readers of a decision you've made. Certainly, your message should not be unnecessarily impersonal. For it would seem stuffy and overbearing if it were.

In general, confine the use of the first person and second person to your transmittal material, where you present your central message. At the same time, avoid the chatty approach of an informal letter or memo.

Remember, you want your readers to comprehend fully the importance of your message. So don't write in a conversational style: "I think you should follow a new procedure." Instead, say "I am sending you the revised procedure."

Or you might choose a more formal approach, omitting the first-person and second-person pronouns entirely: "The following procedure supercedes all previous instructions."

Or you might use the imperative mood: "Follow the revised procedure, attached."

Or you might use the third person ("the department," "the chief engineer," "all employees"): "All engineering department employees should follow the attached procedure, which has been developed by the chief engineer."

Allow common sense to dictate which approach to use with a given audience or message. If you choose your approach carefully, you will often find your writing task eased because you will be speaking naturally yet firmly. Above all, remember to follow any specific writing rules or guidelines mandated by your organization.

3. State Instructions or Requests in the Imperative Mood

Specific instructions or requests should usually be stated in the imperative mood—the language of command. This is the most direct and most economical way to tell people how to get a job done. For example, in directing an employee to analyze a blueprint, it's better to write, "Break down the blueprint into assemblies and subassemblies," than the weaker, awkward form, "The blueprint should be broken down by you." The same is true for most other commands and requests, no matter how simple or complex.

As seen in some of our earlier examples, however, writers occasionally relax a little from the purely authoritative, imperative approach. Especially in simple one-page documents, or at the same time for extremely complex material, commands may sometimes be given in different ways. For instance, you might say "I am asking all engineers to break down the blueprints." Or you might say, "All engineers should break down the blueprints." Note that these sentences are less direct than the example written in the imperative mood.

If you are writing a "Responsibilities" section for the directive or procedure, avoid using the imperative mood in that section. Instead, use the verb form *shall* or *will* and refer to the various parties involved in the third person: "The client shall provide the technical database necessary for the preparation of all licensing documents."

See Section 6 on specification writing for more details on the use of *shall* and *will*.

4. Use the Indicative Mood for Explanations and Precautions

Explanations or precautions to be used in carrying out the directive or procedure are typically placed in the indicative mood—the language of discussion. For example, at the end of the directive for analyzing the blueprints, if the writer were to feel the need for further explanation, he or she might change from the imperative to say, "The distinction between assemblies and subassemblies may not always be clear-cut. Sometimes the breakdown is arbitrary."

However, if a precaution is of paramount importance, state it in the imperative mood. If necessary, underline the precaution and/or type it in capital letters: "DO NOT PROCEED BEYOND THIS STEP WITHOUT AUTHORIZATION FROM THE CHIEF ENGINEER."

5. Write Short, Declarative Sentences

If a sentence contains more than 20 or 25 words, it is often hard to grasp quickly in its entirety. It can contain so much information that the reader's mind can't assimilate all the data. Since a directive or procedure should be written for quick and easy reference, beware of long or involved sentences.

Often the best way to fix an overloaded sentence is to break it into two separate thoughts. Contrast the following two examples:

> Compare, using all available sources, including reports, handbooks, and experimental data, the values that the half-wave potentials obtained for copper and lead in tartrate solutions.

> Compare the values the half-wave potentials obtained for copper and lead in tartrate solutions. Use all available sources, including reports, handbooks, and experimental data.

If you will follow the five guidelines given above, your readers should find the result more interesting and pleasant to read.

PREPARING THE DOCUMENT

Preparation of a written directive or procedure can involve as little as five minutes of effort or as much as five days of work for an experienced technical writer or manager. On the one hand, the typical single-use directive will state a simple command or request and take only a few minutes to write. On the other hand, a complete project procedure or control document might cover a wide range of topics and require several days of information gathering, decision making, and writing, review, and editing.

Considering the wide variability of these kinds of documents, it is useful to know the basic steps common to the preparation of most directives and procedures.

Here is a comprehensive 10-step scheme to follow when you write a directive or procedure. Use it and you'll find it easier to get your point across—regardless of your subject or the volume of your material.

1. Decide What to Put in Writing

In any organization, there are two extremes that writers and managers must avoid: (a) not putting enough information in writing and (b) putting too much information in writing. Indeed, unnecessary directives and restrictive instructions can encumber an organization's operations as thoroughly as a set of vague or confusing oral commands. Over a period of time, a mass of written instructions may add up to red tape and reduce the initiative of managers and technical personnel alike. Some instructions and requests, however, must be put in writing—especially when they are urgent or complicated or when they are part of the contract requirements for the project.

It is essential, therefore, to determine your objectives before you write. *What* is the message? *Why* are you writing? Unless you can answer these questions in depth, you have no basis upon which to decide what to write and how best to write it.

The suggestions given earlier in this section should help you to determine whether a written directive or procedure will be an appropriate medium for your specific message. In particular, check your message against the following factors: (a) purpose or intended result; (b) complexity; (c) urgency; and (d) your prospective readers' current knowledge. In general, if your aim is to get action or to instruct your audience in a complex task, it is advisable to communicate in writing. Otherwise, an oral communication may be equally effective.

We want to emphasize that, used judiciously, written directives and procedures are an effective tool for technical management. But we also wish to caution you against the pitfalls of extensive documentation. Clearly, paperwork is integral to good management; however, nobody benefits by being swamped with paper in order to get a simple task accomplished.

2. Analyze Your Audience for the Proper Angle

Once you decide to write, you must consider your probable readers and determine their main fields of interest. Earlier we saw how you can tailor the language and focus of a directive or procedure for a given class of readers—usually supervisors, engineers, scientists, or support personnel. Analyze your readers before you write and you'll be better prepared to catch their attention and make them act.

When you intend your document to be read primarily by supervisory personnel, give management information—such as work flow, responsibilities, and performance control—the main emphasis. For engineers, emphasize the engineering data. For scientists, focus on the scientific data. For support personnel, emphasize the service aspects.

Always remember to narrow your message on the basis of your reader's experience and knowledge of your subject. Then you'll be able to plan your document for maximum effectiveness.

3. Make a Rough Outline to Speed Your Writing

Draft a few key sentences or phrases, stating the main topics you wish to cover. Start with your central point and then list, randomly if necessary, the various ideas to be incorporated into your directive or procedure. Make certain to list all salient points, including such topics as purpose, scope, steps, responsibilities, and safety precautions. Keep the rough outline brief, but try to write each item in the form of a complete thought; this helps to focus your ideas.

For example, a rough outline for a multiple-page directive covering the control of engineering documentation in a technical firm might be written as follows:

Rough Outline for Directive No. XX

1. *Purpose:* Furnish all project managers with standard practice instructions for control of technical information.
2. *Scope:* All scientific studies and evaluations, environmental impact statements, topical reports, and cost summarizations in the Engineering Department.
3. *Instructions and general information:* Specific directives.

- **3.1.** Compile applicable references, bibliographies, calculation summaries, etc.
- **3.2.** Determine the correct documentation format.
- **3.3.** Identify the sequence of events.
- **3.4.** Review the technical content of relevant studies.
- **3.5.** Establish a control and distribution number.
- **3.6.** Reproduce six drafts of the document for review by management and client.
- **3.7.** Assign an identifying number to each copy.
- **3.8.** Obtain approvals.
- **3.9.** Edit and produce 10 final copies.
- **3.10.** Maintain an up-to-date distribution list.
- **3.11.** Place the originals in a secure file.

If you will be writing a project procedure or a control document involving contract requirements, it is often appropriate to contact the client for additional information and suggestions. By all means, read the contract thoroughly and make a list of all pertinent topics to include in your written procedure. If necessary, contact the client (or the client's organization) to see what—in his or her judgment—needs to be put in writing. Remember to follow any rules your organization has regarding client contacts.

Generally, if you will follow the client's suggestions, you can build confidence in your organization's work, because the client will see you've left nothing to chance. Further, you may be able to prevent potentially disastrous errors or disputes from arising later.

4. Design a Simplified Flowchart for Visual Display

As noted, a flowchart may be used to clarify procedures and to condense information into concise form. Hence, many experienced technical writers and managers make it a rule to develop a simple flowchart for any document that will give a series of instructions or steps. If the chart will aid reader comprehension, it may later be prepared for distribution with the text of the directive or procedure.

To speed your writing with a flowchart, identify the central flow of your subject. For example, you may be covering a sequence of steps, a series of project events or milestones, a data distribution process, or a network of personnel responsibilities. Design your flowchart to display such relationships clearly, using one or more key words for each element in the chart.

5. Expand and Polish Your Outline

At this point, you should be able to rearrange the random list of topics in your rough outline and put them in some logical order. You may also begin to fill in any parts that are missing, such as explanations or descriptions of specific instructions. Once you've expanded the outline to include all the topics and subtopics in a logical arrangement, check your outline against the basic functional format for directives and procedures shown earlier in this section. Reshuffle the expanded list, if necessary, to follow the functional format. For lengthy or complicated documents such as project control procedures, follow the appropriate expanded format.

The process of expanding and polishing your outline may take a few extra minutes, but it will save valuable time and energy as you write. In some organizations, writers of project procedures and technical directives are required to submit a finished outline to top management prior to the actual writing. During review, additional topics or ideas may be suggested. If such a review is required by your organization, be sure to revise your outline according to management's recommendations. Unless you do, there is little chance that the written document will later be approved for distribution.

6. Plan for Any Special Features of Your Message

Occasionally you will have to write to employees who are either too busy or too skeptical to respond to your message in the way that you desire. If you know you're dealing with such a situation, you'll need to take special measures to get action. So don't start writing until you've planned to tackle these problems.

When your message is extremely urgent, you'll need to rely more heavily on attention-getting devices and techniques, such as brief copy and bold headings. You can't expect your audience to read and act on every communication the moment it is received. You have to take positive steps to alert your readers to the urgency of your message.

If you anticipate skepticism in your readers' responses, you'll need to place special emphasis on the reasons for your written communication. Perhaps you'll need to begin your document with the answer to "why" and then proceed to your recommendations. Whatever special problems you anticipate, you'll be better able to handle them by planning in advance.

7. Write Your Directive or Procedure

At this point, if you have followed the preceding steps, you will have finished a large portion of your work. This is the time to gather any last-minute details and assemble all of your material into proper form for writing. Also, you may find that certain ideas or data that you collected earlier may now be set aside.

If your material is voluminous and unwieldy in outline form, you can transfer important items to index cards for easy arrangement. Label each card according to the sequence in which it appears in your outline. Then group the cards by section. Finally, arrange the cards in each group in the order indicated by your outline and desired format.

For a simple one-page directive or procedure, you can usually skip the index card stage and proceed directly to writing. Some effective writers go immediately from a brief outline to the final draft. We recommend that you put the essence down first, however, then edit and rewrite separately.

A first draft is essential for most long or complex documents because it allows you to expand and clarify your ideas later. You can get into writing the first draft quickly without having to worry about your English. Later, as you edit, you can improve your content and grammar, eliminate excess material, and add the finishing touches to your language and style.

Compose your document one section at a time, keeping an eye to your outline. Most good writers draft the body of the directive or procedure first, starting with the "why" and "what" details (purpose, scope, conclusions). Then they write the actual instructions and any additional items (general information, responsibilities) that readers will need to take the desired actions. Finally, they compose the transmittal material, identification items (such as a cover page), and back matter (appendix, illustrations, references).

After you've written the directive or procedure, set it aside for a few hours or a few days, if possible. When you reread the document, you'll be able to examine it from a fresh perspective.

8. Edit Your Document

For most directives and procedures, careful editing is a necessity. You can't submit your document for approval and simply hope your reviewers will check accuracy or eliminate faulty grammar. Instead, criticize and rework your draft. Follow the guidelines given earlier in this section for language and style.

Pay close attention to the accuracy and completeness of the document as a whole. If you find that your entire approach to the subject could be improved, it might be worthwhile to rewrite the first draft. This task is usually a lot less burdensome than it seems. Because you already know what you want to say and you've thought of a better way to say it, rewriting follows naturally. The most difficult work is finished; thus, editing and rewriting can be a rewarding process.

Not every document needs to be rewritten, of course. Occasionally you might turn out a polished draft on the first try. On further scrutiny, however, such a draft might reveal hidden problems. Therefore, check your first draft thoroughly and edit as needed. Remember, effective editing requires much more than simple proofreading. You must mark your copy not only to remove errors of typing, spelling, and grammar but also to delete repetitious and unnecessary words, to improve punctuation, and to enhance the order of your ideas.

Insert headings, if necessary, to make your document more interesting to look at and easier to read. Also check to see that you've made a proper transition from sentence to sentence, topic to topic, and part to part. See the appropriate sections at the back of this book for more details on grammar, usage, and editing. When you've edited your document thoroughly, type it or have it typed.

9. Submit the Document for Review and Approval

In most organizations, a directive or a procedure must be submitted for management approval prior to release. Even if you are a leading partner or executive, you may be required to obtain a clearance from your peers, since your document will affect the operation of the firm. It is your responsibility to secure all management approvals required. And never overlook client review of selected items, if such a requirement is part of the contract obligations.

When you submit your document for review, be sure to include the appropriate signature block or clearance form. After you've obtained the required approvals, in writing, make a copy of the page and place it in your file.

10. Produce the Directive or Procedure

Many technical writers and managers must supervise the production of the documents they write. If this occurs in your case, you can avoid difficulty by following these suggestions:

1. Type the final copy on 8 ½- by 11-inch white paper. Before you reproduce the final draft, proofread it for accuracy. If your directive or procedure is only a few pages long and does not require special typefaces, you can often produce all the finished copies you'll need by using a word processor or copier. Simply collate the copies and staple each completed document in the upper left-hand corner.

2. When the services of a professional typist or publication team are required, mark the final draft for type, spacing, and makeup. Make all changes and corrections in the final draft prior to starting the final typography. Also, make certain that the transmittal material, title page, appendix, and illustrations are supplied with the final draft.

3. Since directives and procedures are frequently in need of change, deletion, or augmentation, some organizations print them on loose-leaf paper. With a loose-leaf notebook, employees can easily update procedure manuals and standard practice instructions. So be sure to specify the correct type of paper to use when you submit the final draft for reproduction.

4. Proofread the reproduction copy, that is, the camera copy, in both the galleys and in assembled form.

5. Schedule production well in advance to ensure distribution when required.

CHECKLIST FOR DIRECTIVES AND PROCEDURES

1. Is your message defined clearly? Does it say exactly what you want your readers to do?
2. Have you analyzed your intended audience to determine: (a) what your readers will be most concerned about; (b) how your document will be used; (c) how much your readers know about the subject; and (d) what problems, if any, to expect in readers' responses?
3. Based on your audience analysis, have you narrowed the message to appeal to your readers' interests?
4. Have you decided how you'll handle any special problems in your reader's reactions? For example, if your message is extremely urgent, have you planned to use special distribution methods or attention-getting devices?
5. Do you know which in-house medium will be most appropriate for your purpose? For example, a single-use directive memo will be best-suited to a simple command or request, whereas a multiple-page procedure is best for a long sequence of instructions.
6. Have you chosen the correct format for your medium? Are all the elements included in the proper sequence?
7. Have your gathered any information needed to write a complete and well-organized directive or procedure?
8. Have you used an outline to help organize your ideas and speed your writing?
9. Have you incorporated all relevant contract requirements, if any, within the text of your document?

10. Have you used terms that all your readers will understand? Have you defined any unusual or unfamiliar words or phrases the first time they appear?
11. Is your coverage complete? For example, have you given any explanations or safety precautions needed?
12. Have you listed any "action" details or instructions?
13. Are all instructions given in the imperative mood?
14. Are your sentences clear and concise? Have you edited your first draft to verify accuracy, eliminate excess, improve grammar, and add style?
15. If your instructions are complex, have you included a flowchart to aid reader comprehension?

SECTION 8
CREATING EFFECTIVE INSTRUCTION MANUALS AND BULLETINS

Instruction manuals and bulletins are the main means of communication between equipment builders and users. When a user of an item needs information on installing, maintaining, or operating that item, a well-written set of instructions offers quick answers to practical problems. Not only do instructions help the user avoid costly errors, they also help potential buyers to evaluate a product before purchasing; for the usefulness of an item depends partly on the quality of the accompanying written instructions. Hence, each manual or bulletin should convey the builder's specific intentions for use and care of the equipment.

To serve these purposes effectively, the manual or bulletin must be detailed, accurate, and clearly written. Specific instructions must be arranged in logical order to allow ready access of key items. Language should be as simple and direct as possible.

FEATURES OF WRITTEN INSTRUCTIONS

Because manufacturing firms intend instruction manuals and bulletins primarily for out-of-company readership, manufacturers often devote special attention to the preparation of product information. Some large technical and scientific organizations have entire departments that write and produce manuals and bulletins, while many smaller firms assign such tasks to a few well-trained writers and engineers. No matter how large your organization is, if you spend much time in industry, you're almost certain to run into an instruction-manual or instruction-bulletin writing job.

Let's take a look at the key features of instruction manuals and bulletins and see how they differ from other forms of technical writing. We'll also examine various strategies that you can use to write instructions quickly, completely, and effectively.

Range of Products Covered

In general, any product requiring care must be accompanied by written instructions for its users. These instructions may be only a few words, such as "insert the program disk." Or they may run to several hundred pages, as do the operating and maintenance instructions for a nuclear submarine. Other products that require written instructions of varying complexity include computer peripherals, audio and video equipment, motor vehicles, private and commercial aircraft, electric motors, washing machines, photographic supplies, and kits for building toys, furniture, boats, motors, and more. There is hardly a product that is used today that is not accompanied by a manual, bulletin, or some other form of documentation to guide the user in installation, operation, or maintenance.

Written instructions are used by consumers, by industry, and by the military. But since many product users are not trained operators or mechanics, the instructions must be written in plain, easy-to-follow language. Incorrect or needlessly complicated instructions can lead to major operating or assembly troubles. Customer complaints, litigation, and a reluctance to buy additional products from the same manufacturer are a few of the problems that poor instruction material can create.

Kinds of Publications

Many names are used to describe written instructions. Thus we have *Trane Refrigeration Manual; Bailey Control Systems Product Instruction—Electronic Differential Pressure Transmitter, Type BQ7 and BQ8, Series 30: The IBM Personal Computer Guide to Operations; Digital Hydraulic Test Unit User's Manual; Instruction Manual for Installing, Operating, and Repairing Electric Motor and Pump Sets; Volkswagen Rabbit Diesel Service Manual; Operating Instructions for the SX-40 Computer-Controlled Stereo Receiver; Instruction Bulletin for Oil-Filter Replacement; Gas Turbine Cogeneration System Maintenance Bulletin,* etc.

Instruction bulletins are brief, tightly organized sets of practical details about a company's products or services. Although some bulletins contain extensive product information, most address a limited number of specific applications and product features. Typically, the bulletin for a particular product will furnish up-to-date details that may not appear in other literature such as the user's manual, sales brochure, or data sheet. In some instances, the instruction bulletin will provide all the information needed for a user to set up and use the equipment properly.

Instruction manuals are usually longer than bulletins. Industrial technical writers seldom use the word *manual* for a written piece shorter than four published pages. Instruction manuals can run to several hundred pages; a few large manuals exceed 1,000 pages in length. For our purposes in this handbook, we will define an *instruction manual* as any set of installation, operation, or maintenance instructions that is four or more published pages in length. Any task shorter than four published pages will be called an *instruction bulletin.*

Note that these are arbitrary definitions set up to form a convenient division between short and long writing tasks. In actual practice, you can choose whatever terms you feel best reflect the form and content of a particular document type. Or you may be required to follow management policy in choosing specific titles.

In some organizations, different names are used interchangeably. For example, a bulletin may sometimes be referred to as an *instruction sheet,* or simply *instructions;* an instruction manual might be called a *handbook, booklet,* or *instruction book.* Occasionally, titles may be dictated by policy or simply by tradition. Within the computer industry, for example, written instructions are almost exclusively referred to as *documentation,* though in many instances the actual publication will bear a more formal title, such as "user's manual" or "handbook of operations."

If you will be preparing instructions for the military, on the other hand, you will have to check with the governing agency or department before you write. Most military departments have specific writing requirements not only for the choice of titles but also for style, language, format, and other features of each set of instructions.

Similarity to Directives and Procedures

Instruction manuals and bulletins share some basic similarities with the in-house directives and project procedures discussed in Section 7. The main similarity between these various forms of technical writing is that all can include specific how-to instructions. Hence, many of the writing suggestions given in Section 7 may also be applied to the preparation of instruction manuals and bulletins.

Keep in mind, however, that most instruction manuals and bulletins are meant for out-of-company readership, as opposed to in-house use. Thus, while a typical directive or project procedure will contain a set of instructions, the document will be concerned primarily with instructing organizational personnel, instead of outsiders who buy and use the organization's products. Further, an instruction manual or bulletin will focus on the practical details of product use, whereas an in-house directive or procedure will concentrate on controlling the firm's personnel.

Scope of Product Information

Probably more instruction manuals and bulletins cover operating and maintenance techniques than any other subject. But a variety of other topics are also covered. For example, a manual may provide complete product information on virtually any kind of component, process, or system. Instructions may include specific how-to details on unpacking, startup, testing, adjustment, safety, troubleshooting, overhaul, and normal and emergency shutdowns. Further, most instruction bulletins and manuals contain illustrations and tables to aid reader comprehension. Among the most common types of illustrations used are line drawings and photographs with perspective, isometric, or exploded views. Instructional publications may also include schematic diagrams, tables of operating data, performance curves—in short, anything that helps tell the reader what to do and how to do it.

Relatively little theory is covered. However, some complex publications do discuss the design and manufacturing principles of a specific product or type of product. *Theory-of-operation manuals,* or *information manuals* (sometimes called *bulletins*) seek to increase the reader's basic understanding of a product or group of products, often without reference to the specific units built by the firm authoring the manual. But since most instruction-manual readers are looking for quick answers to practical questions about a particular unit, theory usually receives less attention than other aspects of the subject.

Other topics popular in nonoperating manuals and bulletins today include selection, application, planning, and management. Since most of these subjects require a good knowledge of engineering or science, the writing is usually done under the supervision of an engineer or scientist. Even if the technical writer is not trained in the subject of the manual, he or she can contribute much in regard to clear writing, effective arrangement, useful illustrations, and rapid production at low cost.

Orientation toward Users

Users of an instruction manual or bulletin may include operators, mechanics, technicians, programmers, scientists, or general consumers. Some may be highly trained—others may never have seen the product before. The writer therefore must strive to understand his or her audience and orient the writing toward that readership. In general, this means you must write in a style that is simple, thorough, and direct.

Awareness of the need for user-oriented instructions has increased in recent years, owing partly to the growing use of computers by people with little or no technical training. As more and more of the general public has begun using computer products, the opportunity for catching flaws in the accompanying instructions has grown. As a result, a number of consumers and industry watchers have complained that the computer industry has not been doing enough to help the public understand its products. The developers of hardware and software have responded to these complaints, making substantial efforts to improve the overall quality of their written instructions.

Remember, always consider the interests of your prospective readers when you write. The instructions you prepare may be used in a home, on a space mission, in an industrial plant, or on an arctic mountain. Your readers will believe and apply every word in the document—as long as your writing is precise and, above all, clear. Instruction writing demands that you continually put yourself in your reader's place. Only then can you anticipate and satisfy the reader's needs.

MAJOR TYPES OF INSTRUCTION MANUALS AND BULLETINS

Today you'll meet two major types of instruction manuals and bulletins—industrial (or commercial) and military. The two types are similar in intent. But the methods for presentation may be different. In writing a military manual or bulletin, you will usually follow a military specification covering the content and format. Few industrial instructions are written to particular specifications, although some organizations do pattern their manuals and bulletins after military specifications. Generally, each manual represents a separate problem and is designed to meet specific requirements of the reader. Since industrial manuals sometimes serve as promotional and marketing aids, they may be elaborately prepared. Use of page tabs, loose-leaf binders, foldouts, multiple colors, and other aids is widespread today.

Military and commercial manuals or bulletins typically contain as many as five different kinds of information: (1) operating instructions, (2) service instructions, (3) overhaul procedures, (4) parts lists, and (5) development equipment. All this information may be combined in a single manual, or it may require two or more volumes. As noted, some technical organizations issue bulletins on specific problems whenever a special need becomes apparent. The current trend in industry, however, is to use a single manual for each product wherever possible. Military practice often requires two or more manuals, each covering a separate function.

Below are typical subjects covered under each of the five divisions of information. You'll also find helpful hints for writing about them. Later in this section we'll expand our discussion to include specific examples from actual documents. Also, we'll look at some other possible elements you might wish to include in a complete instruction manual or bulletin.

Operating Instructions

Specific instructions should be stated in the imperative mood—command language. For example, in giving instructions for operating a circuit tester, do not say, "You should move the TEST switch to the AUTOMATIC position." Instead, say, "Move the TEST switch...." In general, the imperative mood provides the most direct and economical way to tell people how to get the job done.

When writing operating instructions, try to include enough theory and principles of operation to give the operator a fundamental understanding of how to use the equipment intelligently. If you're cramped for space, as you often may be, give greater attention to principles of operation. If the operator wants information on theory, this can often be found in a standard reference book.

Include operating limitations—like maximum and minimum speeds, storage capacity, available memory, pressures, temperatures, etc. Also, you can help operators do a better job if you include a list of precautions they should observe.

Be sure to cover in detail any adjustments, checks, or maintenance or emergency procedures that operators can or should perform. In writing any instructions, keep in mind that operating personnel may not be technical experts. Also, they may not have access to the tools and special equipment needed for major repairs and maintenance. So limit your coverage to routine operating adjustments, unless you are asked to cover more areas.

For some devices and machines, proper warm-up and shutdown procedures are essential operating information. Include all that apply. And be sure to discuss inspection practices, cleaning, preservation, and lubrication methods when these are relevant to the day-to-day task of operating the equipment.

Service Instructions

These instructions usually cover organizational and field service, including preventive maintenance and corrective maintenance. Field servicing is often done by maintenance personnel who are attached to or employed by the using organization or firm. Hence you'll find that field servicing is limited primarily by the tools and maintenance equipment available in the field. Poorly functioning or unserviceable parts and assemblies are usually not repaired in the field. Instead, they are sent to an overhaul depot or shop for repair, rebuilding, or both.

When writing service instructions, be sure to include tests and adjustments coming within the scope of field servicing. Because service instructions and operating procedures are often similar, they may be included in the same manual. Like other types of instructions, service instructions should be stated in the imperative mood.

Overhaul Procedures

The purpose of overhaul instructions is to give detailed information for a major overhaul or complete rebuilding of the equipment. Overhaul shops have more specialized skills and equipment than field service forces.

When writing overhaul instructions for mechanical equipment, try to use exploded views from the parts list. These views help your reader understand the detailed instructions you give on disassembly and reassembly. To make your disassembly instructions easier to understand, number the parts in the exploded views in the disassembly sequence.

Do not omit detailed instructions on how to conduct performance tests of the equipment after overhaul. The only way the operator can check his or her work is by testing the equipment. So be sure to give the operator step-by-step ways to perform the tests.

Parts Lists

The main purpose of a parts list is to locate and identify parts for replacement purposes. Some specifications for instruction manuals require you to identify, describe, and illustrate all parts; others do not require coverage of standard items like common bolts and nuts. But nearly all military specifications require illustration of parts.

Parts of mechanical equipment are usually shown in exploded views. Each part is shown in a disassembled position. But its relative location with respect to the other parts is also shown.

Figure 8.1 shows a typical parts list. The first item in the parts list is the complete assembly. The component parts follow in the order of disassembly. Often, attaching hardware is listed after the part it attaches and is identified as *attaching parts*. When making up the parts list, be sure that you use the part numbers of the manufacturer. Then there will be no trouble in ordering replacements.

Some military agencies require that you show government stock numbers for the part. When this occurs, the manufacturer may be required to submit complete descriptions and drawings of the parts to the government stock-numbering agency.

Indent the part and assembly names in the nomenclature column to show the relative order of assembly. For example, the word *primer* (index no. −5) in Figure 8.1 is indented one space to the right of the fuel primer assembly to show that it is part of that assembly. The fuel primer assembly, part number B-9119, includes the primer. But the primer, part number B-11333, also may be procured separately.

Electronic parts list specifications often do not require exploded views or coverage of standard hardware. Photographs of the chassis with all parts intact are used for identification of functional electronic parts. The reference symbol or part number is shown in the photo or drawing of the equipment. Full parts information can be presented in the parts list. This list is arranged in the order of symbol reference or parts number.

Development Equipment

Instruction manuals for development equipment or prototype units are unique. Usually, only a few copies are needed. Format requirements are generally much

CREATING EFFECTIVE INSTRUCTION MANUALS AND BULLETINS

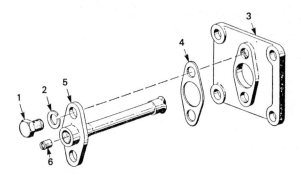

Exploded View of Fuel Primer and Adapter Assembly

Figure & Index No.	Part Number	Nomenclature 1 2 3 4 5 6 7	Units per Assy
2-	C-9058	Fuel Primer and Adapter Assembly	1
-1	A-9120	• Screw — Mach, hex hd, 0.250-28 NF 3	2
-2	A-9121	• Washer — Spring lock, 0.267 in. ID X 0.057 in. thk, stainless steel	2
-3	C-9118	• Plate — Adapter	1
-4	ASD-12219	• Gasket	1
	B-9119	• Primer Assembly — Fuel	1
-5	B-11333	• • Primer	1
-6	A-9123	• • Strainer	1

FIG. 8.1 Typical exploded view and parts list for an instruction manual. (*Courtesy of Product Engineering.*)

less stringent than for manuals covering production items. But this is no reason to slight the writing job. Instruction manuals for development equipment are often expanded to full coverage for production units. So the better your initial writing job, the easier it is to write a finished manual.

Differences in Manuals

Commercial and *industrial* operating, installation, and maintenance manuals have the same objectives as military manuals. Depending on the complexity of the equipment, however, the breakdowns between documents may not be as clear-cut as for military instructions. For example, a commercial user's instruction manual and the repairer's maintenance manual can sometimes be combined into one—especially for a simple piece of equipment. In *military* instructions, operation and maintenance usually are covered in separate manuals. Further, military specifications typically require a separate document for replacement parts; this may be referred to as the *parts catalog, illustrated parts breakdown, supply catalog,* or *standard nomenclature list* (SNL). Commercial and industrial manuals normally include the parts list with the operating and maintenance instructions. For large industrial systems and complex equipment, replacement items may be highlighted in one or more separate bulletins.

WRITING TO SPECIFICATIONS

Military Specifications

If you will be working on a military or industrial manual, bulletin, handbook, or parts list, you should follow the basic rules of good technical writing given throughout this book. In addition to these rules, at least one other factor—the applicable military specification—will govern your writing of military documents; it may also affect certain industrial and commercial manual writing projects.

The *mil-spec,* as it is termed, covers a variety of items like manual content, phases of manual preparation, style, language, understandability, and the arrangement of instructions, chapters, sections, and illustrations. To be acceptable, your material and manuscript must be prepared in strict accord with the applicable specifications. Successful technical writers working the military field—there are thousands of them—have a good knowledge of the specifications in their subject areas.

The way to learn a military specification is to study every item in it. We do not have space in this handbook to reproduce a complete specification. But we will discuss excerpts from one widely used mil-spec. These excerpts will give you an insight into the tasks involved in writing military manuals.

Importance of Specifications in Industry

Even if you never expect to write a military manual, you can learn many useful techniques by studying this and other military writing specifications. Further, several large industrial firms pattern their civilian manuals after military specifications. By studying various mil-specs, you will be better prepared to handle these industrial writing tasks. And if the day ever comes when you must write a military manual, your knowledge of what a specification requires will be helpful.

Sample Specification

Military specification 5474, which is revised periodically, covers the general format of technical manuals of many types. These include operating instructions, organizational maintenance instructions, field maintenance instructions, overhaul depot maintenance, and operation and maintenance support information. Designed specifically for Air Force Procurement, the latest revisions of the specification are indicated by capital letters before and after the identifying number, thus, M-5474A, M-005474B, etc. Whenever you use this or any other military specification, be certain you have the latest revision. If you work with an earlier version, the manuscript may be unacceptable.

Certain *military standards* are commonly considered part of the basic specification. Thus, MIL-M-005474 uses MIL-STD-12, *Abbreviations for Use on Drawings,* MIL-STD-15, *Electrical and Electronic Symbols,* and MIL-STD-16, *Electrical and Electronic Reference Designations.* Therefore, to prepare text and illustrations properly in accord with this specification, you must observe *both* the specification and the standards.

Let's take a quick look at some of the pertinent paragraphs of this specification. (*Warning:* Do not use the following paragraphs for preparation of official military manuals. Refer to a copy of the latest revision of the applicable specification.)

3.4. *The elements of the manual:* Except as otherwise specified, the manual shall have the following elements, in the order indicated:

A. Front matter consisting of the following and arranged in the order indicated:
  1. A title page
  2. A list of effective pages
  3. A table of contents
  4. A list of illustrations
  5. A list of tables
  6. Other front matter required by the detail specification

B. The instructions

C. An index if required by the detail specification. Manuals of 50 pages or less shall not have an index.

3.7.2.1.1. *Accuracy:* The instructions shall be accurate in all respects, whether they cover highly technical subject matter or not. Deficient manuals can be, and have been, the direct cause of poor operation and maintenance; such manuals can result in malfunction, mission abortion, loss of life, and loss of even the most expensive equipment such as highly complex missile systems, aircraft, etc.

3.7.2.1.2. *Understandability:* The instructions shall be so designed, or engineered, as to be readily understandable to the technicians who will operate and maintain the equipment during its service life in the Services, and to permit operation and maintenance by such personnel with optimum facility.

3.7.2.1.2.2. Language and methods of presentation of the type used by an engineer to personnel of virtually equal specialized training are unsatisfactory for manuals prepared to this specification. Instructions calculated to impress others with superior knowledge of the writers and illustrators are also unsatisfactory. The instructions shall be so presented as to bridge the gap between the engineering knowledge that created the equipment and the technicians who will operate and maintain it, becoming the link of communication between the planes of thought of the engineer and the technician.

3.7.2.1.3. *Quick and easy accomplishment:* The instructions shall be so presented as to permit quick and easy accomplishment of the work covered. Accurate and readily understood presentation goes a long way in making this possible. The procuring activity assists in this by setting forth requirements for multiple-volume manuals, and by stipulating in detail specifications the chapters and sections that specific manuals will generally have. In some detail specifications, the procuring activity specifies to some degree what the nature of the main paragraphs will be within sections; usually, the procuring activity specifies the operation or maintenance concept to which the manual will be prepared, and gives somewhat specific instructions as to the type of illustrations required. This specification, in its requirements for text and illustrations, gives other criteria to be used in making the instructions easy to follow in accomplishing the work in a minimum amount of time.

3.7.3. *Media to be used in presenting the instructions:* The instructions shall be presented in the media specified in the following paragraphs, in the proportions indicated.

3.7.3.1. Text is the basic medium in which the instructions shall be presented.

3.7.3.2. The text shall be supported by illustrations to the extent necessary. Illustrations shall also be used when required by the applicable detail specification, to furnish pictorial identification of parts and tools. The minimum number of illustrations essential for such purposes shall be used. Illustrations serving no specific instructional function shall not be used....

3.7.3.3. Combination text-and-illustration presentation (Procedural Illustrations) shall be used in presenting step-by-step procedure where illustration is required to make such procedure easily understood or to permit quick and easy accomplishment. The basic medium (text) preceding each such presentation shall make specific reference to the presentation as being the instructions for accomplishing the procedure portrayed in the presentation.

3.7.5.1.1. *Wording of text:* The text shall be factual, specific, concise, and clearly worded so as to be understandable to relatively inexperienced personnel performing the work on the equipment, yet provide technicians with sufficient information to insure peak performance of the equipment. The sentence form shall be simple and direct, avoiding the obvious and the elementary, and omitting discussions of theory except where essential for practical understanding and application, or as required by the applicable detail specification. Engineering knowledge reflected in the manual shall first be converted into the most easily understood wording possible. Technical phraseology requiring a specialized knowledge shall be avoided, except where no other wording will convey the intended meaning. The prime emphasis shall be placed upon specific steps to be followed, the results that may be expected or desired, and corrective measures that are required when such results are not obtained, rather than on the theoretical aspects of the work.

3.7.5.1.2. *Grammatical person and mode:* The second person imperative shall be used for operation procedures; for example: "Break casing head loose from wheel flange." The third person indicative shall be used for description and discussion; for example: "The torsion link assembly transmits torsional loads from the axle to the shock strut."

3.7.5.1.3. *Nomenclature consistency:* Nomenclature used shall be consistent throughout directly related publications. For example, a part once defined as a "cover" shall not be referred to elsewhere as a "plate." That portion of the complete nomenclature as is used shall agree with the parts list or parts breakdown nomenclature except when reference is made to a name plate or a decalcomania, in which case the exact abbreviation that appears on the name plate or decalcomania may be used.

Specification MIL-M-005474B contains much more information than the few paragraphs presented above. Among the other requirements covered by this specification are: use of abbreviations, notes, cautions, warnings, line drawings, photographs, color illustrations, illustration callouts, tables, manual updates, reproduction copy, type format, and page dimensions.

A number of these requirements are directly related to industrial instruction-manual and instruction-bulletin writing tasks. As an industrial technical writer, you must not be confused by the way the military requirements are stated. You must be able to recognize the specific writing requirements for each new set of product instructions and apply them to your immediate task. This is sometimes difficult to do when you are following an unfamiliar specification or one that contains multiple requirements and long or complicated sentences. To become familiar with a wide variety of instruction requirements, study this and several other military writing specifications.

STRUCTURE AND CONTENT OF INSTRUCTIONS

Many engineers and inexperienced technical writers make the mistake of plunging into the writing of a manual or bulletin without a plan. Because technical instructions are vital to the proper use of most equipment, it makes sense to prepare a preliminary structure, or format, before you write.

If your organization is contracted by a nonmilitary government agency, that agency may insist on manuals or bulletins produced to the format spelled out in a written specification. But for most industrial and commercial contracts that do not involve a formal specification, a separate, detailed writing plan will prove helpful. A good plan will enable you to structure your manual or bulletin around specific characteristics of the equipment you wish to cover. Hence, format may vary from document to document and from subject to subject.

Many large technical organizations have style guides that define one or more preferred formats. If your organization has such a guide, be sure to consult it before you write. A preferred or standard format will usually allow greater speed in planning and writing the document. If necessary, the format can be modified to accommodate special details or features of your subject.

Possible Elements

An instruction manual or bulletin may be made up of few or many elements, depending on specific requirements of the subject and of the intended reader. As the writer, you must decide which elements to include and which to omit, based on what a user normally needs to know in order to use the equipment effectively. For example, an operating manual for a power plant normally requires detailed procedures for handling such emergencies as a runaway reaction, loss of cooling water, fire, leaks, and overheating and for shutting down the process. A bulletin on a set of loudspeakers, on the other hand, might consist of little more than a short explanation of design principles, plus a few paragraphs on wiring the speakers to an amplifier.

To demonstrate the range of contents of technical instructions, we've listed the *possible* elements of a manual or bulletin in Figure 8.2. Note that not all of these elements are required; few, if any, publications would include them all. You should be aware of these elements, however, since this knowledge will help you design each document to meet the user's specific requirements. (For a more detailed listing of typical subject items covered in several types of instructions, see the reminders near the end of this section.)

Let's examine briefly the possible elements of an instruction manual or bulletin. We'll also take a look at some excerpts from typical instructions and see how they are written.

Cover. The information printed on the cover normally includes the title and reference number of the document, the name of the issuing organization, and the organization's logo. Many commercial manuals carry a photograph or line drawing of the product on the cover, as shown in Figure 8.3.

The release date of the current version of the document may also be given, especially when substantial revisions have occurred since the last printing. If the revised instructions correspond to a new version of the product itself, the title often reflects this change. For example, the instruction manual for a valve might be titled *Valve User's Manual;* but the manual for a Valve II would be titled *Valve II User's Manual.* If, on the other hand, the differences between the two versions of the product are minor, a single manual may be prepared; in this case, differences in the instructions for the two models should be noted wherever they appear in the text itself.

Design of the cover is seldom the responsibility of the technical writer; instead, many organizations have a separate production department do this work. You should become familiar with basic cover types, however, since you will then be able to contribute useful suggestions.

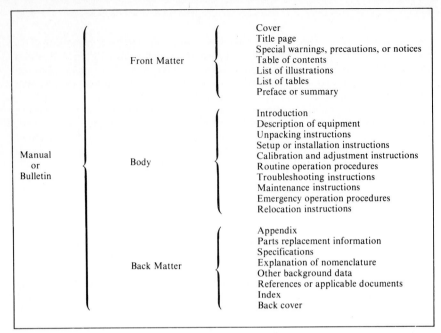

FIG. 8.2 Possible elements of an instruction manual or bulletin.

FIG. 8.3 Cover of an instruction manual. (*Courtesy of Bailey Controls.*)

Instructions may be bound in any of several kinds of covers, ranging from plain paper stock to heavy, rugged plastic or cloth materials. Types of bindings used include metal spiral, plastic comb, loose-leaf, saddle stitch (center-stapled), and perfect binding. The type of cover and binding chosen will depend on many factors, including size, cost, and desired durability of the publication. A large manual designed for field use, for instance, should have a durable cover to resist damage caused by rough handling and weathering. Many field and shop manuals are bound in metal spiral or loose-leaf so they lie flat when opened; this kind of binding enables readers to use both hands for their work.

Bulletins and instruction sheets do not normally require a special cover stock or binding. In most cases, they are printed on inexpensive paper, with the first page serving as both the cover and title page. If desired, holes may be punched in the left margin of the document to allow storage in a loose-leaf binder.

Title Page. The title page is often virtually identical to the cover. Depending on management policy and contract requirements, however, additional information will sometimes be placed on the title page. Such details may include copyright data, publication date, and name of author.

If you must choose the title of a manual or bulletin yourself, keep these guidelines in mind:

1. Choose a short, simple title. You'll make the reader's life easier if you identify your subject concisely. Therefore, equipment should be cited by the name and/or number under which it is normally marketed. Be sure to specify the exact model of the product you are covering. Identify the product in as few words as possible. Also, make certain you choose your title in accordance with company style and contract requirements.

2. Avoid titles that could be confused with those of other publications. Because technicians sometimes need to use a number of different manuals, each manual must be easily distinguishable in order to prevent mix-ups. If necessary, add a short subtitle to distinguish your document from others.

3. For documents that are updated regularly, include the publication date or revision number in the title itself.

Special Warnings, Precautions, or Notices. Occasionally you may need to inform your readers about some vitally important details before they turn to the body of the instructions. For example, if the new document represents a major revision or if using it will involve a substantial safety hazard, be sure to call attention to these facts as early as possible. Legal disclaimers and product warranty information may also be put at the front of the document, depending on management policy and applicable state and federal regulations.

Don't hesitate to highlight general precautions, explanations, or notices by placing them before the body of the instructions. In a manual, put this information on a separate page immediately following the title page, if necessary. In a short bulletin or instruction sheet, place general precautions above the body of the instructions, where they will be easily noticed. Use descriptive, attention-getting headlines. Typical ones include: "Read This First," "Important Notices," "Warning," and "Special Precautions." Underline the words or type them in capitals, if you feel this will help readers avoid serious trouble in using the instructions.

WARNING

ANY SUBSTITUTION OF COMPONENTS MAY IMPAIR INTRINSIC SAFETY.

AVERTISSEMENT

TOUTE SUBSTITUTION DANS LES COMPOSANTS RISQUE DE COMPROMETTRE LA SECURITE INTRINSEQUE.

Signal wiring should not run in conductors or in open trays with power wiring, and should not be run near heavy electrical equipment.

ATTENTION

Le cablage du signal ne doit pas fonctionner sous gaine ni dans des conduits ouvertes aved d'autres cables sous tension, ni ne doit être disposé a proximité de material éléctrique lourd.

FIG. 8.4 Examples of precautions as they appear in a typical instruction manual. (*Courtesy of Bailey Controls.*)

Precautions are sometimes set in bold type or in boxes in the final publication, as shown in Figure 8.4.

Table of Contents. The table of contents lists the various parts of the document—including section and subsection headings, appendix titles, and reference numbers—in sequential order. Usually, any illustrations or tables are also listed here. However, if the document is very long, has a large number of illustrations and tables, or is bound in more than one volume, a separate list of illustrations or list of tables may be required.

Each item in the table of contents should be keyed, by means of numbers, to its exact location in the text. Although page numbers are often used, some technical writers prefer to key each listing by the number of the section and subsection that it covers. This system allows unlimited revision. If only page numbers are used, the table of contents can become meaningless as the document undergoes revision.

Short, simple bulletins and instruction sheets do not require a table of contents. Extremely long or complicated manuals, however, may call for a table of contents plus an alphabetical index. An alphabetically arranged cross-reference index will be particularly helpful to the reader who is familiar with the equipment and who needs specific information. If the index lists all key words and technical terms contained in the text, it will enable the reader to find the desired information quickly without combing through the entire table of contents.

The table of contents is usually the last element of the front matter; hence it is placed immediately before the first page of text in the body of the manual. Some complex manuals include both a main table of contents and a separate contents page at the beginning of each section of instructions to help the reader get to specific subsections more easily. Occasionally the main table of contents may be followed by a short preface or summary. A typical table of contents for an instruction manual is shown in Figure 8.5. This manual covers electronic differential pressure transmitters.

Preface or Summary. In technical instructions, the preface or summary (or foreword—the names are synonymous here) normally consists of a brief statement of content. Thus, a typical preface or summary might include one or more of the following items: (1) general purpose of the instructions; (2) limitations and scope of the information; (3) basic equipment requirements or other data the reader

TABLE OF CONTENTS

Figure List	i
INTRODUCTION	4
DESCRIPTION OF OPERATION	4
UNPACKING	4
CALIBRATION AND ADJUSTMENT	5
Span and Zero Adjustment	6
Elevation/Suppression Adjustment	7
Linearity Adjustment	7
Damping Adjustment	7
Changing Suppressed/Elevated Ranges	9
Readjusting Linearity	9
Span Correction for High Line Pressure	10
INSTALLATION	11
General	11
Mounting	11
Connecting Piping	11
Pressure Piping	11
Wiring	12
Flow Measurement	13
Level Measurement	14
Pressure Applications	20
Hazardous Location	20
TROUBLESHOOTING	22
Checkout of Sensing Element	22
Checkout of Amplifier Assembly Boards	22
Cleaning Isolating Diaphragms	27
Disassembling Transmitter	27
Resassembling Transmitter	28
Interchangeability of Parts	29
Fault Correction Chart	30
SPECIFICATIONS	32
EXPLANATION OF NOMENCLATURE	34
PARTS REPLACEMENT AND DRAWINGS	36

FIG. 8.5 Example of a table of contents for an instruction manual. This manual covers electronic differential pressure transmitters. (*Courtesy of Bailey Controls.*)

must know before proceeding; (4) organization of the document; (5) summary of changes affecting the current edition; (6) general hints or suggestions on using the instructions.

A preface or summary may not be necessary if the first section of your document will be an introduction, as is common; in that case, you may be able to cover the above items in the introduction itself. Short bulletins seldom contain a preface or summary. If the document is long or the equipment you are covering is very complex, however, a preface or summary is suggested; in this case, the introduction may be devoted to a discussion of design and operating concepts.

Location of the preface or summary varies, depending on company style, personal preference, and contract requirements. Often, it is printed on the first page following the title page; but in many instruction manuals the preface or summary follows the remainder of the front matter.

Introduction. The introduction prepares readers for the details that follow in the body of the instructions. Ordinarily, it gives little or no historical background; but it may discuss the design concept and operating principles, especially where complex equipment is involved. The introduction may be short. Often a few tightly organized paragraphs are all that is needed to give the reader a basic understanding of the equipment and its function and to relate this knowledge to correct working procedures.

The following example was taken from an instruction manual for electronic differential pressure transmitters. It illustrates the use of a brief introduction with highly technical material.

Introduction

Type BQ7 and BQ8 differential pressure transmitters measure differential pressures from 0–5″ H_2O (1–1.24 kPa) to 0–1000 psid (0–6900 kPa). A 4–20 mA signal, proportional to the measured pressure, is transmitted. These two-wire units require an external power supply for operation.

Applications for Type BQ7/8 transmitters include measurement of flow using a differential pressure producing device (such as an orifice plate or flow nozzle), and the measurement of liquid levels in tanks. Compatibility of materials of construction and the process fluid being measured must be considered in each application....[1]

Typical items contained in an introduction include: (1) general definition of the type of equipment covered; (2) brief description of the specific model(s) covered, e.g., construction, capacity, range, size and weight; (3) discussion of principal applications; (4) fundamentals of the design concept; and (5) basic operating principles.

If a separate preface or summary is not used, the introduction may also include a discussion of the purpose and content of the instructions. The introduction is the first section of the body of the instructions.

Description. Sometimes a separate section is devoted to a detailed description of the equipment covered in the main text. This section is particularly helpful when covering a complex system or plant. Also, it is recommended when the descriptive material is too long or too unwieldy to fit easily in the introduction. For example, a "Description" section might include the overall process flow diagram for an industrial system or plant; but it would be rare for such a diagram to be presented in the introduction itself.

Ordinarily, this section discusses the construction and arrangement of equipment. In some instances it is used to introduce the reader to the general principles behind specific types of instructions. Thus, in an operating manual, the description section might be titled, "Description of Operation," whereas in a more gen-

---

[1]From Bailey Controls, Babcock & Wilcox Company, *Product Instruction for the Electronic Differential Pressure Transmitter, Type BQ7 and BQ8, Series 30,* by permission.

eral type of manual the description section might be called, "Description of Equipment," or simply, "Description."

Descriptions are usually written in the indicative mood, the language of discussion as opposed to the language of command. The following excerpt is taken from the manual on pressure transmitters quoted earlier. Note that the writer uses the imperative ("Refer to Figure 1") only when instructing the reader to take a specific action. (The figure number in this description refers to a figure in the actual manual; "Figure 1" is reproduced in this handbook as Figure 8.6).

Description of Operation

Refer to Figure 1 [Figure 8.6].

Process pressure is applied to the High and Low process connections. The pressures from the process connections are transmitted through isolating diaphragms and sealed fill fluid to a sensing diaphragm in the center of the sensing element. This sensing diaphragm acts like a spring and deflects in response to the differential pressure across it. This deflection, which is only 0.004 inch (0.102 mm) in either direction for full range of the Transmitter, is detected by fixed capacitor plates on both sides of the sensing diaphragm. The sensing diaphragm is actually the second plate for both of these capacitors, and the change in position of the sensing diaphragm causes a change in both capacitances. The differential capacitance between these two capacitors is electronically converted to a 4-20 mA dc current signal which is proportional to differential pressure.

For direct readings, gage pressure can be measured by connecting to the High input connection and vacuum can be measured by connecting to the Low input pressure connection. The unused connection is vented to atmosphere.

Transmitter circuitry features integrated circuit operational amplifiers along with other active and passive components to convert differential capacitance measured at the sensor to a dc mA signal. An oscillator circuit provides high-frequency excitation to the capacitive sensor.[2]

You can make your manual or bulletin easier to use by inserting subheads where they will help the reader. Use the main subject of a paragraph or other subdivision for the subhead. Try to keep the subheads consistent. For instance, when describing the components of a small office computer system, use only part names as subheads. Thus in a technical manual you might include the following subheads: "Central Processing Unit," "Disk Drives," "Magnetic Tape Drives," "Video Display Terminals," "Keyboards," "Printer," etc. Note that all of these are nouns, and components of a system.

Generally, a description section should be included only when it will aid reader comprehension. If the product is of simple construction and can be described in a few brief sentences, the descriptive material can often be incorporated in the introduction or preface.

Unpacking Instructions. In some manuals and bulletins the first section of instructions covers the removal of the product from its storage container. In general, unpacking instructions need be included only when the equipment is fragile or highly sensitive or when the storage container is very elaborate. Sometimes, however, equipment manufacturers include specific unpacking instructions as a

---

[2]Bailey Controls, Babcock & Wilson Company, *Product Instruction for the Electronic Differential Pressure Transmitter.*

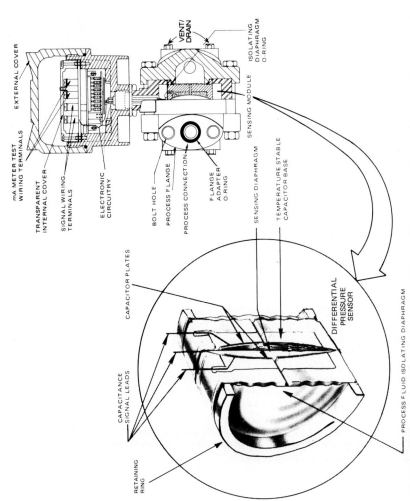

FIG. 8.6 Cutaway views showing typical use of drawings in an instruction manual to acquaint readers with the key components of a product. (*Courtesy of Bailey Controls.*)

means of protecting themselves against potential customer complaints; in this case, the unpacking instructions usually remind users to check for possible damage sustained in transport. For obvious reasons, unpacking instructions are not usually required for large, heavy items like trucks or tractors; but they are often needed with sensitive electronic instruments and measuring devices.

Unpacking instructions should be written in the imperative mood. Whenever possible, the procedure should be broken down into a series of numbered steps. Line drawings may be used where these will help the reader follow the procedure correctly.

The following unpacking procedure illustrates the use of direct, imperative instructions and simple, concise language. Note the use of capitalization to draw the reader's attention to the precaution in step 4.

Unpacking Instructions

1. Check for any obvious damage to carton or contents.
2. Remove all loose packing from carton.
3. Carefully remove transmitter from carton.
4. Before mounting or installing transmitter, check the nameplate (located on electrical housing) to make certain transmitter is suitable for application desired. DO NOT AT ANY TIME EXCEED THE RATINGS LISTED ON NAMEPLATE.

Assembly, Setup, or Installation Instructions. Because proper installation is critical to the functioning of most types of equipment, a "Setup," "Installation," or "Assembly" section is usually included. These instructions are typically divided into several subsections, arranged in chronological order. For example, if you were writing the owner's manual for a microcomputer, you might organize the installation procedure into subsections as follows: (1) cabling; (2) arranging the components; (3) testing the power-on switch; (4) adjusting the keyboard and display. In each subsection you would give step-by-step directions for the specific procedure cited.

Precautions and explanations that apply to the installation procedure as a whole should be given first. To catch the reader's attention, key words and warnings may be set in capitals, in boldface type, and/or underlined. Once the reader is aware that certain precautions must be observed, he or she is then ready to take the steps necessary to install the equipment. In general, these steps are used as the titles of the subsections.

Various types of illustrations may be used, as long as they make the procedure easier for the reader to understand; line drawings are the most common, although photographs, exploded views, and schematic diagrams are sometimes used, depending on the kind of equipment covered.

Installation instructions should be numbered for easy reference. As with other instructions, specific steps should be stated in the imperative mood, as in: "Connect the system's power cord to the system, then to the power supply." Explanations and precautions, on the other hand, are often placed in the indicative mood, for example: "Official certification is voided if the transmitter is not internally grounded."

Where the installation procedure is not complex, adjustment and calibration instructions are sometimes included in this section.

Calibration and Adjustment Instructions. A separate section is devoted to calibration and adjustment when the equipment will require frequent or extensive checking, adjustment, or standardization. This is true for many types of electronic equipment, electrical and mechanical control devices, and measuring instruments.

Figure 8.7 shows a typical illustration from an instruction bulletin on adjustment of a mechanical device; this one is for an automatic starter lock mechanism.

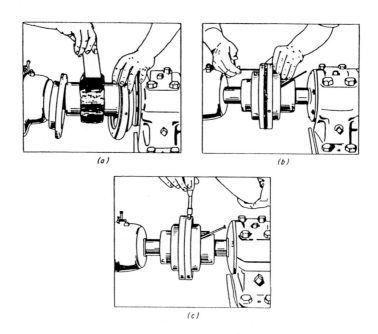

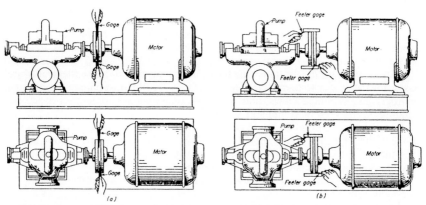

FIG. 8.7 Typical illustration from an instruction bulletin covering adjustment of a product. (*Courtesy of Power Magazine.*)

Operating Procedures. This section of instructions usually begins with a brief introduction to the principles of operation. If there are any general instructions for operating the equipment, these may be placed near the beginning of the section; specific operating instructions are then divided into subsections as needed.

Ideally, titles chosen for the subsections should be in operating terms, for example, "Reading the Manometer," "Measuring Differential Pressures," and "Executing the Overflow Check Program." These headings allow the reader to recognize immediately the techniques that will be used and shows their relationships clearly.

When discussing operating procedures, be sure to choose headings and subheadings carefully. Typical subheads for a mechanical operating procedure might be "Greasing," "Oiling," "Starting," "Stopping," "Inspecting," etc. Do not combine subheads from other sections of the document with those in the operating section, unless they are related. If you must combine terms from two or more sections, such as the operating section and the description section, do it logically, for example, "Oiling the Shaft," "Greasing the Bearings," "Starting the Motor," "Stopping the Motor," "Inspecting the Starter."

If you were preparing the operating section of a computer program user's manual, you wouldn't normally place obscure technical data on program structure in the subheads. Instead, you would choose terms that most users readily recognize. Thus, you might use subheads such as "Starting the Program," "Signing On," "Controlling the Program," "Entering Commands," or "Exiting the Program." Handling your subheads this way helps the reader understand the instructions quickly and makes them easier to use.

For more details on what to include in operating instructions, refer to the list of reminders near the end of this section. Also, review the subsection "Major Types of Instruction Manuals and Bulletins," earlier in this section.

Troubleshooting Instructions. The purpose of troubleshooting instructions is to give concise information for pinpointing and diagnosing equipment malfunctions. Troubleshooting, "fault correction," or "problem determination" instructions find a use in almost every instruction manual, for they tell the reader exactly what to look for when trouble develops. Even if your readers may not always find the exact trouble source in the items you list, they'll have a good chance of finding it through related items. Troubleshooting sections are common in most manuals and bulletins. Some bulletins, in fact, are devoted exclusively to troubleshooting.

The three basic components of a typical troubleshooting section are (1) symptoms, (2) causes, and (3) remedies.

Where the remedy is implied in the cause, specific procedures for the remedy may sometimes be omitted. For example, in instructions for troubleshooting an air compressor, the symptom "failure to deliver air" might be attributed to the cause "valves improperly installed." This implies that the operator should check the valves and install them correctly. Similarly, "suction line blocked—dirty filter" implies that the operator must either remove whatever is blocking the suction line or clean the filter to remedy the trouble. This technique of presenting instructions is simple and economical.

Prepare a list of troubles before writing instructions of this type. List the most common trouble first, the next most common trouble second, etc. This isn't always easy because the exact order is sometimes difficult to determine. The difficulty is further compounded when the troubles and their remedies are complex; in this case, arrange your list as a step-by-step procedure. Give steps in the chronological order that will result in the quickest problem determination and indicate each step in a separate sentence. Where several procedures are involved, divide the entire list into a series of procedures and arrange them in chronological order.

If you are not thoroughly familiar with the equipment you are writing about, get as much expert assistance as possible from people who are. Have an experienced technician or mechanic list every trouble he or she has met in the unit. Try to classify these troubles under a few major causes. Ask the engineers who designed and tested the equipment to do the same. Classify all these troubles. Then have them checked by *both* the technicians and the engineers.

Collect additional data, if possible, from operators and testing laboratories; add these to your list. Send the list out to field engineers, mechanics, and others who know the equipment and its problems. Ask for their suggestions. Include these in the list. Once you've gathered the information and arranged it in logical order, decide on the method of presentation that will be most appropriate for your purposes.

Troubleshooting procedures are usually presented in the form of a chart or table, as shown in Figure 8.8. This allows the reader to find key items quickly. Other techniques—including cartoons and lists—have been used with equal effectiveness. Cartoons are popular in industrial instruction material because they develop reader interest faster than a solid page of type. When cartoons are used, each drawing depicts a typical trouble situation. The drawings are accompanied by concise explanations of possible causes, as Figure 8.9 shows. Often, specific remedies are also included.

Another way of presenting troubleshooting instructions is to list them in outline form, as in the following example:

Troubleshooting—Centrifugal Pumps

If any of the following troubles occur, they may be due to the causes listed below:

A. No water delivered:
 1. Pump not primed.
 2. Speed too low. Check motor voltage.
 3. Air leak on suction.
 4. Discharge head too high
 5. Impeller plugged up.
 6. Suction lift too high.
 7. Wrong direction of rotation.
B. Not enough water delivered:
 1. Air leaks in suction or stuffing box.
 2. Speed too low.
 3. Suction lift too high.
 4. Impeller partially plugged.
 5. Not enough suction head for hot water.
 6. Total head too high.
 7. Pump defects:
  a. Excessive ring clearances.
  b. Damaged Impeller.
 8. Suction not submerged enough.[3]

---

[3]From Ingersoll-Rand Company, "Instructions for Installing and Operating Ingersoll-Rand Centrifugal Pumps."

FAULT CORRECTION CHART

Fault	Probable Cause	Corrective Action
High output	Primary element	Check for restriction at primary element.
	Pressure piping	Check for leaks or blockage.
		Check that blocking valves are fully open.
		Check for entrapped gas in liquid lines and for liquid in dry lines.
		Check that density of fluid in pressure lines is unchanged.
		Check for sediment in transmitter process flanges.
	Transmitter electronics connections	Make sure bayonet connectors are clean, and check the sensor connections.
	Transmitter electronics failure	Determine faulty circuit by trying spare amplifier assembly.
	Sensing element	Refer to "Checkout of the Sensing Element."
Erratic output	Loop wiring	Check for intermittent shorts, open circuits, and multiple grounds.
	Process fluid pulsation	Install dampeners in pressure piping
	Pressure piping	Check for entrapped gas in liquid lines and for liquid in dry lines.
	Transmitter electronics connections	Check for intermittent shorts or open circuits.
		Make sure that bayonet connectors are clean, and check the sensor connections
	Transmitter electronics failure	Determine faulty circuit by trying spare amplifier assembly.

FIG. 8.8 Portion of a typical troubleshooting chart. (*Courtesy of Bailey Controls.*)

AIR-COOLED AIR COMPRESSORS

Locating Troubles

FAILURE TO DELIVER AIR

Suction line blocked—dirty filter
Valves improperly installed
Suction valve unloaders stuck in unloaded position
Strips missing from valves

COMPRESSOR OVERHEATS

Broken valve strips
Wrong direction of rotation
Filter clogged
Discharge pressure higher than rated
Internal leakage
Insufficient lubricating oil

INSUFFICIENT CAPACITY

Excessive leakage in pipelines and fittings and through valves
Discharge pressure higher than rating
Speed incorrect
Filter clogged
Worn piston and rings
Faulty valves
Blown cylinder head gasket
Suction valve unloaders holding strips partially open
Intercooler leaking
Belt slipping

COMPRESSOR OVERLOADS MOTOR

Electrical characteristics of power lines incorrect
Multi-V-Drive belts pulled excessively tight
Discharge pressure higher than rated
Speed greater than rated
Discharge line restricted
Low voltage

INSUFFICIENT PRESSURE

Demand greater than rated capacity of unit
Speed incorrect
Worn rings
Excessive leakage in system and internally

COMPRESSOR KNOCKS

Loose flywheel or pulley
Excessive wrist pin and bushing clearance
Excessive crank pin bearing clearance
Main bearings need adjusting
Loose valve in cylinder
Loose unloader
Excessive end play in motor rotor
Motor rotor shunting back and forth due to belt misalignment or unlevel mounting

FIG. 8.9 Example of a troubleshooting bulletin using cartoons to convey its ideas. (*Courtesy of Worthington Corporation.*)

COMPRESSOR VIBRATES

Not properly secured to foundation
Improper foundation
Piping not supported properly
Shipping blocks not removed under base
Motor rotor out of balance
One cylinder inoperative

UNIT BLOWS FUSES

Fuses too small
Low voltage
Pressure switch differential too narrow
Unit starting against full load
Defective motor
Compressor or motor binding

INTERCOOLER VALVE BLOWS

While Running Unloaded
    Broken or leaking high pressure discharge valve strip
    High pressure unloader leaking air
    Defective or stuck low pressure unloader
    Blown high pressure head gasket

While Running Loaded
    Broken or leaking high pressure suction or discharge valve strip
    High pressure unloader stuck in unloaded position
    Blown high pressure head gasket

EXCESSIVE OIL CONSUMPTION

Oil level too high
Oil too light in viscosity
Too high oil pressure (if force-feed lubricated)
Worn rings or cylinders

RECEIVER SAFETY VALVE BLOWS

Defective safety valve
Safety valve set below cut-out pressure
Defective pressure switch or Trigger valve
Pressure switch or Trigger valve set at too high cut-out pressure
Leak in control line
Inoperative suction unloaders

write us for additional copies or for the complete series

FIG. 8.9 (*Continued*).

The example above shows good organization of symptoms, probable causes, and implied remedies. Although this trouble list does not indicate an actual procedure for the reader to follow, it suggests actions that the reader may take to investigate and solve common problems.

To test the effectiveness of a trouble list or chart, give a copy to an operator or technician and see whether he or she can remedy an equipment problem by reading your instructions. If the reader becomes lost, either your writing is unclear or the causes and remedies are incorrect. If the reader solves the problem without too much difficulty, you've done your job effectively.

Maintenance Instructions. Since proper operation and performance are of top priority to owners of all kinds of equipment, technical writers often pay special attention to the writing of maintenance, or service, instructions. For any given product, one or more sets of maintenance instructions may be prepared, depending on the extent to which expert service will be required.

For example, if it will be possible for operators of the equipment to perform certain maintenance procedures themselves, the writer may include these procedures in the operator's manual, usually placing them near the back of the book. If additional, higher-level maintenance will be necessary, the writer designs a separate manual for trained service technicians; typically, this manual emphasizes procedures only a service technician or mechanic is qualified to perform. In some cases, the service manual is divided into field maintenance instructions and shop maintenance instructions, as we noted earlier. In other cases, only one set of instructions is required. And if the procedures are simple, a single set of instructions may meet the requirements of *both* the operator and the technician and cover all phases of service and repair.

Another way of classifying maintenance instructions is by type of procedure. *Preventive,* or *routine, maintenance* is designed to prevent problems before they develop. *Corrective maintenance* seeks to remedy problems after they have appeared. Most maintenance and service manuals cover both types of procedures. Occasionally, manufacturers issue maintenance bulletins that focus on only one type of procedure.

A set of maintenance instructions may be an extended compilation of step-by-step procedures for upkeep and repair of equipment, or it may be more of a general guide, perhaps only a checklist, for taking care of the product. In general, a complete maintenance manual will provide a wide range of highly detailed procedures, whereas a bulletin (or a maintenance section of an operator's manual) will consist primarily of checklists or charts.

The content of maintenance instructions will depend on the specific requirements of the equipment. Typical subjects include disassembly, lubrication, testing, cleaning, adjustment, and repair. In addition, parts-replacement data and overhaul procedures may be furnished, as long as these are not covered elsewhere in the document.

Here's an example of a corrective maintenance procedure taken from a well-written manual:

The maintenance person should not need to make any mechanical adjustments other than checking brush tension occasionally and replacing brushes correctly. When good commutation cannot be obtained, look over the following mechanical features, trying the commutation after each change, and noting the effect produced:

1. Inspect all connections and be sure that none is loose.

CREATING EFFECTIVE INSTRUCTION MANUALS AND BULLETINS   **8.27**

2. Check the connections and be sure that the commutating field, or any part of it, is not reversed, and that one or more of the main windings are not reversed.
3. Check the brush spacing and alignment. In general, the more accurate the brush spacing, the more uniformly good will be the commutation.
4. Check the mechanical neutral and try shifting the brushes each way from neutral. Very often a slight shift is advantageous. (In shifting brushes, the question of regulation of speed on the motors, and division of load on the generators, should not be overlooked.)
5. Inspect the brushes and see that they move freely in the holder, and that the pigtails do not interfere with any part of the rigging. Look for burning or roughness of the contact surfaces.
6. Check the pressure, and see that the brush fit is good.
7. Inspect the surface of the commutator, and with a canvas wiper, wipe off any blackening. If it is rough or eccentric, causing the brushes to chatter or move in the holders, it should be ground and perhaps turned. (See page 197.)
8. Check the centering of the commutating poles between the main poles, and check both the main and commutating-pole air gap.[4]

Note that the mechanic is told exactly what to do. (The page number in these instructions refers to a page in the actual manual.) On many pages referred to in the manual, both line drawings and halftone illustrations *show* the reader how to do his or her job. So if you want to write effective instructions, you must not only tell the reader what, how, when, and why—*you must also show the reader.*

Clear working drawings and photographs are key elements of successful instructions. In the maintenance section of an operator's manual, illustrations are often limited to a few line drawings and photographs to acquaint the user with important parts of the equipment requiring maintenance. A service technician's manual, on the other hand, usually provides a highly detailed and comprehensive set of illustrations. For example, in addition to the illustrations provided to operators, the service manual may contain: simplified, overall, and unit-by-unit schematic diagrams; flowcharts; exploded views; performance curves; and tables of operating data.

Most complete service manuals also include one or more maintenance charts. Similar in form to troubleshooting charts, these allow quick reference to major topics like lubrication, testing, and adjustment schedules. Further, if the document does not contain a separate troubleshooting section, a troubleshooting chart may be added to the maintenance instructions.

When writing maintenance instructions, be sure to consider your reader's technical abilities. Include only those procedures that your intended reader will be able to accomplish. Also, make certain that the procedures require tools or test equipment that are available to the average reader.

Arrange maintenance procedures by order of importance, with the most important instructions first. In other words, begin the service manual, bulletin, or section with the procedures that are essential to proper operation—these will usually be of the preventive type. Then go on to procedures of lesser importance. If this scheme does not provide a clear-cut arrangement, try placing the most frequently performed procedures first; in the case of mechanical equipment, this usually means that lubrication comes first and parts replacement, last.

Be sure to use the latest product information available, including parts specifications, reference numbers, and other identifying data. Often you can obtain valuable information from technicians, mechanics, equipment owners, and operators.

---

[4]From General Electric Company, "How to Maintain Motors and Generators," with permission.

Write each maintenance procedure as a series of numbered steps, using command language when possible. And don't forget to include any necessary precautions or explanations. For more hints on maintenance instructions, see the list of reminders near the end of this section.

Emergency Operation. The requirements for writing emergency operation procedures are the same as for normal operating instructions, only more stringent. You must be extra careful to ensure that all your information is accurate and complete since a poorly planned set of emergency procedures may lead, directly or indirectly, to a tremendous loss of money or life (or both).

To allow quick reference and to reduce confusion, emergency procedures should be written as clearly and simply as possible. Yet they must provide detailed instructions for handling all foreseeable contingencies, including, where appropriate, emergencies such as a runaway reaction, cooling-water loss, fire, leaks, overheating, and potential loss of electronically stored data. Also, emergency procedures should include instructions for shutting down an ongoing process, with an explanation of alarms and a description of the operation of automatic controls.

Emergency procedures are normally given near the back of the operator's manual, but they may also be issued as a separate manual or bulletin.

Relocation Instructions. This information is given only for products that are subject to frequent moving and that require special handling to prevent damage. Relocation instructions include details on the proper method of preparing the equipment for transport over both long and short distances. In addition, they may include procedures for setting up the equipment at a new location.

Appendix. Material which may prove useful to the reader but which would be inappropriate in the body of the instruction manual may be placed in an appendix. An appendix is almost never used in a short bulletin or instruction sheet. In a manual, an appendix may sometimes be necessary to provide coverage of such items as parts lists, spare-parts lists, specifications, nomenclature, and other background data. These items may be placed in a single appendix or in multiple appendixes, depending on how bulky they are. Alternatively, the name "Appendix" may be dropped entirely; in this case, the final sections of the publication are designated in the same manner as the sections that precede them.

When multiple appendixes are used, each appendix may be designated by a capital letter and a title, as follows:

Appendix A: Parts Replacement and Drawings

Appendix B: Specifications

Appendix C: Explanation of Nomenclature

Other Possible Elements. Everything that follows the body of an instruction manual or bulletin is referred to as *back matter*. This includes not only the appendixes but also the references, an index, and the back cover. In addition, a blank fly sheet may be used as the final page; it also may be used for readers' notes.

Again, not every publication will require all these elements. Most simple manuals, for example, do not carry a separate section for references; instead, applicable documents are cited in the text or in footnotes. Similarly, an index will not always be needed, although it may be helpful to the reader. As a general rule of thumb, an index should be provided in manuals of 50 pages or more in length.

The basic requirements for these items are the same as for most other forms of technical writing. Refer to Section 2 on technical reports for more details on back matter.

Typical Manual Formats

The subject matter and design of instructional materials vary so widely that few generalizations can be made as to what kind of format is most effective. We can, however, describe some features common to many successful manuals and bulletins.

The conventional design for commercial instruction manuals and bulletins is an 8½- by 11-inch page set in two columns. Studies suggest that the two-column design gives improved readability. Another design that is becoming more popular today is a 5½- by 8½-inch page set in one column. The smaller size is handy to use and takes up less room on a desk or workbench. Still another way of enhancing readability is to start each procedure at the top of a new page or column; but this is generally a production concern.

A typical industrial or commercial operating manual follows the format in Figure 8.10. This format includes the basic elements common to most operating manuals, arranged in conventional order.

When your subject is a highly complex one, such as the operation and maintenance of a complete industrial plant, an expanded format may be necessary. There are many ways to expand a basic manual format. One way is to divide the manual into "modules," each devoted to a specific subsystem. With this approach, each module is treated like a self-contained minimanual. Hence, you would plan a module by following the same format as for a larger manual, such as the one shown in Figure 8.10. One advantage of this approach is that it allows you to break up your writing task into a number of more easily manageable phases. And since each module provides all the essential instructions for a single subsystem or group of equipment, you'll be making the reader's job easier too; for your reader can refer quickly to the parts of the system that concern him or her most. Modular design is rarely needed when covering smaller systems like minicomputers, jet engines, or motor vehicles; nor is it very useful within the limitations of a short bulletin or instruction sheet.

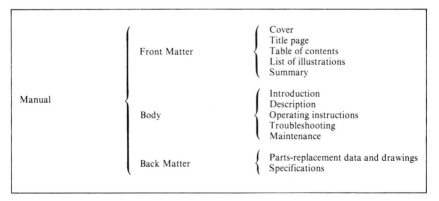

FIG. 8.10 Typical format for an instruction manual.

Bulletin Format

The format for short bulletins, instruction sheets, and user reference cards is simple. As we noted earlier, a bulletin normally focuses on a single topic, such as operation, maintenance, or troubleshooting. In general, ancillary items such as a table of contents, illustration list, preface, appendix, and index are not required. Thus, a typical bulletin might have a cover page and the following four-section format:

Introduction
Description
Instructions
Specifications

Since most types of instructions benefit from graphic material, this elementary bulletin format may be enhanced through the addition of illustrations, tables, and charts wherever they will aid reader comprehension.

HOW TO ORGANIZE THE WRITING JOB

Instructions can rarely be prepared satisfactorily if assigned to individuals in the engineering department in addition to their regular duties. Besides the time involved, close association with the equipment often prevents the objective viewpoint needed in writing for a different level of technical knowledge or aptitude. The task should be performed by skilled technical writers and illustrators headed up by responsible personnel.

The Instruction Development Team

The project engineer or manager is usually responsible for the general task of document preparation. For a small project, the project engineer may work with a writer and an illustrator; rarely does the project engineer do the writing. A large project may involve several writers and illustrators, with the project engineer supervising the work of the various engineering and editorial groups. Occasionally, marketing personnel may also be asked to provide input.

The project engineer must coordinate the flow of information between the various members of the instruction writing team. Since this person is also responsible for scheduling, he or she should specify deadlines for the submittal of the material required. The project engineer may expedite the work by prompting team members via directives, meetings, and telephone calls.

When Does the Writing Begin?

The quality of a manual or bulletin often depends on when it is written—at what stage in the product's evolution the writers are called in. Although there is no agreement on the ideal time, one thing is certain: Before the writing can begin, the product must be clearly conceived by designers. Further, the equipment should be at least partially documented in engineering drawings, flowcharts, schematic diagrams, etc.

Some technical writers and managers believe the task of preparing instructions should be assigned as soon as product development begins. Others warn that if you start too early, you'll later be overwhelmed by engineering changes and new data. In general, the writing begins sometime before the engineering work has been completed.

Writers often work from a rough version of the product and request additional technical data as required. Although this makes the writer's job somewhat like detective work, there is an advantage: Missing information can be obtained while there are still engineering work hours left in the budget. If the project engineer assigns one employee the task of keeping writers up-to-date on product changes, the writing phase of the project can usually be kept well under control.

Ten Prewriting Steps

A comprehensive analysis of the job is a must before starting work. Here is a systematic procedure, broken into ten prewriting steps:

    1. Determine fully the contractual requirements. If you have any questions about what to do or how to do it, ask them at once, not later.

    2. Gather all the available source material. As noted, this may include flowcharts, working drawings, and schematic, block, logic, and other diagrams. Include all applicable specifications. Talk to the product developers. If the product is in the development stage, arrange to receive notification of design changes. Get copies of any existing instructions related to the equipment or its parts. And don't neglect interviews, questionnaires, and other personal sources.

    3. Arrange your material in an orderly manner for easy reference. If some of the material is unwieldy because of odd physical size, note a summary of the data on an index card and add the card to the pile. This card will act as a handy reminder when you're ready to use that particular information.

    4. Read and understand clearly all your source material. Thorough familiarity with the specifications set up for your job is especially important. Regardless of whether the specifications are of commercial or military origin, they can help you avoid costly errors if you know them from the start.

    5. Study the equipment. If you are writing instructions for a machine or a computer program, see if you can operate it yourself, or talk to others who have operated it (or a prototype).

Prepare a systematic breakdown of the assembly and subassembly components. Make certain that your analysis is in keeping with the requirements governing the writing job. Generally, you'll find that it is a good idea to do the parts list first since the parts information is needed for writing the other material. But don't overlook supplementary items such as hardware requirements, accessories, test equipment, special service tools, and connecting cables. A comprehensive component breakdown not only helps establish a presentation sequence but also aids in making a thorough evaluation of the job ahead.

    6. Prepare a preliminary structure for your manual or bulletin. This structure need not be an outline but rather a flowchart or a rough format for the types of information you will present. Follow your writing specifications. Or use the format suggestions in this section as a guide, listing major headings and deleting unnecessary elements as you go along. Once you have a tentative structure, you can fill in the details more easily later.

7. Organize all your information under the major headings. Much of the material may be placed in file folders. Again, items that are hard to handle may be keyed to your structure by means of index cards and short, informative headings such as "Installation: Cabling" or "Operations: Aborting Sort Program." Arranging the information in some logical order—alphabetically, chronologically, or according to significance—will save time and help pinpoint what may be missing.

8. Prepare a detailed outline. Like any other writing job, a good set of instructions cannot be written without an outline. For a short task like a one-page bulletin, a simple outline is sufficient. Thus, you might outline a transistor-replacement bulletin as follows:

Bulletin title: Transistor Replacement
Bulletin number: 7889-B
Issue date: July 15, 19XX
Introduction: Location of transistors in amplifier-stage circuit
Transistor type: PNP junction
Removal: Cut transistor from set base with shears. CAUTION: Do not disturb other circuit elements.
Replacement: Solder transistor to base lug. Connect into circuit. Use care when soldering terminals.
Testing: Operate amplifier; check output.
Illustrations: Circuit diagram; photo of transistor in place.

As the example above shows, a good outline may consist of little more than the major section headings and a few summary statements. For a longer writing task like a 100-page manual, a more highly detailed outline is required; in this case, subheadings should be arranged under the major headings. Then the details belonging under each subhead may be organized into a logical sequence.

9. Check with the person who will approve or disapprove the final manual or bulletin. This person usually will welcome a chance to discuss the document before actual writing starts. These discussions save time, trouble, money, and energy.

10. Establish a schedule and a budget for your writing job. The schedule sets a series of mileposts or checkpoints for evaluating job progress. Assign target dates for the completion of each of the phases so that the final delivery date will be met. Include a contingency factor in the schedule for the unforeseen developments that often happen.

The same principles apply to a budget. Make equitable allocations in hours (or in dollars if cost-accounting service is available) for each operation. Keeping job cards or sheets for each task, on which time is recorded, is a convenient means of checking time expenditures. It cannot be emphasized too strongly that, as with the schedule, contingency allowances must be included in budgeting.

Cost Estimating

If your organization will be building equipment for an outside client (military or commercial), a precise cost estimate may be required. This phase of documentation should be done only by those with extensive knowledge of the factors involved. Historical data on similar projects are the soundest basis for making the estimate.

If the writing task is to be subcontracted to a technical writing service, the service will, of course, furnish a cost estimate. The manufacturers must, however, consider their own costs. There will be liaison with the subcontractor and technical review of the material. Substantial expense may also be incurred for blueprints, photographs, and other material.

In making a cost estimate, the following factors must be determined:

1. What does the source material consist of?
2. What is the end product? What specifications govern it?
3. What are the probable quantities involved?
4. What will the operating conditions be? This includes review procedure, schedule, probability of design changes, and need for travel to military or plant sites.

The best approach to cost estimating is to break down the job into easily recognizable units and estimate the unit costs. Establishing unit costs for typing a page of manuscript or retouching one photograph is a much more manageable proposition than trying to estimate the entire job at once.

Preparing a job breakdown requires a thorough analysis of the job to arrive at the quantitative values for the text and illustrations. An added advantage of this analysis is that it provides an approximate job plan for the execution of the work.

It is advisable to use an estimating form or checklist for compiling costs. Figure 8.11 shows a typical form.

Useful Hints for Writing Manuals and Bulletins

The actual writing is not simple. Good instruction writing is more than skimming through subject matter and setting down points that occur to the writer. Clear writing comes only from much study and careful consideration of the needs of eventual users and from thorough organization of material. See Section 2 on technical reports for more information on outlining; also see the checklist at the end of this section for further suggestions.

The need for you to understand the intended user is fundamental. You must present all the information at (or below) his or her reading level. As a general rule of thumb, the reading level for most instruction manuals and bulletins should be that of a second-year high school student. This provides an extra margin of safety in case some users read at a level below that of the "average" operator or technician. Thus, a typical user also can be expected to have a minimum of technical background and no previous experience with the product. Readers of overhaul instruction manuals can be assumed to have considerably more technical background with other equipment, but none with the equipment covered by the manual you are writing.

Other useful suggestions for preparing instructions include: (1) Use as little introductory material as possible. A manual or a bulletin is meant to be a practical tool, so start with information that is immediately useful to your reader. (2) State a reason for your instructions wherever the "why" of an action is unclear. A plausible explanation, written in the indicative mood (the language of discussion), makes almost any instruction less arbitrary. (3) Give examples, analogies, or

```
                        Estimated Cost Breakdown

        Direct Labor              Hours              Rate              Amount

   Project manager              _____            _____            _____
   Technical writers            _____            _____            _____
   Copy editors                 _____            _____            _____
   Typists                      _____            _____            _____
   Copy check                   _____            _____            _____
   Proofread                    _____            _____            _____
   Art liaison                  _____            _____            _____
   Manuscript production        _____            _____            _____


   Production overhead  (    %)

   Art purchases (explain)

   Consultants

   General service expenses  (    %)

   Other direct costs
       Travel and subsistence
       Illustration reproduction
       Typography
       Printing and binding

   Fixed fee
       or profit  (    %)

   Proposed contract price
```

FIG. 8.11 Typical estimating form.

comparisons when possible—these help the reader to "see" the procedure clearly. (4) Divide the content into major and minor sections, and use headings and subheadings to introduce the various compartments of the document. (5) Choose easy-to-use illustrations—photos, simplified diagrams, and perspective, isometric, or exploded views. Use them singly or in combination to show the reader what to do, and write a caption for each. (6) In the illustrations identify callouts referred to in the text. (7) Tabulate important data for easy reference. (8) Assign an identifying number to each manual or bulletin. (9) Place the date of issue and identifying number on every page of the instructions. This practice helps prevent mix-ups. Also, it does away with the need to use the full title of the piece when referring to that piece in other documentation. (10) Place supporting illustrations or tables as close to each step as possible.

Reviewing and Editing

After the first draft of the manual or bulletin has been typed, it should be reviewed and edited for content, clarity, and conciseness. A lot of time can be saved if the writer makes all necessary improvements *before* submitting a piece for approval or production.

Because reviewing and editing are as vital to the effectiveness of the instructions as the writing itself, these steps must not be rushed or neglected. Try to schedule a review well in advance of the date on which the "final" draft is due; for a short bulletin, schedule at least one day, and for a large manual, preferably a week or more. When you are planning your work schedule, always plan with this review period in mind.

Have your first draft reviewed by the engineers and developers who designed the product. Inform them that they are to check for usefulness, completeness, readability, and technical accuracy. Although good technical reviewers will be as critical as possible, they may need to be reminded that these factors deserve closer scrutiny than literary style. Generally, style and grammar can be checked more quickly by editorial experts than by product designers. Technical reviewers, however, may be asked to mark the copy for any typographical, grammatical, or spelling errors they come across during their review.

After the technical review has been completed, rework your first draft and incorporate the reviewers' recommendations. Then review your draft again. Examine the arrangement of the instructions. See that it follows your outline. If necessary, you can change the sequence. But be sure there are good reasons for doing so. Of course, if you've been asked to follow a commercial or military specification, make certain that the document meets all the applicable requirements. Also check for items such as: consistency of units, terms, and nomenclature; appropriateness of headings and subheadings; logical structure of sections, subsections, and paragraphs; correctness of cross-references, tables, and illustrations; and concreteness of words and phrases.

After the results of the second review have been incorporated into an "approval" draft, proofread the typed copy thoroughly. Submit the polished version to the person who will approve or disapprove the document. Depending on the outcome of this review, a final draft may or may not be required. Refer to the appropriate sections at the back of this handbook for more details on style, grammar, and editing; also see the checklist at the end of this section.

REMINDERS FOR WRITING COMMON TYPES OF INSTRUCTIONS

Product instructions vary so much that it is impossible to set up a typical, detailed outline, as we did for articles. But the following reminders should prove helpful. Each is essentially a listing of important subject items and production needs you'll meet in writing and producing certain types of manuals and bulletins.

To use this list, check the topics you'd like to include. Add any others that are peculiar to the subject or to your field in general. Then you'll have a good outline to guide your writing. And when you write, be clear and direct. Use command language for specific instructions whenever possible.

Items listed under "Description," "Operation," "Theory," and "Maintenance" are common to many instructional publications. So use the list regularly. Keep it on your desk while you write. Delete those topics that do not apply to your writing task.

Description

What is this information?
Purpose of this information.
How to use this information.
Other publications required.
Illustration of equipment.
Identify manufacturer.
Equipment name and number.
Manufacturer's guarantee.
Relate to other models.
Relate to associated equipment.
Modifications available.
If system, what are units?
If in larger system, explain.
What does equipment do?
What goes in?
What happens to it?
What comes out?
Why do it this way?
What is equipment used with?
Where, when, why is it used?
Who uses it?
Advantages of the equipment.
Advertising of product(s).
Advertising of company.
Price.
Reordering.
Order, shipment, contract numbers.
Shipping list.
Shipping damage report.
Reshipping requirements.
Accessories supplied.
Accessories not supplied.
Basic principle of operation.
List characteristics.
List controls.
List connectors.
List cables.
Power requirements.
Weight and dimensions.

CREATING EFFECTIVE INSTRUCTION MANUALS AND BULLETINS 8.37

Operating supplies required.
Installation requirements.
Operator requirements.
Weatherproofing.
Tropicalization.
Warnings and precautions.
If system, repeat for each unit.
Note patent numbers.
Give copyright date.
Notes on government licensing.

Operation

Repeat from preceding as needed.
Write to operator's level.
Operator requirements.
Illustrations of what operator sees.
Identify panel controls.
Unpacking.
Installation requirements.
Installation procedures.
Intercabling drawings, tables, schematics.
Identify by name and number.
Tabulate modes of operation.
When are different modes used?
How it operates, vis-à-vis controls.
Operational block diagrams.
Logic diagrams.
Flow diagrams.
Enough theory for operator.
Preoperational adjustments.
Turning on the equipment.
Connecting: inputs, outputs, accessories.
Standby operation.
Operating the equipment.
Emergency operation.
Turning off the equipment.
Postoperational adjustments.
Keeping log of operation.
Local and remote operation.

Warnings, precautions.
Supplies needed.
Replacing supplies.
Repeat above for each mode.
Define unusual expressions.
Explain unusual procedures.
Refer to technician's adjustments.
Operator's tests for operation.
Operator's maintenance check-chart.
Fuse, adapter, backup tape, circuit board, tube, indicator.
Plug-in-unit replacement table.

Theory

Repeat from preceding as needed.
Write to reader's level.
How it works, not why.
What goes in?
What happens to it?
What comes out?
Compare with familiar equipment.
Differences between models.
Diagrams: systems block, unit block, flow, vector, simplified schematic, complete schematic, primary power distribution.
Oscilloscope wave patterns.
Wave-shape comparison chart.
Signal-flow analysis.
System-operation theory.
Stage-by-stage theory.
Theory: by functions performed.
Theory: output toward input.
Mathematical basis of operation.
Theory: each mode of operation.

Preventive Maintenance

Repeat from preceding as needed.
Write to technician's level.
Technician requirements.
Check-chart for proper operation.

Maintenance check-chart.
Lubrication chart.
Hydraulic maintenance chart.
Periodic checks (hourly, daily, etc., through annual).
How to gain access for maintenance.
Periodic overhaul procedures.
Precautions, warnings.
Special test circuits.
Special test setups.
Safety precautions.
Retropicalization.
Reweatherproofing.
Specify lubricants.
List special tools.
List special equipment.
Fuse, circuit board, disk, tape drive, head, processing unit, crystal, and plug-in-unit locations.
Premaintenance procedures.
Postmaintenance procedures.
Cleaning, decontaminating, demagnetizing, painting, polishing.
Corrosion: rust, water, salt-spray damage; air filter, cable, connector check.
Parts-replacement list and schedule.
Where to buy replacement parts.
Supply replacement schedule.
Where to buy supplies.

Corrective Maintenance

Repeat from preceding as needed.
Write to technician's level.
Technician requirements.
Recommended maintenance approach.
Illustrations of equipment.
Component identification charts, drawings, and photos.
Troubleshooting output to input.
Troubleshooting input to output.
Troubleshooting: for system, by unit, at controls, by symptoms, by error messages, by test points, by test signal.
Signal tracing chart.
Charts: oscilloscope wave-form, voltage, resistance.

Diagrams: servicing block, simplified schematic, overall schematic, unit-by-unit schematic, vector, primary power.
Mechanical exploded views.
Overhaul procedures.
Mechanical disassembly drawings.
Mechanical reassembly drawings.
Accessory equipment required.
Special tools required.
List faults and probable causes.
Special test circuits.
Illustrations showing adjustments.
Electrical, mechanical, hydraulic, pneumatic servicing.
Disassembly, reassembly.
Emergency repair.
Voltages on schematic drawings.
Wave shapes on schematic drawings.
List of component parts.
Parts manufacturers.
Parts ordering.
Repair: cabinet, corrosion, rust, water damage, grounding.

Other Things to Remember

Covering letter.
Get your company's name and address on it.
Get your name on it.
Give date, department.
Type one side, double-spaced.
Drawings and pictures are better than words.
Hold consistent nomenclature.
Hold consistent technical level.
Will it be understood?
Write to reader's level.
Write to reader's interests.
Check security clearance.
Table of contents.
List of illustrations.
List of tables.
Number the pages.
Use short words.
Use simple sentences.

Keep paragraphs brief.
Number sequential steps.
File a reference copy.
Keep a copy for yourself.

CHECKLIST FOR INSTRUCTION MANUALS AND BULLETINS

The following checklist is intended as a guide only. When preparing instructions to a military or commercial specification, the applicable specification requirements must be adhered to in every detail.

Text

1. Use short sentences and known or defined terms.
2. Use short paragraphs and a separate paragraph for each thought.
3. Give specific instructions. Statements such as "Test in accordance with authorized shop procedure" are worthless unless the source of such procedures is given.
4. Put each numbered step in a separate sentence, paragraph, or block of text.
5. Avoid uncommon words, unnecessary technical phraseology, and words peculiar to a single industry.
6. Avoid complicated phrasing. Delete redundant or superfluous words and phrases.
7. Use exact nomenclature. Name each part, component, or subassembly. Do not use vague references such as "the unit."
8. Divide the text by appropriate headings and subheadings.
9. Link discussions with illustrations so that one follows closely on the other.
10. Edit the manual or bulletin, paying close attention to content, grammar, clarity, conciseness, and style.
11. Follow the governing specifications, if any.

Photographs

1. Photographs may be used for equipment views, operational views, and parts identification, as long as they are permitted in applicable specifications.
2. Photographs usually require retouching to retain detail sharpness and contrast.
3. Eliminate superfluous or unattractive backgrounds by cropping, retouching, or opaquing.
4. Use exploded views to identify parts in mechanical assemblies and assembled chassis views for electronic equipment.
5. Identify parts by number, reference symbol, or name on the photo.

6. For identification nomenclature, use copy prepared by typesetting or with mechanical lettering guides or "press-on" type.
7. Before having photos taken, plan each picture and supply the photographer with a rough layout so that each view will cover the required subject matter exactly.

Line Drawings

1. For wiring diagrams and electrical schematics, use consistent practices on symbols, line weights, conventions, and layouts as recommended by the industry or military service involved.
2. Use mechanical schematic drawings to illustrate operating features and maintenance practices. Present hydraulic, pneumatic, and mechanical drive systems in schematic form.
3. Specify plan views to show installation and mounting features, lubrication practices, equipment outlines and dimensions, and panel markings.
4. Simplify the job of explaining operating and maintenance details by liberal use of pictorial drawings. Isometric projection is common, but dimetric and trimetric projection give the subject a more natural appearance. Shade the drawings when it will add realism.
5. Use exploded views in conjunction with parts lists to show assembly and disassembly procedures. Use pictorial drawing technique.
6. In general, limit the use of color to functional purposes such as flow diagrams and color-coded illustrations.

   Note: Pictorial representations of many types of relationships can be produced via computer graphics, by means of dots, lines, curves, etc. Once data has been entered through a keyboard or other device, the desired image is produced on a screen; the image may then be manipulated easily with a device such as a light pen or track ball. This system is especially useful where complex two- and three-dimensional images are involved and in documents that require regular updating. The general guidelines given above apply to both computer graphics and manually produced drawings.

Production

1. Type the manuscript double-spaced on white paper. Leave wide margins, and adhere to precise specifications or company standards, if any, on punctuation, format, style, and abbreviations.
2. Proofread each typed copy for accuracy; usually a rough draft, approval draft, and final draft are necessary.
3. If you are assigned the task of supervising the production of the document, mark the final draft for type, spacing, and makeup. Make all changes and corrections in the manuscript prior to starting final typography.
4. Make certain that the index, contents pages, title pages, appendix, artwork, and other applicable items are submitted with the manuscript.
5. Proofread the reproduction copy—in both galley and rough page form—and check placement of and references to table and illustrations.
6. Schedule printing well in advance to ensure delivery when required.

SECTION 9
WRITING CORPORATE SALES BROCHURES

Sales brochures are today's principal communication tools for technical and scientific firms wishing to present their skills and services to their market. This market may be other technical and scientific firms, industry in general, specialized prospective clients, government (state and/or federal), or the general public. The last market is probably the smallest segment for most technical and scientific firms in business today.

DEFINITION OF A SALES BROCHURE

A sales brochure is a document designed to present the skills, services, or products of a firm to its customers and prospective customers. The brochure is designed and written so it presents the needed information in a concise and useful manner. Color and high-quality paper and binding may be used where a goal of the sales brochure is to attract attention despite competition from other similar brochures. A sales brochure may be as short as 1 page or as long as 500 or more pages. The length of a brochure, its design, and its physical quality all depend on the objectives the firm wishes to achieve with the brochure.

Sales brochures present a unique writing challenge for any technical writer since such brochures can contain history, engineering and technical data, sales copy, and personal biographies of the firm's principals. This section of the handbook shows how to prepare effective sales brochures to meet any requirements—within budget and on schedule. The data provided can be used with any type of sales brochure the writer may be asked to prepare.

THE SALES BROCHURE AS A PROMOTIONAL TOOL

Let's look—quickly—at why there is such a strong need for sales brochures among nearly all technical and scientific firms everywhere in the world today.

Companies Must Promote

Today, more than ever before, competitive marketing programs are usual among engineering and other high-technology concerns. A company's survival depends on its ability to approach a potential customer, to get that person's attention, and to persuade him or her to buy the firm's services or products.

While the sales representative enjoys face-to-face contact and has the best chance to close the sale with the customer, the sales brochure writer has the unique capacity to *presell* the customer *before* the actual sales meeting occurs.

Generally, the customer will request a sales brochure after being referred to your firm by a colleague or business associate, or after seeing an advertisement for your company's products or services. In addition, sales brochures may be picked up at your office, used as part of a direct-mail marketing campaign, or distributed at business meetings and conventions.

The Primary Printed Sales Medium

Brochures emphasize key qualifications and selling points of the high-technology firm, yet they omit extraneous technical detail. Thus the corporate brochure is basically a tool for promoting sales; it is the primary printed sales tool for the majority of service-oriented companies.

Properly employed, the corporate brochure will present the company's image of itself not only in descriptions of the firm's qualifications and in illustrations of its past work but also in the quality and content of the document itself.

WHAT IS A CORPORATE SALES BROCHURE?

Definition of Objectives

The corporate brochure is the basic element of sales literature for organizations offering technical and/or scientific services. Also referred to as a *company brochure* or *professional brochure,* it is the principal document that describes the firm to the marketplace. Corporate brochures are used by a wide variety of technical and scientific firms, including most large design firms; they are also used by engineers, architects, and scientists working as consultants.

The corporate brochure is the *qualifications booklet* that traditionally has been one of the first items the technical sales representative has put into the hands of prospects. It is what you mail to a customer to arouse his or her interest before you make a "cold" sales call. It is the pamphlet you send to a customer who responds to a referral, seeking a concise overview of your organization's capabilities.

The corporate brochure, therefore, is a vital element of the technical promotion process, which often requires as many as six sales calls before a sale is closed. The effective technical administrator uses the document not only as a backup device with actual sales calls but also as a supplement to other media, such as sales letters and advertising.

In sum, the corporate sales brochure is a kind of advertising in which writing and illustrations combine to form a persuasive presentation highlighting the advantages of doing business with your firm. The company brochure may incidentally teach or inform readers about your technical field, but what it must do is *sell*.

General Features

Because it directly reflects the image of the company, the effective sales brochure is usually attractive in appearance and produced in a highly professional manner. Like technical proposals and most other forms of writing described in this handbook, however, corporate brochures vary widely in layout, quality, cost, and other factors. Some firms rarely revise their basic brochures; others update and change the documents frequently, often producing new ones in order to meet the changing requirements of the marketplace.

Despite these variations, most company brochures share a common objective: to describe the performance record of the technical organization in a way that demonstrates the firm's ability to meet the client's requirements. This has led some writers to conclude that brochure copy must rely almost totally on dry recitations of technical and historic data. To be truly effective, however, the company brochure must go beyond this; it must get and hold the reader's attention and persuade him or her to take action. A well-written brochure may move the customer to request a detailed technical proposal, for example, or a meeting with the firm's contract administrators.

Using the Salesperson's Approach

The content of the brochure is dictated largely by the basic information the client needs in order to decide whether or not to retain your firm. As in sales meetings, however, there is room in brochure copy for subjectivity. In fact, some brochure experts firmly believe that a sales brochure cannot be successful unless it is written with the honest approach of an effective salesperson. As a form of advertising, the corporate brochure should not hide the purpose for which it is designed: to sell your organization's services.

Other Types of Sales Brochures

In addition to the corporate brochure, some technical organizations produce sales brochures that are devoted exclusively to one or more products. Like the corporate brochure, the *product brochure* is designed to sell; but here the item being sold is somewhat more tangible. Hence, the product brochure must give specific, detailed technical data—rather than listing qualifications of personnel or seeking to appeal to the customer's emotions to get him or her to buy the firm's technical services.

The typical buyer of an industrial product seeks photographs of the product, specifications, and concise information on such topics as installation, maintenance, and testing procedures. The product brochure does, however, point out the advantages (e.g., durability, reliability, and price) over similar products in the field.

Generally, the layout and reproduction of the product brochure is simpler and less expensive than that used for a corporate sales brochure. Most product brochures are printed on standard 8½- by 11-inch paper, saddle-stitched as required, and prepared with a 3- or 5-hole punch to permit filing in a loose-leaf binder. Length usually ranges from one to eight pages.

Product brochures may be divided into two main categories: single-product brochures and multiple-product brochures.

Cross-use of Brochures and Other Product Literature

In some cases, much of the information and material that is developed for sales brochures can be used in preparing other technical literature—such as advertisements, catalogs, and instruction manuals—and vice versa. Photographs, specifications, and descriptions that are used in a product brochure or instruction manual can, for example, be incorporated into a multiple-product catalog. The product brochure is identical in many respects to the product catalog; it differs mainly in that the product data are presented in a broad, summary fashion rather than in complete detail. For additional guidelines on preparing catalogs or instruction manuals, see the appropriate sections of this handbook.

SIX RULES FOR WRITING EFFECTIVE BROCHURE COPY

Keep the following basic guidelines in mind when planning and writing corporate sales brochures.

1. Present Your Message Clearly and Persuasively

Avoid professional jargon. In rare instances, technical buzzwords may be used, provided that they convey your intended meaning as concisely as less specialized terms can. Even then, you should be sure that they do not make readers less receptive to your message.

2. Be Brief

Because you face busy, skeptical readers, you must steer clear of rambling discourses on management policy, design philosophy, and similar topics which have no place in sales promotion. Choose instead a summary view of the background and history of the firm and its principals. If you are going to write in depth on any particular aspect of the firm, it is best to detail the firm's project experience and technical capabilities.

3. Demonstrate Your Awareness of Client Interests

Honor your potential customer and forget pride of technical accomplishment. Focus instead on cultivating a tone of helpfulness and honesty.

You must not only show the prospective client that you know what he or she requires, you must also prove that your firm can deliver as promised. In this sense, the brochure is a form of unsolicited proposal because it must address the immediate objectives of the client. The principal difference is that the corporate brochure cannot be designed exclusively for one project or one client. It must take a somewhat more general approach, while pinpointing the technical requirements and solutions of exemplary past projects.

Before you write, you must have a "capture" strategy by which you'll look at clients' requirements and see if they "track" with the project experience and qualifications of your firm and its members. By stating these requirements in your own terms—with extraneous trivia stripped away—you will prove your understanding of the client's problems. If you then add interesting descriptions and telling photographs of your firm's solutions to clients' problems on comparable past projects, you will demonstrate that your firm can deliver what it promises.

As with other forms of writing, it is of paramount importance that you know your audience before you write. Are you writing for technical managers? purchasing agents? building owners? several different types of readers at once? Know your readers by their interests, by their job titles and functions, by their industries, by their geographic location, and by their reading levels.

If you will be writing for a large engineering or scientific organization, you will need to gather some of the necessary background information from management, engineering, and marketing. If you will be writing for a firm in which you are solely responsible for all these departments, you may need to look to other sources for help. Any objection to getting some free help from your competitors?

4. Know Your Competition

Obtain and study their brochures. Examine their approach to capturing the client. There may be ways in which you can use this information to your advantage. For example, there must be some way in which your company or service is different from and better than the competition. Find your advantage and tell your readers about it. For more ideas on developing an effective capture strategy, consult the proposal writing section of this handbook.

5. Strive for Straightforwardness in Your Presentation

This is not to contradict our earlier statement that there is room for subjectivity in the sales brochure. On the contrary, it is quite possible to describe your firm's services and experiences in an interesting (if not exciting) way, while emphasizing the benefits of retaining your firm, and being straightforward about its capabilities.

Ideally, the material and layout can be arranged to tell a story. A sales representative should be able to put the brochure on a desk in front of potential clients and leaf through it page by page, giving an organized presentation at the same time. If, on the other hand, the brochure is arranged in a way that makes such a presentation impossible, the client may only glance through it and then drop it into the wastepaper basket.

6. Stick to Nonchanging Information

Omit data that are subject to continuing variation or which may become obsolete in the near future. The cost of producing an attractive brochure is relatively high compared to that of other technical documents; if the entire brochure must be redone to accommodate changing information, the result can be a costly waste of time. Changing information is better dealt with in the form of *customized inserts* or *fact sheets,* which may be placed in the brochure as required (more on these later).

MANAGING THE BROCHURE PREPARATION PROCESS

While the technical writer's responsibility in the brochure writing process is generally confined to copywriting, the manager or principal of the high-technology firm must be aware of the entire process of preparing the corporate brochure. Regardless of your position, it will be to your benefit to know the basic procedures involved. These may include: (1) planning the brochure preparation project (for example, determining project organization, scheduling, and budget considerations); (2) assigning the personnel needed to perform the project as planned; (3) retaining help from outside agencies as needed; (4) overseeing research; (5) directing the creative staff (for example, writers, illustrators, graphic designers); (6) coordinating reviews by editors, engineers, and marketing and legal experts; (7) choosing graphic design and layout; (8) having the document printed; and (9) directing distribution of the final product.

Let's simplify this process by looking at some of its basic elements.

Planning the Brochure

To prepare effective sales-getting brochures, you first must:

1. Decide on the types of clients you wish to attract.
2. Determine your capture strategy.
3. Decide on which information to include.
4. Choose the best way to present the data.

Let's see how we can apply these methods to your brochures.

Decide on the Types of Clients You Wish to Attract. Are you preparing this brochure to appeal to a relatively wide audience—one that includes, for example, public works agencies, private construction firms, and architects? Or do you intend to produce separate brochures for different kinds of readers? A brochure designed to sell your services mainly to public works agencies may take a much different approach from one for architects. Unless you know exactly who your readers will be, you can't determine how best to win potential engagements. So investigate your readers and find out their needs. Only when you know these can you plan an effective brochure that will help sell your services.

Most corporate sales brochures today are written for more than one type of potential client. So we'll give these the most attention.

Determine Your Capture Strategy. How can your brochure grab and hold the prospect's attention? Ask yourself this question before you outline your strategy options. Review the rules for brochure copywriting given earlier in this section.

Tailor your sales pitch to your audience. Decide how you can demonstrate your awareness of client interests, while proving that your firm can deliver as promised. Do you believe that your organization's experience and capabilities closely match the most critical requirements of potential customers? If so, say it. Be ready to demonstrate why this is so.

Decide on Which Information to Include. Again, this ties back to your readers and their needs. Make a list of typical client requirements. Then include answers to as many of these requirements as you can. Do not overlook any of the benefits that may accrue to the client who buys your services.

Make a list of what differentiates your operation from those of your competitors. If yours is the only engineering firm in the state that is run by a president with more than 10 years' experience in the maintenance of air-operated servomotors in multifunction robots, your brochure should not only mention this fact, it should seize on it, glorify it. Make a list of the instances and ways in which this has helped your firm excel in satisfying clients' requirements on past projects. Try to give just the right amount of clear, factual support to your statements of client benefits, and you'll have your readers wondering how they ever got along without you.

Choose the Best Way to Present Your Data. As noted earlier, it is often best if you can arrange your material and layout to tell a story. By "story" we do not mean a narrative of the chronological development of your firm, but a "story" in the sense that the cover and first pages of the brochure *pull* the reader in. By describing the *benefits* of buying from your firm, you can move the reader emotionally—make him or her *feel* something. If you stick to a dry recitation of the *features* of the company, on the other hand, you will not have this emotional effect. The result would probably read more like a specification than an effective brochure.

Later pages of the brochure should sustain and possibly build the reader's interest, while presenting the necessary data to support your statements of client benefits. Forms in which you can present these data may include not only text but also photographs, charts, and tables. Most brochure experts agree that photographs and charts, used judiciously, enhance reader comprehension. Generally it is best to place the more simple and attractive of your graphic aids near the front or middle of the brochure, and to save complex charts and illustrations for the back.

When choosing graphic devices, ask yourself, "What is the reader accustomed to using?" Photographs of facilities? Corporate experience matrices? Block flowcharts? Linear responsibility charts? Include those he or she is likely to be familiar with. Then your reader will find your brochure more interesting and useful.

Remember, the brochure indirectly reflects your company and its services. Write a serviceable brochure, and the users will be inclined to take action—to request a meeting with the company principals or merely to ask for further information. If your brochure can do this, you will have accomplished what you set out to do.

Controlling Costs

Brochures may be either the most productive or the least effective items in your kit of promotion tools. In addition to the obvious requirements that the brochure be written clearly and produced in an attractive manner, the *cost* of preparing the brochure is a factor in determining its effectiveness.

In order to be truly effective, a sales brochure should have a relatively low cost in comparison to the sales it helps to generate. This is not to say you should skimp on brochure preparation. On the contrary, the majority of corporate brochures represent a large investment of time and money. In managing brochure preparation, however, you must pay attention to budget and schedule constraints.

There is no point in paying huge sums of money for the help of a professional graphic designer or outside advertising agency if you expect your sales per potential customer to be very low. In that case, it is better to design the document by drawing on the limited publications skills of in-house staff; then you can put part of the savings toward producing a larger quantity of brochures. This is sound marketing strategy.

Unfortunately, many technical and scientific organizations end up wasting the money they put into an expensive brochure, usually by distributing it arbitrarily to any company or individual who is even remotely interested. In those cases, many of the brochures end up in wastepaper baskets. So you must also plan to distribute your brochure carefully if you want the cost and time invested in it to be worthwhile.

A general rule of thumb of good brochure production is to spend whatever is needed to produce a first-rate sales tool without giving the reader the impression that it is overdone. Not only is an inflated brochure expensive in and of itself but the effect it has on the reader may be the opposite of what is intended. If the client's intended project is a modest one or on a tight budget, or both, the client may—rightly or wrongly—take extravagance in the brochure as a sign of lavishness on the part of the technical staff. And a reputation for excess will not help your firm win clients.

A corporate brochure that bespeaks success, affluence, and enthusiasm can have a positive influence on potential clients, particularly when the competition's brochures are relatively cheap looking. Indeed, it may be helpful to outdo the competition in brochure production. The point is to know your limits.

It is, of course, possible to spend a lot of time and money on a brochure and have it end up looking like a cheap throwaway ad, but that is a different problem. Probably the only way to protect yourself and your firm against this form of waste is to investigate several printers thoroughly before you submit the document for production. Examine samples of their work as they come off the press. Also look at samples of their trimming and bindery work. Finally, be sure to get cost estimates from at least two printers before you decide which one to go with.

Basic Design Ideas

Innovation and creativity are to be encouraged in writing and illustrating, but brochure design is another matter. Here the line between what works and what does not is much more clear-cut. To some extent, the design of the document is dictated by variables such as cost, personal preference, and the amount and kind of data you wish to present. Thus, brochures can consist of a single page, a few pages, or more, up to several large bound volumes. The corporate sales brochures of most professional design firms are at least 10 pages in length.

Sales brochures range in size from the standard 8½- by 11-inch sheet folded to go in a number 10 business envelope to the multipage publication with stiff covers, printed on 11- by 17-inch sheets, folded and saddle-stitched at the center. As we will soon see, it may sometimes be advisable to hire experts from outside your firm to aid in the design effort.

Here are some basic design principles, most of which apply to all printed marketing tools:

1. Watch eye movement in design. Keep readers moving in the right direction through a page or spread.
2. The best design is never noticed.

3. Good design reinforces words and helps comprehension and retention.
4. Capital and lowercase letters are easier and faster to read than all caps.
5. Shorter lines of type are easier to read than longer lines.
6. We look at color before we look at black and white.
7. Use subheads, bullets, dingbats, ballot boxes, and other graphic devices to break up copy and make it more readable.
8. Leading (space between lines of type) helps readability.
9. Black type on white stock reads almost 12 percent faster than reverse type (white type on a black or colored background).
10. Quantity comparisons are better illustrated by bar charts than by line graphs.
11. *Photos:* When there are several pictures on a page, one should dominate all the others. People look at larger objects first. Humans look at other humans first, then objects.
12. *Color:* Color is a very personal thing; it always attracts attention, and it should always help legibility. All colors have certain associations. Red attracts the most attention and gives the illusion of closeness. Blue gives an illusion of distance. Green reminds people of the outdoors, but green and orange are tricky to use because some people are offended by the two colors. Blue seems to be the least offensive, which accounts for its wide use in direct mail.[1]

Page layout is another important design consideration, for which several options are available. Figure 9.1 shows three different page layouts that are popular in corporate brochures.

For descriptions of typical methods of binding publications, see Section 2 on reports.

When contemplating various design options, don't forget that your printer can be an excellent source of advice. Many printers have dealt extensively with sales brochures, so they know what sorts of designs are most popular and effective.

Assignments

The job of preparing the brochure may draw on the knowledge and skills of people in marketing, public relations, personnel, management, and technical functions. While some large technical and scientific organizations employ a full-time public relations writer or advertising copywriter, few firms find it economically feasible to employ someone exclusively for the writing of brochures; the frequency with which brochures are issued is just too limited.

The firms that do well find that the best brochures are produced when input is solicited from several sources. Organizationally, these sources make up a team, much like the teams used to develop written proposals.

In a large organization, the typical team should include a representative for each function relevant to the preparation of the complete document. Hence, a brochure development team might include: a marketing or sales director; engineers or technical specialists needed to ensure that the contents of the brochure are technically correct; one or more technical writers, public relations writers, or advertising copywriters; a photographer, illustrator, and/or graphic designer who will prepare the graphics; and one or more editors who will be responsible for reviewing and editing the document drafts. Production specialists such as a typesetter and printer would round out the team. Figure 9.2 shows eight sources of input for a large corporate sales brochure.

---

[1]Gerre Jones, *How to Market Professional Design Services,* 2d ed., McGraw-Hill, New York, 1983, p. 132.

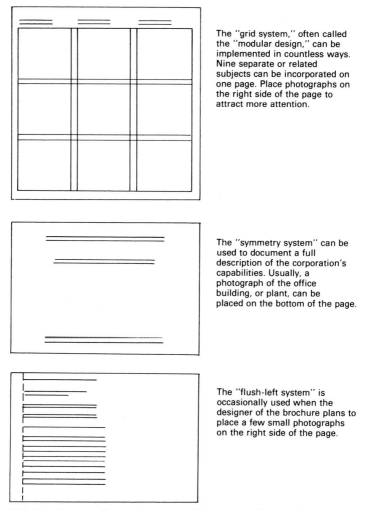

FIG. 9.1 Three popular page layouts used in corporate sales brochures.

Each representative should be assigned the responsibility of obtaining and coordinating the information required from his or her department. Thus, information on how to pitch the firm and its services, selling gimmicks, market analyses, etc., can come from the marketing representative. The mechanics of producing the brochure, such as typesetting and printing, are normally contracted externally. In a small firm, however, the secretarial staff might be assigned to produce the brochure on a high-quality electronic typewriter or word processor.

Companies that employ most or all of these talents and own certain capabilities—such as typesetting—are fortunate. Creative talent can be related to the firm and its services more readily. Further, original writing and promotion can be

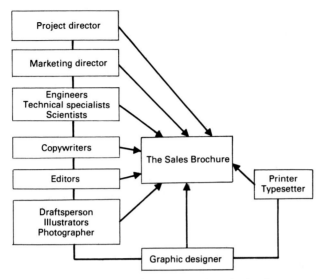

FIG. 9.2 Typical sources of input in preparing sales brochures.

achieved at a relatively low cost. In general, the capabilities of an internal team should be utilized to the fullest extent possible.

If the team is drawn together as soon as the general assignments are made and if it spends a few hours developing an outline detailing exactly what will be included and who will assemble each element, the process of preparing the brochure becomes both more manageable and more efficient. In the event that the technical or scientific organization does not have all the necessary talents and capabilities in-house, outside assistance can be sought.

Outside Assistance

Experts from outside the firm who can assist in developing winning brochures generally come from either of two categories: agencies and freelancers. Specialized agencies include public relations firms, advertising agencies, graphic design companies, photographers, typographers, and printers.

Some public relations consultants work only on press releases and media gatherings, while others offer complete services for all aspects of PR, from writing and producing brochures to holding press conferences. When seeking help in brochure preparation, you may want to inquire at various PR firms about the range of their services and request cost estimates for brochures.

Advertising agencies are usually oriented more toward graphics and ads, while graphics firms provide help in developing layouts, choosing color schemes and paper stock, and working with printers.

Professional photographic services may be needed for your brochure, particularly if there is no photography expert on your staff. Generally, it is good practice to take at least three representative photographs of each major project in which your firm is engaged; if possible, these should be taken during the actual engagement. Some of these photos may be able to be used directly in a brochure, provided that they are of high quality and that they support your message clearly. Other photos taken at the time of a specific engagement may suggest additional photographic ideas to use in the brochure later. Again, those that you choose to put in the brochure should drive home your point about certain benefits of your organization's capabilities.

For many firms, a printer offers one of the most economical sources of outside brochure services, often including typesetting. If necessary, a typesetting firm may also be employed.

Freelancers are generally available for many aspects of brochure preparation, from graphic design and type specification to preparation of an entire brochure. In fact, some marketing experts recommend the use of outside brochure consultants over in-house talent. The writer and other specialists assigned by the brochure consultants come to the project fresh, open-minded, and objective—in that they may be working for your firm for the first time. It's up to you to decide whether these forms of outside help will be worth the added expense.

Gathering Background Information

The task of researching the corporate brochure generally requires that the persons responsible for providing background data look *outside* to the clients' requirements and *inside* to the firm's overall capabilities. Although much of the background information may be useful in determining capture strategy and other general features of your basic presentation, little or none of this material will end up in the brochure in its original form. The work of skilled researchers is nonetheless vital to the development of an effective brochure.

Typical research efforts can be summed up roughly as follows:

1. Develop a comprehensive list of client requirements. Who are your typical clients? What are their technical or scientific needs? In seeking to answer these questions, it may be worthwhile to examine certain publications issued by the client; these may include both external documents such as clients' own brochures, catalogs, statements of work, or requests for proposals, and in-house documents—if you can obtain them—such as reports, procurement directives, and memoranda relating to specific projects or management objectives.

2. Try to determine what your potential clients look for in the following major areas:

       Technical capabilities
       Budgetary considerations
       Management expertise
       Staffing characteristics
       Professional qualifications
       Scheduling considerations
       Innovation (technical or of project design)
       Reliability (of project planning)

3. Examine your firm and its services (or products) to determine how your company can satisfy the clients' basic desires and requirements. Investigate past projects to detail how they "track" with clients' needs.

4. Determine your firm's overall capabilities. Obtain documentation of personnel responsibilities and work performed on past projects.

5. List professional qualifications, educational background, and special accomplishments of the members of your organization. Examine licenses, certificates, résumés, awards, etc.

6. Study documentation of your firm's in-house technical procedures, standards, design philosophy, management and personnel policy, and the like. Investigate in-house manuals, project procedures, directives, memoranda, etc.

7. Study the competition's promotional literature, including corporate and product sales brochures, catalogs, and press releases.

STRUCTURE AND CONTENTS

A five- or six-section format is standard for the corporate brochure. These sections are:

1. Foreword (optional)
2. Introduction
3. Descriptions of the firm's principals
4. Services available
5. Projects
6. Clients

Front Matter

Unlike annual reports, professional brochures do not usually require a table of contents or a letter from the firm's president. A brief introduction, including a summary of the company's main objectives and some pertinent historical data, is sufficient but not entirely necessary. When a brief foreword seems appropriate, it is usually written by the organization's chairperson or president.

Descriptions of Company Principals

Personnel bios should usually be limited to the company's principals and officers, accompanied by photos of one or two of the most prominent individuals. Operating personnel are usually not included in the main body of the brochure, for they are not as certain to stay with the firm for a long period of time as are the company heads. Further, the special expertise of individual employees may not be relevant to every client's needs. For this reason, many firms choose a form of binding that permits them to alter brochures to fit changing requirements. Other options are available for dealing with these sorts of demands, as we'll see shortly.

Description of Services

The next item that appears in most brochures is a description of the firm's services. Services can be summarized and/or accompanied by photos of outstanding projects. If desired, services can first be listed in summary fashion and then described in depth with a more thorough accompanying narrative. Or a combination can be used. For instance, basic services can be listed and then broken out and described in detail in independent subsections.

Project Experience

Most firms follow the description of services by listing specific, exemplary projects in which they have been involved. Like the services listing, the project section can be emphasized with lots of detail, if needed, or it can be presented as a short summary. A good rule of thumb is to follow the example set by the federal government: in questionnaires and RFPs, federal agencies usually ask for a listing of 10 recent projects similar to the ones proposed. Thus, in a brochure, a reasonable estimate would be to list 10 to 50 recent projects, giving the type of work performed, plus location, name of client, date of completion, and perhaps a paragraph or two of description. Remember not to bog down your readers with too much material.

Clients

For a firm that does not have a large list of impressive clients, a separate client listing can be especially valuable. If your firm has worked on several projects for a few large clients, listing the projects under the clients' names can also be helpful.

Photographs

The role of photos in professional brochures can hardly be overemphasized. When a potential client says, during or before an actual sales meeting, "Let's see what you've done," the client means exactly that. Photos, rather than words, sometimes make the strongest impression.

Optional Materials

Two valuable tools that you may incorporate into your overall presentation are customized inserts and fact sheets. *Customized inserts* are inexpensively produced copies of information pertinent to a specific client or project. Examples of these are bios of key personnel appropriate to a particular promotion. These can be produced quickly by your secretarial staff as needed. If you design the brochure cover to include a pocket (usually on the inside back cover), you can insert these sheets without destroying the continuity of the presentation. Other data that are subject to frequent change are also candidates for customized inserts.

In addition, you can hand out *fact sheets* of one to four pages on specific projects. These facts sheets may come in handy when you want to focus on an especially pertinent project, or on a relevant project that has been completed after the formal brochure is produced. Fact sheets might include photos of the project, together with a detailed cost breakdown, and other vital data.

By putting major projects on individual fact sheets, you can choose those projects that are most appropriate to a particular prospect. Fact sheets are useful in avoiding the waste that may result when diverse examples are bound in a single package or brochure. If the potential client is interested in a highly specialized air-conditioning system, you are wasting your money and his or her time if you provide too many examples of a simple refrigeration unit or other system in which there is no interest. Used properly, customized inserts and fact sheets can be powerful sales tools.

WRITING THE BROCHURE COPY

The process of writing the corporate sales brochure will vary from company to company, depending on such factors as the complexity of the organization and the number of people involved. However, the general writing task can be summarized in seven basic steps: (1) Collect data and illustrations. (2) Make an outline. (3) Write a rough draft. (4) Check your facts. (5) Submit the draft for editorial review. (6) Polish the rough draft. (7) Send copy and art to the production department.

Data and Illustrations

Get these from the engineering, sales, and advertising departments. Try to get the latest material available. If your material is not up-to-date, you will probably have to change it before the brochure is printed. With every advance in the production process, changes made to correct errors in data or to update data become more expensive and often lead to wasted effort, lost time, and misunderstandings.

Don't overlook clients as potential sources of illustrations. Many of your clients will be glad to supply photos of your completed projects. Or they will allow you to visit their facilities to take photos of the projects.

Insist on being supplied all relevant data. True, you may use only a tenth of it. But having the information on hand will help you write your text.

Make an Outline

You'll be lost without one. Study other brochures to see how they are written. Remember to examine your competitors' publications. Some of your best ideas may result from reading the writing of others; you'll not only learn from their successes but also from their mistakes.

When preparing your outline, try to imagine exactly what the typical reader will be seeking. Arrange the outline so the needed information is given in the sequence in which the potential client will want it.

Study your outline and revise it until you are satisfied it is the best you can prepare. Key the outline to the illustrations and other examples you will use. For more hints on outline preparation, see Section 2 on reports.

Write a Rough Draft

When you write, keep in mind the rules for good brochure copy given earlier in this section.

Do not discourage your reader with pages upon pages of unrelenting print. Break up your text with illustrations, lists, and other aids to readability. Relate the illustrations to the text. Refer to each illustration if possible. Use short, active-voice captions that distinguish your firm from the competition.

Check Your Facts

Once your rough draft is finished, make at least one copy and underline any facts that need to be verified. Do not rely on your own feelings about the accuracy of the data; many writers who felt sure of their facts have later found themselves to be wrong, often with disastrous results. Generally, you can save time and avoid embarrassment by checking facts yourself, even though the document may later be subject to a number of thorough reviews.

Submit the Draft for Review

A technical or scientific sales brochure is usually subject to some or all of the following reviews:

  Management review, to ensure that it reflects the desired company image
  Engineering review, to verify that it is technically accurate
  Marketing review, to ensure that it satisfies marketing objectives
  Editorial review, to verify that it is written and organized correctly
  Legal review, to make certain it does not contain any hidden liabilities, such as use of a client's proprietary data

Some of the reviewers will probably have comments. Listen to them. Then make the changes that they request. If necessary, double-check with the main editor or project director to settle any sticky questions.

Polish the Draft

Use the hints given in Sections 2, 3, 4, and 5 of this handbook. Then you'll be certain that your copy is as good as you can make it.

Send to Production

Don't skimp on production—but don't overdo production either. In general, you'll get the best results if you leave the final design and production decisions up to an experienced graphic designer, printer, and/or publicity agency. Customized inserts and fact sheets, on the other hand, can usually be produced in the office by a competent typist.

GENERAL STEPS IN BROCHURE DEVELOPMENT

Figure 9.3 shows the general steps in evolution of brochure copy.

Figure 9.4 is an example of a typical document control sheet, which may be used for both corporate brochures and other publications. In some firms, this type of control sheet is required to help the manager or brochure director keep track of the status of all functions. The control sheet accompanies the main document as it is circulated for review, editing, revision, and production.

CHECKLIST FOR BROCHURE WRITING

Capture Strategy

1. Have you set realistic goals for use of the brochure? Don't expect your readers to change basic buying habits. Don't try to sell your services or product through the brochure alone if a more complex sales campaign is required.
2. Does the brochure address the specific interests of your intended readership?
3. Will it capture the attention of prospective customers and make them want to read on?
4. Does it prove your understanding of client requirements?
5. Does it show the benefits of buying your firm's services or products rather than those of the competition?
6. Does it give sufficient evidence to show that your firm will live up to the promises it makes?
7. Does it use clear-cut requirements and desires of the prospective client in order to persuade him or her to take action (e.g., place an order, request a sales meeting, or ask for more information)?

Organization and Design

1. Does the overall organization of the material make it easy for the typical reader to use? Don't write one long, generalized brochure for all your clients if several, more specialized documents will be required. Can some of the information be presented more effectively in the form of a separate brochure or in customized fact sheets or other inserts?
2. Is the overall layout simple—i.e., will it promote readability while it also grabs attention? Avoid clutter.
3. Are graphics, captions, narrative, lists, charts, and other elements arranged to support your points as effectively as possible?
4. Do the graphics stimulate reader interest? Photographs generally capture more

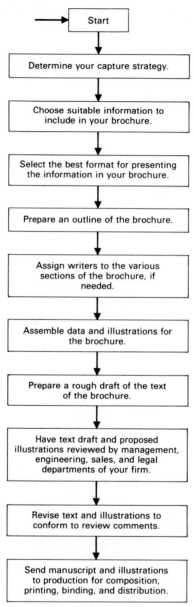

FIG. 9.3 Sequence in writing a typical sales brochure.

```
1. Document title _____

   Writing assignment number _____

2. Required publication date _____

3. Writing to start _____

4. Writing to finish _____

5. Copyediting to start _____

6. Copyediting to finish _____

7. Typesetting to start _____

8. Typesetting to finish _____

9. Remarks _____
   _____
   _____
   _____
```

FIG. 9.4 Typical document control form for preparing sales brochures and similar written items.

interest than other forms of graphics. People look at graphics, particularly photographs, before they read the text.

5. Have you used captions to introduce readers to your firm's main features and strengths? People drawn to the graphics read captions.
6. Do your major benefits (i.e., selling points) stand out? Use headers, special typefaces, large letters, white space, and contrast to increase the impact of the actual writing.
7. Have you checked with someone experienced in brochure design to discuss the layout, number of pages, binding, trim size, colors, type, etc.?

Structure and Content

1. Does the format follow the traditional structure for brochures presented earlier?
2. *Introduction:* Is it limited to a brief summary of information that will be interesting and useful to the typical reader (e.g., pertinent historical data, important design objectives)? Don't waste your readers' time with lengthy explanations of management philosophy or other peripheral matters. If a foreword is included, apply the same rules to it.

3. *Members of the Organization:* Are descriptions brief and limited to company principals and officers? Is pertinent experience summarized effectively?
4. *Services:* Are overall capabilities described, with special emphasis on those features that will be most important to your customers? Have you shown ways in which your firm is unique? Have you addressed not only the customer's typical requirements but also the benefits to be obtained through buying from your company? Is the information presented in such a way as to allow quick and easy reference to important details?
5. *Projects:* Have you summarized the performance record of your firm in a way that demonstrates its ability to meet the client's requirements? Are telling photos used? Is the description concise?
6. *Clients:* Are important clients highlighted to enhance your firm's overall experience profile?

Miscellaneous

1. Does the brochure project the desired company image?
2. Have all technical facts been checked for accuracy?
3. Is proprietary information and other sensitive data protected? Have approvals been obtained for use of special information from appropriate members of the firm?
4. Are data on your services or products summarized into short, concise statements?
5. Are special features highlighted by bullets or other graphic aids?
6. Are lengthy or voluminous data summarized in the form of an easy-to-read chart or list?
7. Is technical jargon avoided? Tell the story in direct, plain language.
8. Does the brochure tells the reader clearly the next step to take (see item 7 under "Capture Strategy")? Does it make responding easy?
9. Have all necessary cover data been included, e.g., brochure title and company name and address on the front cover with a list of branch offices and the company name, address, and telephone numbers on back cover?

SECTION 10
HOW TO WRITE BETTER CATALOGS AND ADVERTISING

If you do much technical writing, you're almost certain to run into industrial catalog and advertising jobs. Many of these will come to you by indirect routes. For instance, the sales department may find one of its catalogs doesn't give enough information. So the manager comes to you for more data. Within minutes you're telling the sales manager how the catalog should be written. Or worse yet, you're rewriting a paragraph for him.

With ads the approach is often different. You are asked to check an ad. Just the technical content, you know. Technically the ad is correct. But from the writing angle, it's for nuclear scientists—instead of the auto mechanics who'll be reading it. So you get out a pencil and begin making suggestions. From then on, every important ad the company publishes will wind up on your desk just before press time.

Writing catalogs, bulletins, and ads is a challenging task worthy of your writing skills. And this area of technical writing can be lucrative—if you're good. Let's take a quick look at how catalogs and ads differ from other kinds of technical writing. We'll look at catalogs first.

CATALOGS SHOULD INFORM USERS AND MOTIVATE SALES

To paraphrase a well-known catalog collection: *Buyers seeking sellers use catalogs; sellers seeking buyers use advertising.* Keep these two principles in mind and you'll write better industrial copy with far less pain.

Catalogs are used by many potential buyers. Every time a person reaches for a catalog, a sale hangs in the balance. If the catalog contains the information he or she wants, there's a good chance the sale will be made. If the catalog doesn't have the data, you're lost. For busy engineers and scientists want a quick answer. The catalog must give enough data for the reader to decide if he or she wants to know more about your product. A skimpy, poorly prepared catalog implies an inferior product.

WHAT IS A CATALOG?

The simplest answer is that a catalog is a basic source of information about products. It must tell the user what you have for sale. And it must help the user to decide if your product can help him or her. Anson A. MacLaren defines a catalog as *a form of advertising which wraps up in a single package the full details about a product or group of products to give information quickly and easily so a salesperson need not be called until an order is to be placed.*

Users refer to a catalog to get answers to basic questions *before* the salesperson is called. So you must write your catalog to give a user the answers he or she wants. How can you do this? It isn't easy, but it can be done.

PLAN YOUR CATALOG BEFORE YOU WRITE

To write effective sales-getting catalogs you must:

1. Decide the purpose of the catalog.
2. Determine who will use it.
3. Decide what information to include.
4. Choose the best way to present data.

Let's see how we can apply these methods to your catalogs.

Decide the Purpose

Why is this catalog being written? Ask yourself this question before you put a word on paper. Is this to be (a) a mail-order catalog, (b) a single-product catalog, (c) a distributor's or jobber's catalog, (d) a condensed catalog, or (e) a prefiled catalog? The answers to these questions will tell you the purpose of your catalog.

Most technical writers in industry today are called on to write the single-product catalog, condensed catalog, or prefiled catalog. So we'll give these the most attention.

Determine the Users

A catalog for an auto mechanic has a much different approach from one for a research scientist. Unless you know exactly who your readers will be, you can't decide what approach to use. So investigate your readers. What is their average education? Do they read regularly? What is their favorite reading matter? What do they want from a catalog? Prices? Design data? Dimensions? Weights? Only when you know the answers to these questions can you plan an effective catalog that will help sell your product.

Decide on the Information to Be Included

This ties back to your reader and his or her needs. Make a list of these needs. Then include answers to as many of these needs as you can. Don't overlook items like finish, color, materials, sizes, capacities, allowable loadings, speed, range, fuel consumption, models available, discounts, dealers' addresses, and phone numbers.

Choose the Best Way to Present Data

What is your reader accustomed to using? Charts? Tables? Circuit diagrams? Include those he or she is familiar with so that your catalog will be easier to use and more helpful.

Remember—the catalog indirectly reflects your company and its products. Write a serviceable catalog and the users will buy your product. Don't use extravagant colors, complicated foldouts, or cheap emotional appeals. If your reader wants these, he or she won't turn to a catalog but instead will buy a publication that provides these better than any catalog ever could. Catalogs are not designed for leisure reading—they're tools, just like a hammer or saw.

WRITING THE CATALOG

There are six steps in writing effective catalogs: (1) Collect your data and illustrations. (2) Make an outline. (3) Write a rough draft. (4) Check your facts. (5) Polish the rough draft. (6) Send copy and art to the production department.

Data and Illustrations

Get these from the engineering, sales, and advertising departments. Try to get the latest data and illustrations. For if you get old ones now, you'll have to change them before the catalog is published. Such changes lead to wasted effort, lost time, and misunderstandings. And don't overlook customers as potential sources of illustrations. Many of your customers will be glad to supply photos of your product in use. Or they'll allow you to visit their plant to take photos of your product in use.

Don't cut costs on illustrations. Get the best. Remember—you're after sales. The better the appearance of your catalog, the greater your chances of making the important sale.

Insist on being supplied *all* data related to the product. True, you may use only a tenth of it. But having the data on hand will help you write your text. This will show up in the greater assurance your writing reflects.

Make an Outline

You'll be lost without one. Study other catalogs to see how they're written. And don't overlook your competitors! Some of your best ideas may result from reading poorly executed catalogs. (Not your own, of course.)

Here is the outline of a well-prepared 12-page catalog on valves for high-pressure steam. Note the sequence of the various items:

Yarway Welbond Valves for High Pressure and Temperatures

1. Front cover (1 page)
   a. Title of catalog
   b. Three photos of valve
   c. Catalog table of contents
   d. Company name; address of home office; telephone number
   e. Catalog number; date of issue
2. Valve details (2 pages)
   a. Materials
   b. Packing
   c. Handwheels
   d. Seat, disk
   e. Body, yoke, nozzle
   f. Valve uses (over 12 listed and described)
   g. Special valve features—materials, seat, stem, packing, accessible parts
3. Valve selection (2 pages)
   a. Range of pressure and sizes
   b. Dimensions and weights
   c. Prices
   d. Installation and maintenance notes
   e. How to select proper valves
   f. Selection table
   g. Pressure rating chart
4. Typical installations (2 pages)
   a. Ten photos of typical installations, with captions
   b. List of 37 representative users of these valves
5. Valves for 1,500 lb/in^2 (1 page)
   a. Cross-section drawing
   b. Parts and materials list
   c. Text on typical uses, ways to order
   d. Dimension drawings (two views)
   e. Tabulation of valve figure number, dimensions, weights, stem rise
6. Valves for 2,500 lb/in^2 (1 page)
   a. Cross-section drawing
   b. Parts and materials list
   c. Text on typical uses, ways to order
   d. Dimension drawings (two views)
   e. Tabulation of valve figure number, dimensions, weights, stem rise
7. Valves for water-column shut-off (2 pages)
   a. Three views of valve (valve open, lock engaged; valve closed, lock disengaged; dimensions)
   b. Valve uses; code requirements
   c. Tabulation of weights, dimensions, figure number
   d. Two installation views and descriptive text
   e. Related equipment built by firm (two photos)
8. Back cover (1 page)
   a. Related equipment for high-pressure boilers
   b. Seatless valves (two views)
   c. Hard-seat valves (two views)
   d. Unit-tandem valves (two views)
   e. Descriptive text about each valve; numbers of catalogs covering each valve

f. Listing of other Yarway equipment
g. Company name and address
h. List of cities and countries in which sales representatives are located

When preparing your outline, try to visualize the information the typical user will be seeking. Arrange the outline so the needed information is given in the sequence the user will need it. Though there are no fixed rules, many engineers prefer to see the information presented in the following order: (a) catalog contents; (b) product uses; (c) materials of construction; (d) product features; (e) product ratings, capacities, loads, etc.; (f) selection procedure; (g) product dimensions, weights, color, finish, etc.; (h) product prices; (i) ordering procedure; (j) extract of a typical specification for ordering the product; (k) typical users; (l) sales representatives; (m) home-office information; and (n) other related products.

Study your outline and revise it until you're satisfied it's the best you can prepare. Indicate where you'll use illustrations. Specify which illustration will be used where. It's amazing how quickly you can forget which photo you intended to use on page 2, or page 19.

Write a Rough Draft of Your Catalog

When you write, keep these rules in mind. Be specific. Be concise. Be concrete. Leave the adjectives at home the day you write your catalog. Give enough information to the catalog user to invite further action on his or her part. Relate the illustrations to the text. Refer to each illustration if possible. Use short, specific captions for the illustrations.

Don't discourage your reader with pages of formidable text. Break up your text with illustrations, tabulations, and lists of applications of the product.

Be sure your reader knows exactly what the catalog covers. Keep the text short enough to encourage reading. But give the needed facts. Remember—catalog users are looking for a quick, exact answer. Forget the history of your company, how long the firm spent testing the product, and how many generations have used the product. The salesperson can supply this information if the customer wants it.

Make all detail drawings large. Be sure the user can read the dimensions. And give those important limiting dimensions—like how much clearance is needed for installation and how close two units can be placed. Don't overlook weights, colors, finishes, materials, and special considerations. They're extremely important to the user of your catalog.

Never overlook product sizes, capacities, figure numbers, models, and other variations. Chances are that your product will exactly fill the special needs of certain customers. If it does, you've got a sale.

And show illustrations of related products. But use judgment in this. Don't show auto tires in a catalog on radiator valves. There is hardly ever a customer who needs both at the same time. Instead, include items like radiators, radiator traps, and piping in a catalog on radiator valves. These are items the user might logically need, besides the valves which he or she was originally looking for in your catalog.

If at all possible, include examples of product selection. When it's necessary for your reader to compute capacity, size, rating, or some other factor, give him or her several worked-out examples showing how to make the computation. The catalogs that get the most use in engineering offices are those that save the user time.

Use tabulations, charts, and formulas, in that order, to present selection and product physical data. Arrange the data so that they can be read easily, without a need of squinting.

If your product has limitations—and almost all do—state exactly what the limitations are. Then your users will make fewer mistakes, and there will be less dissatisfaction with your product. Though catalog writing is less restrained than the style used in articles, papers, and books, watch the superlatives. They can get you into trouble.

Check Your Facts

Once your rough draft is finished, type it neatly and collect all your illustrations. Send the copy and art to the engineering and sales departments for checking. Never begin final production of a catalog until you've done this and received approval of the content.

Engineering and sales will probably have comments. Listen to them. Then make the changes they request. But don't let the sales department inject too many adjectives or superlatives into your copy. The back-slapping that gets into many catalogs annoys most engineers. Why? Because they are usually modest people who state the facts and go on from there. So be clear; be brief; give the facts. That's enough for most catalog users.

Polish the Rough Draft

Use the hints given elsewhere in this handbook. Then you'll be certain that your copy is as good as you can make it.

Send to Production

Don't skimp on catalog production. Use the best paper, printer, and binding your budget will stand. Remember—your catalog subtly reflects your company. So settle for only the best. You'll never regret it. In general, you'll get better results if you deal with an experienced advertising or publicity agency when producing catalogs. Good industrial agencies have long experience and can advise you what to do.

CATALOGS ARE IMPORTANT

Don't ever ignore catalogs. They're one of the strongest links between your firm and the purchasing public. And since most purchasers today are well-trained technical personnel, the importance of good catalogs will continue to grow.

In a survey of a committee of 100 prominent engineers, *Consulting Engineer* magazine made these findings:

> Somewhere between a half and two-thirds of the catalogs received in the mail or directly from salesmen are filed in the wastebasket. This is because too many catalogs

contain more sales story than pertinent technical data. The consulting engineer does not want the two combined.

Some manufacturers seem to have a tendency to include not only promotional material but also maintenance and operation manuals and perhaps histories of their companies or detailed reports of the preliminary research required for development of their product. Most of this is superfluous from the consulting engineer's point of view. The primary concern is preparing good specifications for the client.

Contents of a Good Catalog

A good catalog should contain, as a minimum: A brief description of the product, pointing out its special features or characteristics. Complete technical data including ratings and detailed dimensions. Clear statements as to the product's limitations as well as its proper applications. Prices wherever possible.

Generally speaking, consulting engineers prefer tabular data to charts, and prefer charts to formulas. Some charts are quite useful, but they are seldom as easy to read or as accurate as tabular data. Formulas should be avoided except when absolutely essential.

The Committee emphasized most strongly the need for adequate dimensional drawings. It seems that the majority of manufacturers fail to put all of the dimensions required on their product drawings, and they frequently forget to show required clearances. Nothing is more frustrating to the engineer than to find missing from the catalog drawing one of the dimensions needed.

The engineer also likes to have complete information on the physical and chemical properties of all materials and complete ratings on all equipment. This is particularly important in connection with new products.

One other interesting point from this important survey will be of use to you when you prepare catalogs:

Engineers, as might be expected, prefer catalogs in which specification data is referred to as "engineering data." They strongly object to bulletins in which all technical material is labeled "architects' data." The consulting engineer is particularly offended when the manufacturer addresses itself to architects in connection with products that are obviously within the specification area of engineers. However, the engineer will forgive this if the manufacturers will make a real effort to provide better technical information.

INDUSTRIAL ADVERTISING—A QUICK LOOK

As a technical writer, you'll seldom have to write an ad from scratch. Instead, your job will be *rewriting* ads. Sometimes this may mean changing only one word, comma, or colon. At other times you may have to recast whole sentences or paragraphs. But with the high proficiency of industrial ad agencies today, you'll seldom have to rewrite an entire ad. Hence our coverage of industrial advertising here will be extremely brief.

Purpose of Industrial Advertising

Study a few good industrial ads and you'll find that most aim at one or more of the following: (1) finding new markets for established products, or markets for new products; (2) helping spread news of product uses; (3) helping customers get better use from products already purchased; (4) inviting inquiries about new or established products; (5) helping keep the firm ahead of its competitors; and (6) reducing the number of complaints to the service department.

It's almost impossible for a single ad to serve all these purposes. So you'll find the range and scope of industrial advertising a constant challenge.

Almost every industrial ad is composed of three elements—copy, illustrations, and layout. By *copy* we mean the words—headline and text—in the ad. (At times *copy* is used to mean the entire ad—text, illustrations, and layout.) And since this is a handbook on writing, we'll neglect the mechanics of preparing an advertising layout, for the art director has more responsibility for the layout than the copywriter.

SIX RULES FOR WRITING GOOD COPY

Industrial advertising is big business. *Business Week* says, "There are about 7,000 periodicals published in the United States. Of these, more than 2,000 are business papers, a group that is thriving as American industry expands." So don't ever slight an industrial ad. It deserves every bit of your skill, ingenuity, and effort. Good advertising has been one of the main forces behind the rise of many outstanding industrial firms in the United States today.

There is no magic formula for writing good copy. But the following rules, if followed sensibly, should help you turn out good copy.

1. Know Your Readers

You can't write convincing ad copy unless you know who will be reading it. So before you start to write, find out where the ad will run. Then write your copy to fit the readers.

The copy you write for *Business Week* certainly would not be suitable for a technical ad in *Water & Sewage Works*. These two magazines have two different kinds of readers, with different interests. If you want readership and action from your ad, you must pitch it to the reader's job interests.

Where can you learn who reads a given publication? There are a number of sources of such information. The publication itself will supply a comprehensive analysis. Standard Rate and Data Service is another excellent source of readership data. And your ad agency (if you're employed by an industrial firm) is a reliable source of information about typical readers, their interests, needs, and preferences.

Study the readership data you obtain. Try to form an accurate mental picture of the man or woman you'll aim your ad at. If you can't seem to visualize your typical reader, get out and meet as many actual readers as you can. Visit them on the job. Make notes on how they talk, what their problems are, the atmosphere in which they work. Go back to your office and study your notes. Your typical reader will come to life. Then write your copy for him or her.

2. Make Your Ad Say Something

Readers of business magazines are busy people. But at the same time they're anxious to learn. If your ad speaks the reader's language and tells him or her something that will help on the job, he or she will read your copy. But if your ad lacks facts, it won't be read. It isn't worth the reader's time.

So before you start to write, have something to say. If you don't know what to say, stop and find out. Empty ads waste money, magazine space, and valuable time.

What will your ad say? Any of a thousand things. But if you've read many industrial ads, you'll find most stress an advantage of some kind. Typical advantages popular today are (a) saving of time, money, effort, space, weight, labor, etc., and (b) better product quality, service, life, reliability, and performance. Here are typical ad headlines which say different but specific things:

Reduced voltage starters...easy on the motor...easy on the line.

Solve difficult word processing problems.

Silicone-insulated motors help keep production moving.

700 percent longer service with these Crane valves.

Maintaining *maximum* flow in serum lines with *minimum* pressure drop.

New PC system steps up capacity.

Again: Make your ad say something. You'll get better readership, action. If you don't know what to say, find out. Keep in mind that the usual reader reads his or her magazine *at* work. Say something he or she wants to hear. This will ensure the readership you deserve.

3. Keep Your Message Simple

Few of your readers are Ph.D.s. Most are working people, struggling to do the best job possible. Speak in terms they can understand at a glance, like these:

Microcomputers cut costs.

These Reliance meters will not corrode.

Underwater, underground—this hose loves its salt diet.

Emphasize the product. There's no doubt what the three ads above cover—microcomputers, motors, hose. If the reader is interested in any of these, you can be sure your ad will be read.

Don't deal in complicated symbols. If you're writing an ad about industrial hoses, talk about hoses—not symbols like fluid highways and liquid conduits. These are one step removed from hoses. And your reader wants information on hoses—regardless of how prosaic this word may seem.

So decide what you want to say—then say it in as simple and as direct a manner as possible. And be sure your message relates to the reader's job. If he or she wants to read about golf clubs or skin-diving equipment, he or she will buy a magazine covering these. In a business magazine he or she wants, and expects to get, business information.

Make your headline forceful, interesting, attention-getting. Don't go highbrow—stick to the simple message, like:

Now, a PC giant in capacity...yet moderate in size

Pumping molten chemicals? Write Taber.

A package for pollution control

4. Tell Your Story with Pictures

Show your reader what you mean. Give a photo of the unit, a chart to compute capacity, a cutaway drawing showing how the device works. Pictures get your story across to the reader faster, save words, attract more attention. And if you're using photos, try to include installation views of the product. Engineers and scientists prefer installation shots. There's much to be learned from how the other person piped up a unit or braced a foundation. Avoid the symbol type of photo—dogs, cats, lions, seals, etc. They mean little. Instead, show your product. That's what your reader is interested in seeing.

Next to photos, charts and graphs are most popular. Try to use a chart or graph that gives information of value to the reader. A pretty but unusable chart means little. Present a chart or graph and show the reader how it will solve a problem for *him or her.*

Cutaway views have many applications—machines, buildings, ships, planes, and various equipment. But don't clutter the cutaway with type that obstructs the construction details. Never forget that your readers are scientifically trained. They want to learn how the other designer, engineer, chemist, or geologist solved a problem. Show them—use cutaways.

Steer clear of frills, scrolls, and all other fancy designs around the illustrations in your ad. Use a simple rectangular or square photo, drawing, or chart. The scientific mind is geometric-conscious. So leave the frills for consumer ads and concentrate on the simple illustration that shows your product clearly.

5. Be Helpful—It Never Hurts

Your readers want help with the problems they meet on the job. If you help, they'll buy your product. Take a look at a few ads stressing help (only the headline and first paragraphs are given):

Why Celanese Chose These Mixers for Low-Pressure Polyethylene
  How can mechanical mixers help you give the touch of success to an important new process?
    Celanese Corporation of America faced this question when its Plastics Division designed a plant to produce 100,000 lb/day of....

Here is another ad designed to help the reader solve his or her problems:

B. F. Goodrich report:
  No holes, no leaks, no repairs—liquid alum stored safely in rubber.
  Problem: This paper manufacturer wanted to cut costs by switching from powdered to liquid alum in their pulp mixture. But getting a tank to hold this corrosive solution was a problem. Wood tanks shrink, often leak. Lead-lined tanks require frequent repairs.
  What was done: B. F. Goodrich engineers recommended Triflex rubber-lining for the two 5,000-gallon steel storage tanks....

Aiming at the major problems of many engineers helps get your message across:

  Get product uniformity faster and at less cost...with NETTCO Engineered Agitation.

  Increase productivity, lower power costs, and minimize maintenance requirements...with process-rated Model WT minicomputers.

When you offer help, be sure your claims for the product are fair. Don't exaggerate! You'll only cause more grief and trouble. State what your product will do. Then stop.

6. Keep Your Ad Professional

Speak the language of your readers. Don't introduce unnecessary chatter or unrelated information. State your claims, and document each claim with simple, unassailable facts. Then your readers will be more likely to ask for more information. This can eventually lead to the sale of the product or service you're aiming at.

INDUSTRIAL AD HEADLINES

If you're going to catch a reader, you'll probably do it with an interesting headline. For almost everything in an ad depends on an effective headline.

Two rules will help you write better headlines: (1) Strive for complete clarity in the headline. (2) Tie the headline in with the illustrations in the ad.

Clarity is of utmost importance because business-paper readers are busy people. They haven't time to spend trying to decipher a trick headline. So state your facts clearly and interestingly, the way these examples do:

> Modern, efficient NINEMILE POINT plant uses modern, efficient turbine oil: GULFCREST.
>
> "Big Station" combustion controls pay off for boiler plant.
>
> The blow-off valve trend on "package" boilers is YARWAY SEATLESS.

Tying your headline in with the illustrations adds to the effectiveness of both. In the first of the three ads above, two views of the plant named are shown. In the second ad three views of boilers and their control are shown. The third ad features seven illustrations of the product mentioned in the headline. Make your headline and illustrations work together. Your ad will have greater unity, more punch, and fuller meaning.

SECTION 11
WRITING TECHNICAL AND SCIENTIFIC BOOKS

In Section 1 you learned the many advantages you secure from writing good technical material. Review these advantages now by reading Chapter 1 again. Pay particular attention to the information on books. The technical book, since it is far longer than a technical article, is a much more difficult writing task. But the rewards are in proportion to the extra work required. A good technical book has a long life, contributes much to the knowledge of your field, and can be an important factor in your personal finances. Also, the prestige you acquire from a well-written book can do much to further your career.

KINDS OF TECHNICAL BOOKS

What kind of books do engineers, scientists, and technicians write? There are many types, but they can be roughly classed in seven categories: (1) home-study texts, (2) technical training texts, (3) college texts, (4) industrial reference books, (5) advanced engineering books, (6) engineering monographs, and (7) handbooks. Let's take a look at each type.

1. Home-Study Texts

These are designed for use by men and women who wish to increase their knowledge by spare-time study. Since the student usually has no instructor other than the book, you must be extremely careful in this type of writing. You must put yourself in the reader's place and try to visualize what questions will occur to him or her as he or she uses your text. These questions should be answered in the explanatory portion of your book.

Two prime requirements of the home-study type of book are (a) clear writing and (b) enough illustrative and study problems to allow the reader to get a firm grasp of the subject matter. Of course clear, easily understood writing is essential in every technical book, but as an author of the home-study type, you must take extra precautions to make yourself understood. If you do not, you may find yourself swamped by letters from users of your book!

Illustrative problems show how to perform the steps described in the text of your book. Study problems give the reader opportunity for practice and self-testing. You should include the answers to the study problems so that the reader can check his or her work. Avoid trick problems because your reader has no one to turn to when he or she meets a problem that is confusing or misleading.

To get a better concept of this type of book, refer to the two following examples: W. S. LaLonde, *Professional Engineers' Examination Questions and Answers*, McGraw-Hill Book Company, New York, and T. C. Power, *Practical Shop Mathematics*, McGraw-Hill Book Company, New York.

2. Technical Training Texts

These may or may not be more advanced in subject matter than books written for home-study use. The usual text of this type is prepared for study by candidates for training for technician-level jobs in industry. These are usually high school graduates having the fundamental training in physics and chemistry given in secondary schools. So you can use a higher-level approach in the technical training text than you can in the usual home-study text.

Good technical training texts are characterized by clear, simple writing, combined with careful matching of the text and illustrations. These characteristics are desirable in every technical book; however, the extra effort you make to produce an outstanding technical training text is appreciated by all your readers.

Two examples of technical training texts are *Automotive Engines*, W. H. Crouse, McGraw-Hill Book Company, and *Basic Television*, Bernard Grob, McGraw-Hill Book Company.

3. College Texts

These are most frequently written by teachers in engineering schools and by part-time faculty members. Relatively few engineers and technicians in industry can write a text specifically for use in colleges. But some of the books written by engineers in industry are adopted as college texts. Most books of this type that are adopted are also of interest to engineers in the field and are usually written for them. College adoption comes later. Keep these facts in mind when you think about writing a text for college use.

Occasionally college texts are coauthored by an engineer on the teaching staff and an engineer in industry. Each contributes his or her particular talent and knowledge to produce a well-balanced coverage. Collaborating with a teacher is probably the best way for you to author a college text if you are not a faculty member yourself.

Two typical examples of outstanding college texts are *Engineering Drawing*, T. E. French and C. J. Vierck, McGraw-Hill Book Company, and *Water Supply and Sewerage*, E. W. Steel, McGraw-Hill Book Company.

4. Industrial Reference Books

These are written primarily for engineers, scientists, and technicians working in a particular field. They may also be used by students, salespersons, technicians, and others interested in that field. If you study a number of industrial reference books, you will see that this type puts somewhat more emphasis on applications than on theory. This is because the person in the field is most interested in solution of problems related to his or her job. Theoretical discussions are of little use unless he or she can see some relation between them and daily work.

Two examples of industrial reference books are *Pump Application Engineering,* T. G. Hicks and T. Edwards, McGraw-Hill Book Company, and *Water Treatment for HVAC and Potable Water Systems,* Richard Blake McGraw-Hill Book Company.

5. Advanced Engineering Books

These differ from industrial reference books in a number of ways. In general, the advanced engineering book is more theoretical. It is directed at a higher audience level and is likely to have a mathematical approach. The subject matter generally deals more with design and engineering considerations than with the problems of installation, operation, and maintenance. There are, of course, some variations in these characteristics, but they are typical of many advanced engineering books today. Keep these in mind when you think of writing an advanced engineering book.

Two typical examples of this group are *Solar Heating Design Process,* by J. F. Kreider, McGraw-Hill Book Company, and *Modern MOS Technology,* by D. G. Ong, McGraw-Hill Book Company.

6. Engineering Monographs

These books usually cover a limited area of a particular field. Instead of giving a broad, limited-depth view of a field, as some industrial reference books do, the usual engineering monograph delves deeply, but over a restricted area. A mathematical approach is commonly used. The mathematics may be more rigorous than in an advanced engineering book.

The Engineering Society Monographs, sponsored by AIEE, ASCE, AIME, and ASME, are probably the best examples of books of this type. Two titles available today in this series are *Hydraulic Transients,* G. R. Rich, McGraw-Hill Book Company, 1951, and *Theory of Elasticity,* S.P. Timoshenk and J. N. Goodier, McGraw-Hill Book Company. The engineering monograph has a long life because its approach is basic.

7. Handbooks

These are not "written" in the same sense as most other technical books are. Instead, there are a number of contributors, often more than 50, each of whom writes a portion or section of the book. These sections are submitted to the editor of the handbook, who works the material into a unified coverage of the subject matter. To achieve this, the editor may restyle some of the material submitted but he or she never alters its factual content. When the handbook editor is a specialist in some phase of a field, as is often the case, he or she may write one or more sections of the handbook.

In many ways, editing a handbook is the ultimate in technical authorship. You must have long experience in your profession and a high degree of writing skill before you can qualify for handbook editorship. A wide acquaintance with the outstanding members of the profession covered by the handbook is another necessity if a large number of contributors is contemplated.

Two examples of modern engineering handbooks are T. Baumeister, *Marks' Standard Handbook for Mechanical Engineers,* McGraw-Hill Book Company, and *Chemical Engineers' Handbook,* John Perry, McGraw-Hill Book Company. Most engineering handbooks have a long life because their approach is fundamental.

WHO USES BOOKS

Though a rough classification of typical books written by engineers, scientists, and technicians is easily set up, as we have done here, it is extremely difficult to predict the final users of a given book. For example a home-study text might find use in technical training and college courses. Many handbooks are used in college courses. Industrial reference books, advanced engineering books, and monographs are used up and down the scale of readership. Book use depends on the needs of the user.

So you should regard the classification given above merely as a device to aid you in understanding the range and scope of technical books written today. You can control what goes into your book, but that's where your jurisdiction ends. Users are free to choose books that appeal to them, and the variety of their choice is often surprising.

WHAT MAKES A GOOD BOOK

Deciding to write a technical book can be one of the most important decisions in your career, for a well-written book can, to repeat, bring you prestige, recognition, a feeling of accomplishment, and financial returns of generous proportions. Not only that, a good technical book is a permanent contribution to the literature of your profession. As such, it stands for all to see—the bench mark of one engineer or scientist who had the energy and ability to do something to help his fellows in the daily problems of engineering or science.

But such returns are not easily won. Writing a good technical book is a challenging and arduous task. It takes enthusiasm, skill, energy, and a sincere desire to help others. That it can be done is proven every time a good book is published. Now let's take a look at the factors that go into the writing of an outstanding technical book.

An outstanding technical book usually meets all of the following five requirements: (1) It answers the specific questions of its intended readers. (2) It takes a comprehensive view of the subject covered. (3) It is scrupulously accurate in text and illustration. (4) It takes progressive steps from the known to the unknown—from the reader's viewpoint. (5) It is written in a clear, simple, grammatical manner, organized to present the subject matter logically and effectively. An outstanding technical book is not only a work of science; it also approaches a work of art. The following paragraphs cover many of the important problems you meet in writing a technical book of any kind.

WORDS OF CAUTION

Before you decide to write any technical book, consider the implications of your decision. Experience indicates that the average author of a technical book spends at least one year writing it. Why? Consider this. Most technical books have between 10 and 20 chapters. You will find that if you write your book in your spare time that the average chapter requires about one month. So you can see that with 10 or 15 chapters you will spend approximately one year actually writing your book.

In terms of words and illustrations, the average technical book presents a major challenge to any writer. For example, a book containing 300 published pages will have 50,000 to 100,000 words, depending upon the number of illustrations. The usual technical book published today contains about one illustration per page. In a 300-page book you are likely to find between 200 and 300 illustrations.

Since you must either draw or obtain each illustration from an outside source, it is easy to see that you can spend much time just securing illustrations. If you type your manuscript so that it has an average of 12 words per line, which is usually the case, you will find that you must have between 4,000 and 8,000 lines of typewritten material for the text of your book. And since the usual double-spaced typewritten page contains 25 lines, your manuscript will consist of somewhere between 160 and 400 typewritten pages. Many authors find that just the physical preparation of their manuscript is a major undertaking.

Besides the physical preparation of the text and illustrations, the writer of a technical book must have the necessary know-how. Since it is impossible for you to know everything about the subject matter of your book, you will probably have to do some research. This, of course, takes time. And you will have to secure permission from publishers to use copyrighted material assembled in the course of your research. To secure such permission, you must write the copyright owner. This, too, takes time.

Besides all the work that you put into the book, you must be ready to accept and use the criticisms of the reviewers of your manuscript. Some reviewers are extremely critical of the material they read. While such criticisms may upset the new author, experienced authors learn to disregard the bite in the critic's remarks. The experienced writer takes all the critic's worthwhile points and tries to address them in the manuscript. Some authors find it difficult to learn to do this without becoming upset.

HOW YOU CAN WRITE YOUR BOOK

There are four ways of writing your book. These are: (1) as the sole author, (2) with one coauthor, (3) with more than one coauthor, and (4) with contributors. Let's take a quick look at each of these schemes.

One Author

Here you write the entire book yourself. You may secure help from others, but the major writing task is yours. You must also secure the illustrations yourself, although your associates may be kind enough to supply some for your book.

When writing a book by yourself, you must meet and solve a large number of minor problems. The first problem you run into when thinking of a book is: What type shall it be? Shall it be a home-study text, shall it be an industrial reference book, or shall it be a textbook? Although a publisher can give you a little help with this decision, you will find that you must take most of the responsibility yourself. Additional decisions that you will have to make during the writing of your book include such things as chapter length, the number and types of illustrations to use, and the style of writing. As you can see, all these decisions require time and energy.

Although the effort required in writing a book by yourself is a major task, there is the consoling thought that the returns come to you alone. You acquire any prestige that the book commands; you receive the royalties from the sale of the book. With any other arrangement you must share both the prestige and (usually) the royalties.

Before deciding to author a book by yourself, think carefully. You should have had some previous writing experience—the more the better. You should be able to construct a good outline, and you should be willing to follow the advice of books like the one you are reading, as well as the advice that your publisher may offer you. While the strong individualist may appear to be a romantic character, the successful technical author today is the man or woman who is willing to listen to and take advice from others in the publishing business.

Two Coauthors

In this arrangement you work with one other person in producing your book. The exact scheme used depends upon the individual situation. In many cases the work will be split equally between the two authors. For example, one will write the first half of the book while the second writes the other half. Each secures his or her own illustrations. Another scheme teams an author having a great amount of practical know-how but relatively little writing talent with an experienced writer with less technical knowledge. Here the experienced person gains the advantage of having a well-qualified writer prepare the material.

For two authors, coauthorship can be an extremely successful undertaking. But the temperament of one author must mesh with that of the other. Otherwise you may find that one of the coauthors is doing much more than his or her share of the work. This can lead to misunderstanding and difficulty.

Often two authors share the royalties from the book equally. Other arrangements, of course, are possible. For example, one author may receive 20 percent and the other 80 percent of the royalties. The exact split does not concern the publisher. But every publisher is interested in seeing that both coauthors are satisfied with any arrangement that is decided upon.

If you are considering authoring a book with someone else, the best advice that you can be given is to select your coauthor with as much care as you would a wife or a husband. Some engineers and scientists overlook the important personal aspects of coauthorship. In forgetting that their coauthor is a human being, they fail to anticipate the problems that may arise. So before becoming a coauthor or asking someone else to become a coauthor, think carefully. Once you decide to coauthor a book with someone, make it perfectly clear exactly what you expect of him or her and what you will contribute.

To obtain the maximum benefit from a dual-authorship arrangement, you should set up a schedule of deadlines for each coauthor. Arrange and understand how your correspondence or other communications will be carried out. Keep informed at all times of the progress of your coauthor and be careful to inform him or her of exactly what you have accomplished. If two people are completely honest with each other, joint authorship can be an extremely pleasant and rewarding experience.

More than Two Authors

Much the same arrangement is used with three or more authors as with two. But instead of the work's being divided between two people, three or more do the work. Once again each of the authors must be carefully chosen, and definite arrangements must be made for the amount of work to be done and the manner of communication between the authors. If you carefully plan how the work will be divided and how the communication will be carried on, a major task in the writing of your book will be solved.

Here again the personalities of the authors involved are crucial. They must all have a common aim. If they do not, the book may turn out to be a failure. If possible, one of the authors should assume the responsibility of seeing that the other authors deliver their material on time and that it is of satisfactory quality. Often the person who originates the idea for the book fulfills this function.

Contributors

The contributed book is usually a handbook. As we said above, each contributor writes a section or chapter. The section assigned any contributor is one with which he or she is completely familiar. A chief editor correlates the work of all the contributors. Instead of sharing in the royalties, as do the coauthors under the two previous arrangements, a contributor is usually paid a flat fee for his or her work by the editor. This fee is based on the number of published pages in the contributor's section.

Reference books and text books are also sometimes written by a number of contributors. The plan is the same as for a handbook, but on a smaller scale. All the recommendations made above in connection with coauthorship apply to the contributed book, whether handbook, textbook, or reference book.

QUALIFICATIONS FOR TECHNICAL BOOK AUTHORSHIP

No two authors of technical books are exactly the same, of course. However, they all have some characteristics in common. These are: (1) thorough knowledge of a specific subject, (2) a strong desire to help or inform others about their subject, (3) ability to write clearly, and (4) a sense of organization—that is, the ability to see and plan a major project in an effective and logical way.

No doubt many books have been written by authors who lack one or more of these characteristics. But the outstanding technical author usually has at least three of them. So before you decide to write a technical book, give some thought to your own characteristics. If you find that you more or less fill the bill, you can be reasonably sure of being successful.

It may seem strange that in this section concerning technical authorship there are many remarks of a personal nature. But, after all, the writing of a book is much more than just putting facts on paper. Though the task requires primarily the use of your talents and abilities, your entire personality becomes involved. Unless you have the needed temperament as well as the needed skills, you may find it is impossible to complete your book. And since failure to complete a book is always a major disappointment, it is best that you have the complete situation in your mind before deciding to go ahead with a technical book.

FINDING IDEAS FOR TECHNICAL BOOKS

Finding an idea that can be developed into an outstanding technical book is difficult. An idea for a book usually develops slowly. Article ideas and technical paper ideas often come quickly. With a little experience you should be able to develop an article idea within a matter of hours after you first think of it. But it may take you many months or even years to develop the ideas for an outstanding technical book. Let's see how some book ideas are found and developed.

Probably the most common incentives for the writing of engineering and scientific books are (1) the discovery that there are no books covering a particular subject and (2) the realization that the existing books in a field are outdated.

Either of these is sufficient to start you thinking about writing a book on a given subject. But before starting to write a book for either of these reasons, give the project careful thought. Also, read the remainder of this section.

Book ideas sometimes develop from other types of writing you may do. For example, if you write a series of articles on a given subject, you may suddenly discover that you have the nucleus of a book. Or if you prepare several technical papers on your specialty, you may find that they are suitable for use as the basis of a technical book. Or you may write lectures, class notes, catalogs, bulletins, or reports that can be worked into a book.

When considering any of these sources of book ideas, you must remember that a good technical book is far more comprehensive than an article, technical paper, lecture, catalog, report, or bulletin. A technical book has far greater depth of coverage and scope of subject matter than the usual short technical piece.

A technical book, to be of real value to its readers, must cover a fairly wide portion of its subject. This is important. While the subject itself may be a narrow one—for example, the subject of pumps—you must cover a sufficient number of types of pumps to make the book of use to a wide range of readers. Some beginning writers make the mistake of limiting the scope of coverage of a book. This reduces the potential audience for the book and may make the cost of publication almost prohibitive.

Let's assume you have an idea for a book—the first book you've attempted. What do you do next? Here is one procedure that has worked successfully for a large number of technical authors throughout the world. Study this procedure carefully and see if you can apply it to your idea. With only slight alterations it should be usable for any of the seven types of books we discussed in the previous section.

DEVELOPING BOOK IDEAS

The first step to take after you get an idea for a book is to classify the book under one of the headings in the previous chapter. Unless you know what type of book you are trying to write, it will be extremely difficult to do a good job.

So consider your idea carefully. Shall your book be a home-study text? Shall it be an industrial reference book? Shall it be a college text? Or shall it be an advanced engineering text? Decide which category your book will fill.

Once you have decided the type of book you plan to write, your next step is to secure from four to six representative examples of this type. You can secure these from a good library or you can purchase them. Read each of them carefully. Note the style that each author uses in his or her writing. Observe how many chapters the author uses. Study the subheads, paragraphing, tabulations, illustrative examples, and student exercise material in each of the books.

The idea in studying examples in this way is not to imitate the works of others. Rather, it is to stimulate your own ideas. Once you have written a technical book, it will not be necessary for you to study the work of others. But for your first book it is wise for you to learn as much as possible about how other people work. Never overlook this step in developing a book idea. You can learn a great deal from the work of other authors.

Note that the books you study need not have the same subject matter as you plan to cover in your book. But if there is a book on the same or a similar subject as yours, there is no harm in seeing how the author arranged his or her subject matter. While studying these books, you should also observe the number and type of illustrations the author used, and their sources.

In your examination of books you will find some authors whose work you admire. Other books may not appeal to you. Discard the books that you feel are not up to the level that you would like to attain. Note that you are not examining the books for subject matter; you are attempting to learn the way a book is put together from the standpoint of chapters, paragraphs, illustrations, etc.

You will find that most well-written technical books follow a logical line of thought from the known to the unknown. You will also observe that the writer is extremely careful to consider his or her readers with every word he or she chooses, every sentence he or she writes, and every chapter he or she develops. So do not skip this extremely important phase of developing your idea into an outstanding book.

YOU MUST OUTLINE

Never attempt to write a technical book without a comprehensive outline. While some books have been written without an outline, the resulting product is seldom outstanding. The shortest road to misery and pain in writing a technical book is to attempt the task without an outline.

Even experienced writers sometimes have difficulty writing an article without an outline; writing a first book without one will almost surely prove an unmanageable task.

You must remember that just the physical size of a book in terms of manuscript pages and illustrations is enormous compared with any other writing you may have done. This massive bulk can quickly go astray unless you have it under adequate control. And your best control is a good outline.

There are two steps in outlining a book. These are: (1) Prepare a chapter outline. (2) Prepare an outline of each chapter in the book.

Think of the first step as giving you an overall view of your entire book. Once you have prepared this outline, you will be able to judge if the balance of the various phases of the subject matter is suitable for your readers. If your plan devotes too much space to one phase of your subject, this will probably be apparent in the overall chapter outline.

While preparing the overall chapter outline, note alongside the name of each chapter how many printed pages you would like to devote to that chapter. "But," you say, "I do not know how many pages that chapter should have." If you don't know, guess. Ask yourself how important this chapter is to the reader and to the subject matter you cover. If the chapter is extremely important, it deserves more space than a less important chapter. If you are introducing difficult concepts, you are likely to need more space than if you are reviewing some material that the reader is probably familiar with.

As you recall, earlier we noted that writing a book requires that you make a series of small decisions. This overall chapter outline and the selection of the number of pages for each chapter is an excellent example of the decisions that will constantly face you while you are writing your book. If you do not start to make the decisions early in your book, you will have an extremely difficult time with the text.

Once you have finished your overall chapter outline, you should begin your individual chapter outline. As you proceed with this outline, you will probably find it necessary to change the number of published pages you tentatively assigned to certain chapters in the overall chapter outline. Do not be afraid to change the numbers. When you are finished with your individual chapter outline, you will go back and add up the various page numbers. For the usual technical book published by a commercial publisher today, the total number of printed pages should run somewhere between 100 and 400.

While shorter technical books are occasionally published today, and also longer ones, your first effort should be within the range that publishers prefer. Today this is 100 to 400 printed pages.

When preparing your individual chapter outline, it is wise to make it as specific as possible. If you skip work now, you will have more to do later. But if you make a complete outline now, you will be free later to concentrate on writing. You will be free of the burden of worry lest you have left out something important. And consequently you will write a better book.

In a word, the outline is designed to save you work. So do not regard it as a nuisance. Actually, a good outline allows you to write a better book in shorter time with less effort. Of course you will probably make some minor changes in your outline as you write your book. This should cause you no concern. But if you feel you must make major changes in your outline, you should stop writing and review the entire project, because this usually indicates that there is an unbalance in the division of your subject matter. Do not hesitate to revamp your outline if necessary. The effort required is usually slight compared to rewriting a major portion of your book.

TYPICAL OUTLINES

Two typical outlines of published books are given below. The first is for an industrial reference book and the second is for an engineering handbook. While neither of these outlines could be adapted to a book you might write, they are both extremely useful for study. They illustrate many points we have discussed in the text of this chapter.

The subject matter of the industrial reference book was such that it could be divided into three major sections. Thus when studying the overall chapter outline, you will note that the book is divided into three parts. This was done because dividing the book into parts made it easier to achieve a unity within each part. This is extremely important, and you should consider the possibility of dividing your book into parts. Not only does the division into parts contribute to unity; you will find that the divisions aid your thinking and act as inventory-taking points during your writing.

Since writing a technical book is a task involving a large number of words and illustrations, you can often help yourself by setting up a series of small deadlines to act as mileposts. Both part and chapter divisions provide such mileposts for you. Reaching any milepost gives you a sense of accomplishment. It also means that you are nearer the end of your task. And don't think you won't be glad when you come to the end of the book, no matter how much of a pleasure it has been to write it.

Reference Book Outline

The following outline was prepared before any writing was done on the book. Note that the estimated number of printed pages in the overall chapter outline was 400. The actual number of published pages was 432. This illustrates how closely you can control the published length of a book if you prepare an accurate and well-organized outline.

Overall Chapter Outline
Pump Selection and Application

Part 1 *Pump Classes and Types*	
Chapter 1. Centrifugal Pumps	24 pages
Chapter 2. Rotary Pumps	7 pages
Chapter 3. Reciprocating Pumps	28 pages
Part 2 *Factors in Pump Selection*	
Chapter 4. Head on a Pump	25 pages
Chapter 5. Pump Capacity	29 pages
Chapter 6. Liquid Handled	31 pages
Chapter 7. Piping Systems	48 pages
Chapter 8. Drives for Pumps	20 pages
Chapter 9. Pump Selection	14 pages
Chapter 10. System Economics	25 pages
Part 3 *Pump Application*	
Chapter 11. Power Services	15 pages
Chapter 12. Nuclear Energy	10 pages
Chapter 13. Petroleum Industry	12 pages
Chapter 14. Chemical Industries	15 pages
Chapter 15. Paper, Textiles	11 pages
Chapter 16. Food Processing	9 pages
Chapter 17. Water Supply	14 pages
Chapter 18. Sewage and Sump	10 pages
Chapter 19. Air Conditioning	10 pages
Chapter 20. Irrigation	9 pages
Chapter 21. Mining, Construction	8 pages
Chapter 22. Marine Services	11 pages
Chapter 23. Hydraulic Services	10 pages
Chapter 24. Iron, Steel	5 pages
Total page count	400 pages

This outline does not contain any information about the number of illustrations to be used in the book. As you recall, however, earlier we noted that many modern technical books have about one illustration per page. This 432-page book has 412 illustrations. The author of this book planned to use about an illustration per page, and the results were reasonably accurate.

To learn what the usual preface of a book contains, read several technical books. You will find that the good preface tells *who* you've written the book for, *why* you've written it, and *what* the reader will get from your book. And the preface is the place where you thank anyone who has helped you with your book.

Handbook Outline

The following outline is for a major engineering handbook. As you can see, the overall section outline of this handbook is relatively short. The individual chapter outlines are quite comprehensive and too detailed to include here. Study this outline and note how good a concept of the finished book you can acquire from it. This outline proved itself to be an excellent device to control the writing of the book.

Overall Section Outline
The Standard Handbook of Lubrication Engineering

PREFACE: The preface theme will serve as an introduction to the lubrication-engineering profession. There will be a preview of the handbook's scope, plus a general discussion of the four-part structural outline:

Part   I. Lubrication Principles
Part II. General Lubrication Engineering Practice
Part III. Lubrication of Specific Equipment
Part IV. Lubrication in Specific Industries

GET YOURSELF UNDER CONTRACT

Once you are confident that your outline is satisfactory, the next step toward publication of your book is to secure a contract from the publisher of your choice. The contract, also called a *Memorandum of Agreement,* is a document drawn up between publisher and author. What the document does essentially is to set up a partnership between the author and the publisher.

By the terms of the contract, you as author agree to supply a complete manuscript that is technically accurate and in good physical condition. Most contracts for technical books set a date when you must deliver your manuscript. Many publishers like to receive the manuscript within about one year after the contract is signed.

The publisher, as one of the two partners to the agreement, promises to manufacture your manuscript in book form, to promote the book through the normal distribution facilities, and to pay you royalties on all copies that are sold. So the contract adds up to this partnership: You supply your technical know-how and the completed manuscript, along with the illustrations to be used in the book. Also, you secure all necessary permissions for use of copyrighted material that appears in the book. The publisher, on his or her side, supplies the manufacturing facilities, editing skill, design talent, sales facilities, and the financial investment required to produce and market your book. No financial investment of any kind is necessary on your part if your book is published by a reputable publisher. Of course, the publisher cannot assume any of the costs that you incur in the preparation of your manuscript. These costs include items like typing, securing illustrations, or travel to sites important to your book.

The advantage to you of being under contract is that you are assured that your book will be published if it is of suitable quality. Never attempt to write an entire technical book without first getting yourself under contract to a suitable publisher. For after you get the entire book written, you may find that it is unpublishable. Or you may find that extensive revisions are necessary before publication is possible. So every experienced technical writer gets himself or herself under contract just as soon as he or she can after he or she conceives of an idea for a book. How do you get a contract? Let's see.

Most book publishers today require three items from you before they will consider your project. These are: (1) a detailed outline of each chapter in the book, as well as an overall outline covering all the chapters in the book, (2) two chapters in finished form as they will appear in the book, and (3) a preface in which you tell whom you have written the book for, why you have written the book, and what the reader will get from the book. With this material in hand the publisher can go about making the decision as to whether he or she wishes to go ahead with your project. If he or she decides to publish your book, he or she will immediately offer you a contract.

If you are really serious about writing your book, you will make the three items listed above as good as you possibly can. For usually it is upon these that your project will either stand or fall. If your outline is poorly organized, if your chapters are hurriedly written, or if your preface rambles, your book does not have a very good chance of acceptance. So take every precaution possible to do as good a job as you can on this sample material.

Make your outlines as logical as possible, and see that they include every essential entry. Write your two sample chapters with utmost care. Include all the illustrations you intend to use in these chapters in the book. For your samples choose, say, Chapters 1 and 10 rather than Chapters 1 and 2. Or you could choose Chapter 2 and some later chapter, say 8 or 10 or 15. See that your preface states adequately why you are writing your book. If you cannot express the reasons for writing your book, examine your thinking once again. For if you cannot tell the publisher why you are writing the book, there is really little reason for him or her to invest thousands of dollars in producing it.

Once again, you can see that writing a technical book is not a simple task. It involves the expenditure of a great deal of time and energy on your part. Since there is little point to wasting time, you should be completely certain that you have a genuine desire to write the book and the necessary talent to complete the project. If you find that you do not have these requirements, it is best to discard the idea early. For if you do not, you may find the entire experience a disheartening one.

Your outline gives the publisher the scope and organization of your book. The completed sample chapters show him or her how well you can express yourself about your subject. They also indicate to the publisher the level of the audience you are aiming at. The preface tells whom you've written the book for, why you've written it, and what your readers will get from the book.

Once the publisher receives your sample material, he or she sends it to outside critics and reviewers to analyze the material and to suggest how it can be improved, if this should be necessary. Suggestions that reviewers make are usually aimed at making your book more understandable or at tailoring it for a broader audience.

The publisher's decision on your book is based on the opinions of the reviewers and on his or her own study of the market and the manuscript. If his or her decision is favorable, he or she will enter into an agreement with you to publish your book. He or she will offer you a contract which you sign, retaining a copy for yourself and returning one to him or her. The terms of typical book contracts are outlined in Section 1. Review these terms now.

Once the publisher agrees to publish your book, and sometimes even before, he or she will assign an editor to work with you. The editor stands ready to advise, help, cajole, or humor you, according to your needs. Since the book editor is usually a person with long experience in the field, he or she can be of major assistance to you. For besides answering questions on manuscript preparation, copyright laws, illustration sizes, and hundreds of other routine matters, the editor can advise you on the broader aspects of your task. Thus, if your outline needs some changes, if you wish to have an unbiased review of your ideas or manuscript, or if you need some helpful encouragement, the editor is ready to come to your assistance.

Behind every book published there is at least one author and one editor. The closeness of their association varies. Some authors must be led by the hand. Others are heard from by their editors only twice—once when the book is proposed and contracted for, and once when the manuscript is delivered. But regardless of the amount of attention or help you need, the editor will always be a sympathetic and understanding friend.

After an editor has been assigned to you and you have signed your contract with the publisher, you proceed to finish your manuscript. While working on your manuscript, you can keep in touch with your editor, telling him or her of your progress. If any questions arise on the handling of your material, he or she is ready to answer them.

On completion of your manuscript, you submit it to your editor. The manuscript is again reviewed and a report is prepared for you. When you read it, you may wonder why the reviewer didn't write the book himself or herself instead of making unflattering comments and remarks about the manuscript. But you'll find when you reach this stage in your book writing that most outside appraisal is usually constructive. The comments will call your attention to points that may be troublesome to your readers. If necessary, you may be asked to revise your manuscript to bring it into line with the reviewer's suggestions.

The procedure given in this section for outlining a book can be used for any type of technical book. In addition to the seven types listed here, which are probably the more familiar ones to you, many technical writers today author company books and industry-sponsored manuals. While both these types can be outlined as we have discussed here, you will find that there are certain differences, because not all the requirements of a regular commercial publisher must be met. But these differences are usually minor. When writing a book that is sponsored by a company or industry, you will find that the nearer you can make the book approach a standard type, the better reception it will have. And since all sponsored books are written for the largest distribution possible, you should make every effort to have the text and illustrations conform to what the technical reading public wants.

GET READY TO WRITE

Once a publisher has put you under contract to write your book, you are ready to begin the long journey between the sample material and the finished manuscript. Right at the start you should recognize that writing a book is not a simple task. But if you approach it with the right attitude of mind and apply some of the hints given in this section, you will find the task somewhat easier. All the remarks in this section are useful for any type of technical book. So study them well because they will save you much misery and pain.

Before starting to write, review your outline. It will probably have been in the publisher's hands for some time. And if you have not looked at it during this time, you will have a fresh approach when you see it again. Try to be very critical while you look it over. If any questions occur to you, note them immediately in the margin of the outline. Go through the entire outline quickly, marking down any thoughts that come to mind. Then once again go over the entire outline, this time slowly. Make your evaluation as objective as possible.

Make any changes that you find necessary. When the outline appears satisfactory, you are ready to proceed with your writing.

TWO IMPORTANT STEPS BEFORE YOU WRITE

With a good outline and a great deal of enthusiasm, you need only two other tools before you start writing. These are: (1) a writing schedule and (2) a progress notebook. Let's see what each of these is and how it will help you with your writing task.

Writing Schedule

Trying to write a book without a schedule is much like taking a long trip without a timetable. You have no idea where you will be on a given date. So if you try to check your progress after a few days or weeks, you have nothing to tell where you should be. A writing schedule will relieve you of wasting time trying to figure where you should be.

To set up a writing schedule, study your outline. You already have 2 chapters completed. So subtract the number 2 from the total number of chapters in the book. Say that you plan to have 12 chapters. Subtracting 2 leaves 10. If you assume that it will take you 1 month to write each chapter, the total writing time will be 10 months. But one of your chapters may be a lot longer than others. Plan on taking more than a month for this chapter, say, 2 months. Then your total writing time will be 11 months.

Now if you have had a good deal of writing experience, you will know how closely your ordinary writing approaches finished copy. If your rough draft requires only minor changes, then you can plan on another month or two for polishing the finished manuscript. But if you are relatively new to the technical writing business, you should plan on a longer period for rewriting. Assume that you allow 4 months for rewriting the first draft. With these time intervals, the experienced writer as a rule will spend 12 to 13 months on his or her manuscript, while the newer writer will spend 14 to 15 months on his.

But you say, "I do not know how long it will take me to write a chapter. What do I do then?" Well, since you don't know, you must make some assumptions. Guess at how long it will take you to write each chapter. Allow a shorter time for the chapters that cover the material you are most familiar with. Allot a longer period for the chapters needing research or other special work.

Do not allow a writing schedule to frighten you. The schedule is nothing more than a series of check points. If you write rapidly and beat the deadline for a given chapter, go right ahead to the next one. The main purpose of a writing schedule is to help you see your way to the end of the project. If you try to write a book without a writing schedule, you may find that somewhere near the middle you begin to lose enthusiasm. This is because you feel you have done a great deal of work and the end of your task is still far off in the future. But with a writing schedule, when you reach the middle of your book, you will know that the end is only a few months away. This thought will act as an incentive to you to push ahead to that magical last chapter.

Don't underestimate the value of a writing schedule. It will keep you moving ahead toward the completion of your book. And it will add to your sense of accomplishment as you finish each chapter. You can even play a little game with yourself. If in comparing the actual amount of writing time you require with your scheduled time you find yourself ahead, you can imagine yourself as picking up days or weeks. This will allow more time for polishing your manuscript.

A schedule also helps you direct your energy output. As Figure 11.1 shows, a schedule keeps your writing effort at a high level throughout the writing period. This means that you will waste less time worrying about meeting the final deadline.

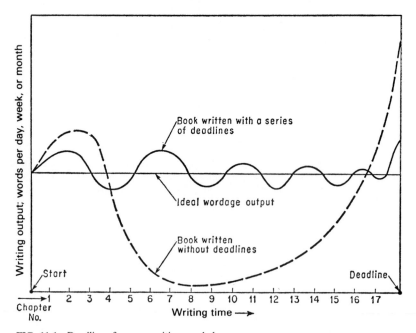

FIG. 11.1 Deadlines for your writing can help.

Another advantage of a writing schedule is that it helps you deliver your manuscript on time to the publisher. Since publishers have to plan well ahead of time on the books they will publish in the future, they are very appreciative when an author delivers his or her manuscript on the promised date. When you come around with another book, the publisher will recall how promptly you met your deadline. This will incline him or her to regard your next project favorably. So do not start writing until you have prepared a realistic schedule.

Use a Progress Notebook

Writing a technical book is a full-time mental activity. You will find that you think of your book throughout your working day, and into the evening. During this time many thoughts will occur to you about the book, some of them extremely valuable. Unless you make a note of them immediately, you may never be able to recall them.

Get yourself a 3- by 6-inch notebook that you can carry with you at all times. Number the pages in the notebook from the beginning to the end.

On the first page of your notebook make a list of the names and numbers of the chapters of your book. On the next page enter the time schedule you have set up for your writing. Include the specific date when you expect to finish each chapter. Next, enter the heading "Titles." On this page you can enter any good title ideas that occur to you. If you are a typical author, you will start with 1 title for your book and by the time you are finished, you will have thought of at least 12 others. Enter each of these on this page so when the final choice must be made, you will have a list of the good candidates as well as the bad ones.

Devote the next few pages of your notebook to the heading "General Data." In these pages you will enter assorted facts and general thoughts that occur to you while you are writing the book. For example, you might enter the date on which you first got the idea for the book, the date on which you signed the contract, and the date on which you started to write. Other entries might include the names of magazines in which you think your book should be advertised, the names of companies or people who could help you with research or illustrations, and any other information pertinent to the writing of your book and to its production and marketing.

The next heading in your notebook should be "Preface." Assign a few pages to it. In these pages make notes of information to be included in the preface of your book. List the names of people whom you wish to thank or otherwise acknowledge. If there are a great many acknowledgments to be made, you may find it advisable to set up a special "Credits" page in your notebook. Then there will be no chance of your overlooking an important person or firm when you come to write your preface. All this may sound unnecessary, but you will be amazed at how easy it is to overlook an important credit. With this special page for credits in your notebook, you can enter all the credits you want to remember as they occur to you and they will be there to be readily picked out when you need them.

Skip a few pages and make the next entry in your notebook. This will be "Chapter 1." Skip a few more pages and enter "Chapter 2." Proceed this way until you have listed all the chapters in your book. Leave four to six pages for each chapter.

Now go back to Chapter 1 in your notebook and enter the outline for this chapter. Also enter any special thoughts you may have concerning Chapter 1. Do the same for all the other chapters in your book. You will now have in easy-to-use form a complete outline of your book, as well as a place to enter any thoughts you may have concerning any part of it.

Now as you go about your business, attend meetings, ride trains or planes, or do anything else, you have an easy way of recording every thought that comes to you about your book during the moments when your mind is free. Write these thoughts in your notebook as soon as they occur to you. For instance, if you learn that a certain firm would be an excellent source of an illustration for Chapter 8, make a note of this on the page devoted to that chapter. Or if you find that a slight change in the outline of Chapter 8 is necessary, make the change at once. Don't wait until the evening or some other time. You will probably have forgotten the thought by then, and you will waste time trying to recover it.

The whole idea of this notebook is based on the enormous amount of work and thinking you will do in producing your technical book. If you can keep your manuscript under control at all times, your writing will benefit and you will not have to work as hard as if you let your energy scatter. So don't ever try to write a technical book without the aid of a progress notebook and a writing schedule. These two tools can mean the difference between an outstanding book easily prepared and a poor one that took twice as much time as necessary. With your schedule and notebook set up, you are ready to begin the serious task of writing your book. Let's see how you can do this without wasting time and energy.

WRITE, WRITE, WRITE

The only way to write your book is to put it into the form of words. And the only way to get the words on paper is by writing. Use a typewriter, a pen, or a pencil. But write.

Try to estimate how many words there will be in your book. Let's say that you find you will have about 60,000 words in the book. Set up a word schedule for each day. If you are an experienced writer, you may be able to produce 1,000 words a day. As a beginner, 500 words may be your limit. But no matter which class you are in, decide how many words you will write each day you work on the book. Then try to meet the goal you have set up.

When you sit down to write, don't worry about the quality of your English. Just be sure of two things: (1) that your technical facts are completely accurate and (2) that you meet your wordage goal for that day. With accurate facts and a steady progression toward completion of your manuscript, you will find that your writing improves as you go along. If you worry and fret about verbs, nouns, infinitives, and other grammatical stumbling blocks, you will slow your writing. Then you may fall behind your overall schedule, and this can lead to discouragement and a loss of enthusiasm.

As the pile of completed manuscript pages rises, your enthusiasm and interest will grow. So write and write. Remember that if your technical facts are accurate and you have a good subject, you can always have a specialist in English rewrite your manuscript if your English is extremely poor. Your main contributions in any technical book are the facts you present and the manner in which you present them. So get your facts straight. Then write a little each day.

You need not necessarily start writing with Chapter 1. If Chapter 6 is the easiest one in the book for you to do, start with it. For the more chapters you complete early in your schedule, the easier the book will be. Many writers experience a letdown somewhere near the middle of their work. This is less apt to occur if the first half of the book has been easy going as the result of tackling the familiar chapters first. Or at least the letdown will not be serious. Another advantage of starting with the easier chapters is that it gives you more time to think over the difficult chapters. While you are thinking about the tough chapters, you will probably get some useful ideas. Enter these in your notebook and they will be waiting for you when you reach the chapters you've been sidestepping.

Use Headings

Headings are used in technical books, textbooks, and handbooks as a convenience to the reader. They point up the main theme and outline the development of thought.

You will notice that in this book two forms of headings predominate. The one in italic and all caps, centered, on a separate line, divides the section into its main parts. An example is the *WRITE, WRITE, WRITE* which appears above. The second type of heading we use is in boldface, capital and lowercase, flush left on a line by itself. As you have gathered, the cap and lowercase words are subheads under the main headings dividing the chapter into sections. You, of course, need not use the same scheme. Choose a scheme that fits your book.

You can follow a main heading with a subhead, and this in turn may be followed by subsidiary headings. For the sake of clarity and smooth reading, restrict yourself to two or at the most three values of headings; too many guideposts may cause your reader to lose sight of the main thought in the chapter. And in the interests of attractive design, try to avoid a heading at the start of a chapter where it will follow directly below the chapter title and create an awkward space above the text.

Use center headings for major chapter breaks, and for "Problems," "Bibliography," etc. The prevailing head throughout the book should be a side head. When typing your manuscript, follow a consistent style of indicating the value of the headings, as shown in Figure 11.2..

It is extremely important that you make the relative values of your headings clear. For if the headings are marked for type incorrectly by the publisher's copy editor, considerable resetting may be required later. When a heading printed on a separate line is changed in the galley proof to be run in at the beginning of a paragraph, usually the whole paragraph has to be reset. If you have any questions about the proper setup for headings, list all the headings in outline form for the guidance of the copy editor.

Although the numbering of headings may look needlessly precise, it is helpful in textbooks, advanced technical material, and in handbooks. Use either consec-

---

MAIN HEADING: CAPITALIZED AND CENTERED

Subheading: Separate-line, Flush with Margin, Underlined, Main Words Capitalized

Subsidiary Heading: Paragraphed, Underlined, Main Words Capitalized. The text follows in the same paragraph.

FIG. 11.2 Use consistent headings throughout your book. (*Courtesy of McGraw-Hill Book Company.*)

utive numbers throughout the book or double numbers by chapter, giving the chapter number first—for example, Section 3-6 for the sixth section in Chapter 3. Double-numbering of headings, tables, illustrations, equations, etc., is recommended for textbooks and reference books because it simplifies cross-reference and reduces the amount of renumbering when a revised edition is prepared. Handbooks are almost always double-numbered because they are frequently revised, and this system make the revision easier. Do not use complicated numbering systems with triple or quadruple numbers. They are cumbersome and are likely to be confusing to the reader.

Give enough thought to your headings and subheads to make them suitable for your particular book. Well-organized headings can help your thinking and can make your book much easier for the reader to use. So the time you spend developing headings is well worth while.

Chapters

A chapter should cover an important phase of your subject. It should have unity within itself. To achieve unity within a chapter, you should limit its content to exactly what is covered by the chapter heading. For instance, if the name of your chapter is "Centrifugal Pumps," do not include any other types of pumps, except as they are related to centrifugal pumps. (And when referring to other types, be sure to state how centrifugal pumps are related to the others; not how the others are related to centrifugal pumps.) In a chapter dealing with an elementary phase of your subject, do not bring in advanced concepts, except to refer to them if they are covered in a later chapter.

By planning your chapters so that each is a well-organized unit, you make your writing task easier. While writing any given chapter, you can forget about the others since you know that they will adequately cover their subject matter.

How long should a chapter be? Only you can answer this question for your book. There are no fixed rules. But if you study a number of technical books, you will find that relatively few contain chapters that are less than two pages long. This is because you seldom find a phase of a technical subject that can be covered in two printed pages.

If you can possibly do so, arrange to have a short chapter followed by a longer one. More important, try to follow a long chapter with a shorter one. This arrangement of your subject matter will make your book more appealing to readers. Many readers become bored with a book when they find that it has a repetitious pattern of chapter length. By following a long chapter with a short one, you give the reader a sense of accomplishment when he or she finishes the short one in a smaller interval of time.

In an illustrated technical book, you will find that you can use your photographs and drawings to vary the chapter pattern. Thus in a chapter that is illustrated mostly with photographs, the reader may begin to wonder what the interior or hookup of the illustrated devices looks like. Perhaps you can satisfy his or her curiosity in the next chapter by using a number of line drawings showing these details. In this way the reader obtains a clearer concept of the subject, and his or her interest is maintained at a higher level.

As in all other types of writing, you must think of your reader first when planning your chapters. For remember that although writing your book may give you a great deal of pleasure, you are still writing it to be read by others. So avoid trick schemes when planning your chapters. Instead aim at the utmost clarity you can achieve. Only when you strive constantly for clarity will you feel that you have done your best for the reader. After all, if you don't do your best, you might as well not write the book. Never forget that the easier your book is to understand and use, the more readers and appreciation it will attract.

Work Your Illustrations

Make your illustrations carry their share of the load. Never use an illustration just because it is pretty or it appeals to you. Choose and use only those illustrations that show your reader an important point.

Try to refer to every illustration in your text. Nothing confuses a reader more than to see some illustrations in a book without any reference in the text. Since the caption or legend for an illustration must be relatively short, you generally must include some additional data about the illustration in the text of your book. If you do not have any references in your text to the illustrations, the reader is at a loss as to just how he or she should relate the illustrations to the text.

By referring to your illustrations in your text, you not only make the illustrations carry their share of the load but add to the unity of your book, because all the elements are neatly tied together. Your reader will show appreciation by using your book more than another book that is poorly put together.

Every time you choose an illustration, remember that the making and handling of illustrations greatly increase the manufacturing costs of your book. These, in turn, increase the price of your book. So before deciding to include an illustration, ask yourself, "*Is this illustration necessary?*" If the answer is yes, then include the illustration in your book. But if you are at all in doubt about the value of the illustration, put it aside and think further about it. Eliminate illustrations that do not appear absolutely necessary for the understanding of your subject matter.

In considering each illustration, ask yourself the following questions:

1. Is it informative? Does it enhance or clarify the text?
2. Is it pertinent to a main theme or merely to a side issue?
3. Is it appropriate in period, content, and style to the text it illustrates?
4. Is it clear, interesting, up-to-date?
5. Does it duplicate some other illustrations you have chosen? Or does it illustrate something you have described so graphically that the reader can visualize it without seeing an illustration?
6. Is it of good quality for reproduction?
7. Can it be reduced to page size or smaller and still be clear?
8. Can it be set vertically on the page so the reader will not have to turn the book to look at it?
9. Does it include unnecessary details or areas? If so, could it be made more pertinent by cropping or retouching?

By applying these questions to each illustration, you can determine if it is suitable for your book. To find exactly how the publisher wants you to submit illustrations, contact your editor. He or she will have specific instructions as to how you should submit drawings and photographs. Follow these instructions exactly. For in doing exactly as the publisher requests, you save him or her time and money. Your book will probably be published sooner, and its price will be lower than if you make your own rules as you go along.

Captions

Number your captions, or legends, to correspond to the illustrations. Submit two copies of the captions, typed double-spaced in list form. Make your captions concise and pertinent; do not use the captions as a catchall for information you have neglected to supply in the text. Be sure that your captions follow the style of your manuscript, rather than the varying styles of the sources from which you secured them. Figure 11.3 shows several different arrangements for typing illustration captions.

Numbering Illustrations

Either number your illustrations consecutively throughout the book or double-number them by chapter. Refer to your illustrations in the text by figure number, not by page number. Correct page numbers cannot be supplied until all the pages have been made up by the printer.

Double-numbering is preferred by many publishers today, and most technical books use this system. It means that you refer to your illustrations with two numbers. The first number is the chapter number. It is followed by a hyphen or a period and a second number, denoting the number of the illustration within the

---

Fig. 3-5. Diagrammatic representation of concept formation. (*From D. M. Johnson, Essentials of Psychology, McGraw-Hill Book Company Inc., New York, 1986. By permission of the publishers.*)

Fig. 3-5. Diagrammatic representation of concept formation. (*From D. M. Johnson, Essentials of Psychology, McGraw-Hill Book Company Inc., New York, 1986. By permission of the publishers.*)

Fig. 4-5. Interior structure of space vehicle: (*a*) front view; (*b*) rear view. (*Drawn by James Brown.*)

Trend of newspaper and magazine advertising in 52 cities by five major classifications, 1928 to 1988. (*Figures from Media Records, Inc.*)

Fig. 266. Digital and analog oscilloscopes. (*Courtesy of Dr. and Mrs. F. M. Mac Farland.*)

Fig. 350. Arrangement of seats and cargo bays in supersonic military aircraft. (*After Viollet-le-Duc.*)

FIG. 11.3 Typical legends for books. (*Courtesy of McGraw-Hill Book Company.*)

chapter. For example, "Figure 3.5" refers to the fifth illustration in Chapter 3. The double-numbering system has the advantage that if you add or delete illustrations, you need change the illustration numbers and references only within the chapter. Also, it is easier to insert new illustrations when you are revising your book at a later date.

Finding Illustrations

You can obtain illustrations from the public relations department of business and industrial firms, from government agencies, service organizations, libraries, schools, museums, commercial photographers, and publishers of periodicals and books. Ink drawings and glossy photographic prints or halftones usually give best results in the printed book.

The printed page of a book or magazine can sometimes be used as copy for a line cut; but it will not reproduce as well as original copy. Never use a printed halftone as copy to make another halftone, for a second pattern of black dots will be superimposed on the previous one and the result will be a cross pattern, making an unattractive illustration.

You must secure permission in writing for the use of illustrations from any source, whether they are reproduced exactly or modified. Include credit to the source of the illustration in the caption of the illustration. If the wording of the credit is specified by the source, call this to the attention of your publisher.

If you take an illustration directly from another book, introduce the caption for this illustration by "From." If you redraw an illustration from another book, use the word "After." But if you base a new illustration on an existing illustration in a book, use the words "Adapted from."

Summaries

You can help your reader by giving a summary at the end of each chapter. You need not use more than one or two paragraphs to do this. Write your summary by going through the entire chapter and picking out the key thoughts. Include only these in your summary. Never try to rewrite the chapter in your summary. You will only bore your reader and waste his or her time.

Problems or Questions

Include problems or questions at the end of your chapters if your material is such that your readers will benefit from solving problems or answering questions. Once again *you* must decide if problems or questions should be included in your book. As you know, most textbooks have either problems or questions, or both. Probably the only type of modern technical book that does not include either problems or questions is the handbook. We shall discuss the handbook in greater detail later in this section.

If you include problems in your book, be sure to solve them and check each solution several times before inserting it in your manuscript. If your book is a college text, you may want to omit the answers to problems. Or you may wish to include only every other answer. You would be wise to follow the recommendations of your publisher in this matter.

Answers to questions require more space and are generally not included. But if you wish to include some of the answers to the questions, try to reduce the answers to their essentials. Again, the decision rests with you. If you are in doubt, check with your editor to find out what the publisher's policy is.

HANDBOOKS

A handbook is a documented volume of proven facts written in a practical, concise form. It comprehensively covers a broad subject area. The handbook is the basic, authoritative reference work for the practicing engineer, scientist, technician, and student.

There are definite basic differences between a handbook and a textbook. A textbook deals with rudiments, reasoning, theory, problems, and mathematical concepts. It usually contains only enough practical treatment to clinch the theory. A handbook contains working information, and only enough theory to explain basic practice. The handbook condenses theory and hypotheses to ultimate terseness to give more space to compiled data and tables designed to meet the needs of the engineer and scientist.

If you take on the job of being the editor of a handbook, your best course to follow is to work closely with the editor assigned to you by the publisher. Your editor will probably have had long experience in this field, and he or she can advise you on how most efficiently to produce the book.

In editing a handbook your biggest task is not writing—rather, it is the selection of suitable contributors. The ideal contributor is a person who has achieved some prominence in his or her sphere. He or she has probably written technical papers on his or her speciality and may have written articles for technical magazines. Often he or she is the author of a book dealing with a particular field. And he or she is usually an active worker on committees of technical or engineering societies. You may find this type of contributor working in industry or in a research laboratory or teaching in a college or university.

Another type of contributor not to be overlooked is the person who is a level or two below the top people in the field. He or she probably is not known outside of his or her chosen area but is considered by his or her associates and those working in the field as a person with demonstrated ability and future promise.

Experience shows that people of both types are quite willing to contribute to handbooks, whereas they might not write for another type of publication. As editor-in-chief, your initial approach to your contributors will be in person, by telephone, or by letter. In your first contact with a prospective contributor, explain the scope of the handbook, the reasons for its preparation, and the fact that the handbook has been placed under contract by a recognized publisher. You should also indicate the name and scope of the section you would like the contributor to write, and the tentative deadline for his or her manuscript.

Your first letter to a prospective contributor is merely an exploratory one, opening the way for more detailed correspondence later. If your publisher has had much previous experience with handbooks, he or she will probably be able to supply you with typical letters you can use as a guide in preparing your own. Examine the sample letters carefully and then write your own. It is not desirable in your first letter to include a copy of the entire outline of the handbook, or mention the honorarium you expect to pay.

The follow-up letter must be tailored to fit the particular situation. In the follow-up letter acknowledging the acceptance of the writing assignment by the contributor, you give additional information and specific instructions to guide him or her in preparing his or her manuscript. Other types of letters are used to follow-up a tentative acceptance or a weak refusal. At this stage of your project it is well for you to keep in close contact with your publisher. Often a letter or call from the publisher to a prospective contributor is very helpful.

It is extremely important that you keep in touch with each contributor as he or she develops his or her section. It is wise for you to work with the publisher's handbook editor in setting up a series of letters covering the entire production of the handbook. You should set up these letters at the time the first invitations to potential contributors are issued. If you set up a basic set of letters in advance, you can use them as needed, with appropriate modifications.

A typical series of letters might include the following: (1) invitation to potential contributor to write a section; (2) follow-up letter to potential contributor invited by telephone or in person; (3) follow-up letter to contributor accepting your original invitation; (4) follow-up letter to weak refusal or tentative acceptance; (5) acknowledgment and instructions to those agreeing to become contributors to your handbook (this should be done by both the editor-in-chief and publisher); (6) a series of letters at regular intervals giving each contributor the names of other contributors, special information, progress reports, etc.; (7) notification of approaching deadline—one month in advance; (8) acknowledgment of the manuscript; (9) a follow-up letter two weeks after the manuscript is due if it has not been received (discuss overdue manuscripts as soon as possible with your publisher so that a suitable course of action can be taken to expedite the manuscripts); and (10) return of manuscript to the contributor after it has been reviewed if further work is needed on it.

It is very important that you as editor-in-chief set up some system for keeping track of the status of each contributor's manuscript: when follow-ups should be made, deadlines, etc. Some handbook editors prefer to use index cards with a card for each contributor; others find it more practical to use a loose-leaf notebook and assign a page to each contributor. Still others find that a large wall chart gives them the information they need at a glance.

Contributors' Compensation

Handbook contributors are paid by you as editor-in-chief of the handbook. The compensation is in the form of a cash payment, an honorarium, as well as a complimentary copy of the handbook. The budget for these payments is discussed and worked out with the handbook editor of the publisher in advance of contacting possible contributors. Usually the publisher advances funds for these payments, against the editor-in-chief's future royalties. In certain cases the publisher may advance funds for stenographic and editorial help.

Because of the number and caliber of the contributors involved in a major handbook, it is economically impossible to pay each a sufficient sum to compensate him or her for the time, effort, and experience he or she gives to the project. Obviously, then, there must be a stronger reason than the money involved to make contributors willing to give their time and effort to write sections for a handbook. Long experience in developing handbooks shows that contributors are motivated chiefly by the following factors:

The top-flight person will contribute a section for the simple reason that if he or she does not become associated with the project, he or she will be conspicuous by his or her absence. If the book is a basic handbook and there is every reason to believe that it may become the bible in its field, this alone will attract many top-flight contributors. Another reason that you should not overlook is that the handbook gives the leaders in any field a good opportunity to make a lasting contribution to the literature of their fields. To the person on the way up, the biggest lure is that the handbook, by bracketing him or her with the top names in the field, could well be the means to more prestige, a better job, more pay, etc. That this is sound reasoning has been proven time and time again.

Information for Contributors

With the contributor's acceptance in hand, the editor-in-chief should furnish him or her with detailed information on how to proceed with the actual writing of his or her section. At this point it is well to stress again to the contributor the objectives of the handbook, its potential audience, and the basic underlying philosophy behind it. The contributor should be given an overall outline of the entire handbook so that he or she may see how his or her section fits into the picture. He or she should also be given a skeleton outline of his or her section with an invitation to suggest improvements in it.

The editor-in-chief should emphasize to the contributors the following general points that should guide them in preparing their individual manuscripts: general descriptive matter and information, history of the art and its development, and like material should be condensed to the shortest possible form without omitting essentials or sacrificing accuracy. In general, the fundamentals of engineering theory should be presented in the briefest form, but no reasoning or hypothesis should be included, nor the derivation of formulas. The last belongs to a textbook or treatise; a handbook should present results alone. In short, it is to be assumed that the biggest user of the handbook will be an engineer, scientist, or technician having some previous acquaintance with engineering or scientific theory and practice.

Point out to all contributors that in selecting data for use in the handbook, each contributor is expected to exercise critical judgment and choose only the best, and to sum up the results of experience to indicate what constitutes standard practice under typical conditions. When authorities differ as to what is preferred practice, the contributor should state the case fairly for both sides, regardless of his or her own opinion or convictions. In the case of unsettled controversies, or where a rapidly changing art has not yet evolved a dominant or ruling practice, conclusions should not be attempted but the case presented as it stands.

The physical preparation of the manuscript cannot be stressed too often by the editor-in-chief. The contributors must be told in exact detail just how many manuscript pages are expected and in what form they should be prepared. As editor-in-chief, you must understand the importance of and assume the responsibility for proper manuscript preparation. A good manuscript will go through production more smoothly and save both the publisher and the editor-in-chief time and money in author's alterations. The publisher's editing staff will be able to read and mark the manuscript more accurately, and the printer will therefore make fewer typesetting errors. If the manuscript is not acceptably prepared, the publisher will have the manuscript retyped at the editor-in-chief's expense. Each contributor should also be warned to keep a carbon of his or her manuscript.

All contributors should be given a firm date for submitting the first drafts of their sections. This is usually covered in the editor-in-chief's letter acknowledging the acceptance of a contributor to write a certain section. Three or four months is the usual time allowed for writing a section. Contributors are expected to get permission to use material copyrighted by other publications.

The Manuscript

As the individual manuscripts are received, you should examine them for style, content, and length. Do as careful an editing job as possible. It is your responsibility to be sure that the manuscript covers the area assigned, is concisely written, and subscribes to the overall philosophy of the book. While it is hoped that the contributors and editor-in-chief will give some thought to spelling, grammar, standard abbreviations, and mechanical details of this sort, it is not expected that a disproportionate amount of time be spent on this phase. The publisher's own copy editors will do the major part of this job. This, of course, is not meant to imply that they will or can rewrite any manuscript. It is the responsibility of the contributor and editor-in-chief to furnish a clearly written and understandable manuscript. The editor-in-chief must stress and see that each contributor conforms to the following so that the final manuscript submitted to the publisher is:

1. *Typed on one side of the sheet:* The ribbon copy should be submitted; a photocopy is unsatisfactory and must be retyped before it is submitted to the publisher.

2. *Double-spaced:* This includes footnotes and bibliographies. Tear sheets from other periodicals and copy printed on both sides of the sheet should not be submitted.

3. *Numbered consecutively:* The manuscript may be numbered consecutively through its entirety or submitted by section with each section consecutively numbered.

4. *Separated from the illustrations:* Two copies of the legends or captions for the illustrations are necessary and should be submitted with the manuscript.

With a vigorous system of contributor communication and control, the average handbook manuscript can be completed and submitted to the publisher within 14 to 18 months.

Manuscript Review

In addition to the editor-in-chief's review of the various sections, it is usually well to get a critical appraisal from another source. The publisher makes arrangements for this review using a staff of outside advisers. These reviews are usually obtained as each manuscript is received by the editor-in-chief, rather than when the complete manuscript is in hand. This section-by-section review permits any necessary rewriting, adding, or deleting of material without holding up the entire project once the entire handbook manuscript is completed and ready for production. If the editor-in-chief has any specific queries on any section, these can be sent on to the reviewers for their comments as well.

After the review, the manuscript is then sent back to the contributor for final polishing in line with the editor-in-chief's and reviewers' comments. (Under certain circumstances the editor-in-chief can make the necessary changes without returning the manuscript to the contributor.) A rigid deadline one month later is usually set for resubmitting the manuscript. In some cases the first draft can be considered the final draft. If a contributed section is not up to par and it would be embarrassing or difficult for the editor-in-chief to so inform the contributor, the publisher usually takes over this diplomatic assignment.

Illustrations

Most handbooks, because of their length, are printed on Bible stock, which does not take photographs clearly. For this reason it is urged that line cuts be used exclusively. It is hoped that contributors will furnish inked illustrations ready for reproduction, but if this is not possible, the publisher will have the illustrations redrawn from pencil sketches. Of course, these sketches should be clear enough to permit the drafters to work from them.

IT'S UP TO YOU

Use of the hints given in this section should be a big help in the writing of your book. The hints should also be of real value to you if you are editing a handbook. But you must remember that in the long run you, and only you, can write the book.

When you reach the last page of your manuscript, you will probably find that you will have to reread the entire text at least once. While you are rereading the manuscript, change the wording here and there as necessary; check the illustrations against the text to see that your references are correct, and check all tables. Make every effort to see that your manuscript is in the best possible condition.

SECTION 12
HOW TO PREPARE LETTERS AND RESUMES

EFFECTIVE LETTER WRITING

The preparation of business or transmittal letters must always be treated with serious concern and must be carefully planned, with attention to such details as the final typing and the stationery (see Figure 12.1). All letters to people outside the firm not only create a lasting impression but also represent a summation of the firm's competence. When writing a specific transmittal letter associated with a request for proposal or qualification, every effort must be employed to avoid repetition of the firm's technical competence or of information already given in the firm's sales brochure.

In the event that a transmittal letter is specifically used for the purpose of providing an assessment of a technology or a progress report of an analysis, it is important to provide a conclusive abstract of the report as an attachment. Use the letter to communicate the progress.

In the case of preparing a transmittal letter with a request for proposal or a qualification, the first paragraph should merely state, for example, "We are pleased to submit this summary of our qualifications for the subject studies relating to the need for cogeneration in the three government installations in Kansas City." This statement should lead the procurement specialist at the client's office into the abstract and the proposal submitted.

You should briefly state what is being transmitted for the client's consideration and close the letter with a statement which places your firm in the light of having the special competence to execute the work efficiently and within the cost constraints.

Your closing statement might be worded thusly: "We are most interested in working with you on the development of a master plan to convert one or more of the government installations. We believe that our capabilities and technical experience can contribute significantly to a successful project."

There are hundreds of situations where transmittal letters are used by professionals of all types and descriptions. It is virtually impossible to provide instructions for the majority of situations. What is important, however, is that transmittal letters be brief and accurately worded and that they deliver the message to support the enclosures being submitted, if any.

The signer of these transmittal letters should always be the person or executive responsible for the client's relations on the proposal being submitted

> **THE QUALITY OF PAPER IS OF MAJOR IMPORTANCE**
>
> Transmittal letters are exceptionally important in reflecting a company's image; thus the quality of paper must always be superior. In examining the cost of business or professional correspondence, your concern should be with the "cost of writing a letter," rather than simply the cost per thousand sheets of letterhead paper.
>
> Valuable time, considerable thought, dictation, and transcription are all important parts of the cost of the letter, as are the design and printing of the letterhead, the postage, and general office overhead.
>
> In observing the total costs, the actual cost of the letterhead and envelope paper takes on a different significance. Estimates show the following breakdown:
>
> | | |
> |---|---|
> | Dictator's time | 22.8% |
> | Secretarial time | 28.5% |
> | Nonproductive labor | 7.7% |
> | Fixed charges | 25.7% |
> | Mailing cost | 6.7% |
> | Filing cost | 5.0% |
> | Materials cost | 3.6% |
>
> Should the total cost of the average professional letter be $25, the materials will account for only 3.6 percent of this total, or approximately 90 cents. These estimates could vary with the changing economy, but the percentages would be relatively the same.
>
> Business and professional stationery must do more than just deliver a specific typewritten message. The letter as a whole represents the writer and his or her firm. The stationery must portray both the character and the standing of the firm in the business world. Using the finest-quality paper helps convey the proper image of the firm. Thus, the total effectiveness of the letter is enhanced, and the small added cost of using prestige paper becomes a sensible investment.
>
> When dealing with a vendor representative for the selection and purchasing of the firm's stationery, request an abridged dictionary of paper terms. These lists of definitions will assist you immeasurably in making the proper selection of your firm's stationery. Crane & Company of Dalton, Massachusetts, provides an excellent brochure with actual specimens to assist sales executives, project managers, and engineers or scientists in the selection of letterheads, envelopes, and stationery for business letters.
>
> Your main objective must be to sell your idea to the prospective client effectively. You have taken extreme care in presenting your ideas, and in presenting your firm's capabilities in an accurate manner; to do so you used the most qualified method of transmitting your company's intentions. The other half of the decision now rests with your client's impression of your company.

FIG. 12.1 The quality of paper is of major importance. (*Courtesy of Crane & Company.*)

and all future business. For convenience, submit your business card with your transmittal letter, should you be the executive-in-charge.

TEN GUIDELINES FOR IMPROVING BUSINESS CORRESPONDENCE

Listed below are brief guidelines you should follow to improve the effectiveness of business letters. These guidelines are applicable to transmittal letters, response letters to complaints, and general communications with clients.

HOW TO PREPARE LETTERS AND RESUMES   **12.3**

The majority of business letters can be made more pertinent, more potent, and more persuasive by using a planned approach. Before you write or dictate your next business letter, refer to the guidelines listed below:

- *Be friendly:* There's no sense of making enemies through your letters. Start your letter by thanking your correspondent for any help you've been given, for materials or information sent, etc. A friendly opening gets results.
- *Be polite:* Ignoring common courtesy can turn people against you or your firm. So be polite, even if your letter demands some type of performance from the addressee. Politeness can win people to your way of thinking.
- *Be concise:* Write your letter like a telegram where you're paying for each word. Conciseness ensures greater attention to every thought in your letter. With such attention to the contents of your letter, results are more certain.
- *Be accurate:* Remember that the recipient of your letter will probably act on its contents. So be certain your facts are completely accurate. Incorrect information can lead to many problems. So—above all—be accurate.
- *Be specific:* "Tie down" every bit of information in your letter so there's *no* doubt as to your meaning. Repeat words, if necessary, to remove any doubt as to your exact meaning. Name names—don't allow vagueness to cause errors.
- *Be detailed:* Tell what your recipient needs to know. Use names, numbers, specifications, and titles to detail exactly what you mean in your letter. There's almost no way in which you can be too detailed where complex matters are covered.
- *Be aware:* Know your reader and what he or she may or may not know about the content of your letter. Take time, and words, to inform and bring your reader up-to-date on the subject matter discussed in your letter.
- *Be cooperative:* You won't make much progress toward the goal of your letter if you're uncooperative. So try to give the reader of your letter what he or she needs to make the goal reachable without enormous amounts of negotiating.
- *Be sincere:* Your letter's goal is to accomplish a certain business purpose. So take every step you can in your letter to be sincere and convince your reader of your dedication to getting the results you both seek.
- *Be serious:* Humor is *not* appropriate for a business letter. Stick to the subject of your letter. Say what you must say. Stop when you've finished your statement. Save jokes and anecdotes for the telephone or personal meetings.

Figure 12.2 is a sample of a transmittal letter sent to a client with a qualification document. In this particular qualification document, the firm submitted their expertise in plant conversion for implementing cogeneration. Since the firm developed a good reputation with successes in several previous projects, the qualification document contained a list of completed projects, résumés of key personnel assigned to these projects, and the normal capability definition of their firm. As a result, the firm's vice president was not required to provide a repetition of proof but merely to express his confidence in the firm's capabilities in the particular technology—cogeneration.

Figures 12.3 and 12.4 are sample letters that respond to different situations. They are provided in this section for your review and reference.

> H&V Engineering Services
> 1234 Palo Alto Drive
> Dallas, TX 75247
>
> January 10, 19XX
>
> Mr. Robert Doe
> Senior Contracts Specialist
> Division of Contracts and Procurement
> Brookhaven National Laboratory
> Upton, L.I. NY 11973
>
> Dear Mr. Doe:
>   Subject: Request for Technical Proposal No. RFTP-1
>   for the Feasibility Study and Cost Estimate
>   to Introduce Cogeneration of Facility XXX
>
>   We are pleased to submit this technical proposal for the subject services. This proposal fully responds to the contents in your RFTP-1 dated December 15, 19XX. We propose to perform the work described in the proposal for the cost estimate submitted in the attachment.
>
>   With special regard to the restrictions imposed by the Fuel Use Act of 1978 and requests made to convert power plants for oil to alternate fuels, we have experience in identical work and a proven track record on similar projects.
>
>   We appreciate the opportunity given to submit this proposal, and we look forward to working on this very important assignment.
>
>   In the event that we can provide you additional answers to questions you may have, please call me on (714) XXX-XXXX.
>
>   Very truly yours,
>
>   Alan Smith
>   Vice President, Marketing Division
>
> AS/avs
> Enclosure

FIG. 12.2 Sample of a transmittal letter sent to a client with a qualification document.

HOW TO PREPARE PROFESSIONAL RÉSUMÉS

This section provides a number of methods and examples for preparing effective professional résumés. This section also demonstrates the varied usages of abbreviated and full-length types of résumés. The point to be noted here is that in every case, or application, *résumés* must accurately describe the technical competence whether in the short or long format. Generally, the rule is that a résumé should not exceed two typewritten pages.

    In the establishment of a firm's capability to qualify for a particular project or technology, it is necessary to present certain key personnel who have the expertise in that technology required in the proposed project. The summary of the key technical personnel's competence must be presented in the Standard GSA Form 255. See Figure 12.5 for an example. This portion of the form is specifically used

H&V Engineering Services
1234 Palo Alto Drive
Dallas, TX 75247

January 19, 19XX

Mr. Warren Smith, Project Manager
Propellant Tank Computer System
Electronics Inc.
Fairfax County, FL 32888

Dear Mr. Smith:
    Subject: Discrepancy Report No. FFF-1 Reporting
              Failures of External Cabling

    We have thoroughly reviewed your Discrepancy Report No. FFF-1 and physically examined the external cabling returned with the subject report.

    Recognizing the schedule of events at the field site, we have shipped today, via express mail, a duplicate set of identical cabling so you can continue your countdown procedure without delay. To avoid damage to each connector stationary sleeve, and to protect the grounding of each connector, we have added on each electrical wiring connector on the replaced cabling a shrunk polyethylene boot.

    Our failure analysis results show that cabling rotating sleeves are severely bent out of round, which has disallowed mating with a receiving plug. Also, there is evidence that these connectors have been dragging on a rough surface, such as concrete or steel. We believe that the polyethylene boots will assist in obtaining longer wear on the cabling connectors. We have changed the design drawing and manufacturing methods to include these protector boots on all future cabling connectors delivered to the field site. A copy of the failure analysis report is attached for your review.

    We regret that this problem caused you inconvenience. Should we be able to be of further assistance, please call me on (714) XXX-XXXX.

                      Very truly yours,

                      John Thomas, Manager
                      Customer Service

JT/avs

  cc: T. Evans, Engineering
         J. Doe, Manufacturing
  cc: A. Woe, Cost Accounting

FIG. 12.3 Sample transmittal letter.

to describe the key personnel's technical capabilities. The examples shown are typical of the personnel engaged in an architect-engineer firm. Appropriately, you should use the same personnel if you are awarded the proposal resulting from the submittal of the qualification document. The demonstration of the key personnel's competence should qualify you in receiving a request for proposal (RFP) for a study.

```
ABC ENGINEERS INC.
     8910 Beveret Drive
     Houston, TX  77000

                                                February 3, 19XX

Mr. Harold Reeves
Senior Contracts Administrator
Electric Power Research Institute (EPRI)
Post Office Box 8888
Palo Alto, CA  70672

Dear Mr. Reeves:

   As you requested in our telephone conversation today, below is the estimate for
converting the three fusion-fission draft reports in accordance with the EPRI Style
Guide. This transmittal supersedes the previously submitted estimate of August 6,
19XX, to Mr. Ira Helms of your operation. Your request for an executive summary
will be completed by March 30, 19XX, as promised.

   A.  Edit, reformat, and type three reports to the           $30,000
       EPRI Style Guide and print 10 copies:
   B.  Prepare an executive summary report on the
       Fusion-Fission Project and print 10 copies:             $20,000

       Total cost estimate:                                    $50,000

   The actual costs for preparing an executive summary were not shown in our
previous estimate to Mr. Helms, since it was not a requirement at that time. We are
pleased to fulfill this new requirement and shall transmit 10 copies of all reports via
express mail by March 30, 19XX.

   Thank you for this new assignment. Should you have any questions, please advise
us.

                         Very truly yours,



                         Mark Stellman
                         Project Manager
CE/avs
```

FIG. 12.4 Sample transmittal letter.

You will note that the firm is submitting capabilities to qualify for a cogeneration project; therefore, the firm's executives selected the most qualified personnel.

How to Prepare Résumés for Gaining Employment Interviews

A well-prepared résumé is the key element for obtaining an interview and a good position. The majority of employment-hires are generally introduced with a résumé. Most candidates seeking a position receive employment interviews from an initial résumé contact. In many instances, an engineer or scientist is deprived of the opportunity to interview for a position because the résumé submitted lacks sufficient information to merit further consultation. It is important, therefore,

Brief Resume of Key Persons, Specialists and Individual Consultants Anticipated for this Project

a. Name & Title:
Joseph R. Seamans, Consulting Engineer

b. Project Assignment:
Project Engineering Manager

c. Name of firm with which associated:
Brown Engineers & Associates

d. Years experience: With This Firm 10 **With Other Firms** 2

e. Education: Degree(s)/Year/Specialization
Drexel University, BSME, 1965
Drexel University, MSME, 1967
Drexel University, MBA, 1974

f. Active Registration   Year First Registered Discipline
1968, Pennsylvania   Mechanical

g. Other Experience and Qualifications relevant to the proposed project:
Mr. Seamans has extensive experience in evaluating equipment reliability, long term economics for steam power plant cycles with improved efficiency. He has served as project coordinator for numerous studies to evaluate the technical and economic feasibility of advanced (high pressure, high temperature) Rankine steam power plant cycles. He also served as leader of standardization program on preparing evaluation characteristics for a gas turbine project. Mr. Seamans prepared the balance-of-plant specifications, plant layout, conceptual design for a new electrical generating plant. Also, he served a project coordinator of a fluidized bed boilers project and provided innovations in the design and equipment selections.

a. Name & Title:
Ronald T. Martinson, Consulting Engineer

b. Project Assignment:
Project Manager

c. Name of Firm with which associated:
Brown Engineers & Associates

d. Years experience: With This Firm 11 **With Other Firms** 0

e. Education: Degree(s)/Years/Specialization
Massachusetts Institute of Technology, BSME, 1970
University of Delaware, MSME, 1972
Drexel University, MBA, 1974

f. Active Registration: year First Registered/ Discipline
1973, Pennsylvania   Mechanical

g. Other Experience and Qualifications relevant to the proposed project:
Mr. Martinson's expertise is in the technical and economic evaluation of cogeneration plant. He served as project engineering manager of several project studies, conceptual designs, and cogeneration applications for utilities and government agencies. Numerous plants have been converted to district heating and cogeneration modes following the studies submitted under his direction. Also, Mr. Martinson has developed economic factors, such as fixed charged rates, discount rates, and levelization factor for various economic evaluations. He performed a number of capital and total generating cost evaluations for nuclear, fossil and plants using alternate energy sources. Mr. Martinson has served as project engineering manager of the previous seven projects involving applications of cogeneration.

FIG. 12.5 Brief resumes of key persons in Standard GSA Form No. 255.

12.7

that the engineer or scientist approach the task with the understanding that preparing a résumé could be considered a sound investment for the future.

There are numerous professionals who elect to quickly "throw" a résumé together and take their chances on the outcome. These professionals have no way of knowing about the position they may have missed because they were not invited for an interview. So take the time to prepare your résumé well, and use the format that follows, adjusting it as necessary to your individual circumstances. In any event, you should have your résumé typed by a qualified secretary.

Described below are some firm recommendations on an acceptable résumé format. In addition, Figure 12.6 gives an example of a résumé used in applying for an engineering position. Should you be an engineer, scientist, accountant, or sales executive, the format should not be different, except for your professional experience.

Heading. Every résumé should list the candidate's full name, home address, city, state, zip code, and telephone number, including area code. Obviously, this information is necessary to contact you in the event further consultations are required.

Objective. One of the most important facets of a résumé is your "career objective." The career objective must properly be presented to allow the recruiter to evaluate your objective and consider you within the framework of the available jobs. If you want to be considered for a management position as well as a senior staff position, you should make this career plan known in your career objective statement. This type of objective shows the recruiter that you have ambition and that you are willing to work on staff as well as to manage.

Education. Each degree and the year it was achieved should be listed with the highest degree being noted first. Candidates possessing advanced degrees should note the subject matter of their thesis. Also, every additional postgraduate course which has been completed should be listed, including any company-sponsored courses.

Professional Experience. The professional experience of a candidate should always be presented in reverse chronological order, with the most recent position being stated first. A prospective employer is particularly interested in a candidate's present experience because it might be related to the present job openings. Should a candidate be employed with a company for several years, the experience must be divided into specific assignments so that it reflects increased proficiency, promotions, diversifications, and depth. In these descriptions of assignments, be careful to use action verbs, e.g., *designed, performed, managed,* should these functions be truthful. Do not list the cut-off date of your current assignment if you are presently employed; i.e., describe your current assignment as, for example, "1984 to present."

References. On the initial written contact with a company, it might prove to be detrimental if the names of references are used; be sure to advise your references of your intention and the nature of the work you are seeking. It is strongly suggested that references be provided only upon request. When listing previous employers, it is not necessary to note the names of supervisors for each position unless such information is specifically requested on the application.

```
                        GEORGE C. SCOTT
                          342 Elkridge Drive
                     Voorhees Township, NJ 08043
                       Telephone (609) 712-XXXX


OBJECTIVE          Responsible position in the interpretation
                   of onsite meteorological and air-quality
                   data. Interested in a management position
                   in the meteorological division.

EDUCATION          M.S. Environmental Sciences, Drexel
                   University, 1978
                   B.S. Meteorology, Pennsylvania State
                   University, 1969

EMPLOYMENT HISTORY

1974-Present       GENERAL ELECTRONICS CORPORATION Senior Meteorologist
                   Performed climatological and severe storm analyses, cooling-
                   tower plume impact, and solid deposition prediction, ground-
                   level and stack emission-diffusion modeling, including data
                   processing and analysis. Responsible for the research for and
                   preparation of numerous study reports containing results of
                   climatological analyses and impact emissions from power plant
                   cooling towers.

1969-1973          STONE & WEBSTER ELECTRONICS Meteorologist
                   Responsible for field monitoring of meteorological variables and
                   air pollutants including their vertical and horizontal gradients.
                   Designed systems for the acquisition of meteorological data for
                   the above urban experiments for Commonwealth of
                   Pennsylvania Air Monitoring System. Conducted measurement
                   and analysis of CO, NOx, nonmethane hydrocarbon, oxident, and
                   meteorological data taken between facilities and the densely
                   populated Tremont Apartment Complex.

PERSONAL           Married                U.S. citizen
                   Height 6 feet, 2 inches        Weight 180 pounds
                            Secret Clearance


REFERENCES         Furnished upon request.

PROFESSIONAL       American Meteorological Society (AMS), President of the Greater
MEMBERSHIP         Delaware Valley Branch
```

FIG. 12.6 Sample résumé for applying for a new position.

Personal Information. Many advisors on résumé preparation believe that information in a résumé should be limited to marital status, security clearance, citizenship, and height and weight.

Publications. This particular facet must be treated with a good deal of judgment. List only those publications or papers that have been presented at technical or scientific meetings and that are relevant to the position for which you are applying. A separate list of all your publications should be made available and presented to the technical interviewer at the time of a plant visit.

Cover Letter with the Résumé. When applying for a position, you must include a cover letter with the résumé. This letter should be addressed to the employment manager or to the industrial relations representative whose name appears in the advertisement. The letter should state the type of position for which you are applying, including the dates you will be available for interview, location preference, and salary information. In your cover letter, it is strongly suggested that you emphasize the experience in your résumé which you feel would qualify you for a particular position.

The cover letter which accompanies the résumé is extremely important. Figure 12.7 provides a sample of a cover letter for your reference. Of course, should your professional experience be different from the one shown in the example, you should alter your cover letter accordingly.

Résumé Formats Used in Requests for Proposals

The résumés used in requests for proposals (RFP) are different in composition and format. For instance, the "objective" contained in the previous type of résumé is not appropriate for résumés submitted with a proposal. The other fac-

---

342 Elkridge Drive
Voorhees Township, NJ 08043
December 1, 19XX

Mr. Frederick Smith, Sr.
Employment Manager
Computers Corporation
Route 41
Cherry Hill, NJ 08034

Dear Mr. Smith:

    Enclosed is my résumé for your consideration for the position of meteorologist which you advertised in the Sunday Philadelphia Inquirer.

    I sincerely believe that my scientific experience in environmental sciences for the past several years, and the data I provided numerous clients during my career, qualify me for this position. My present salary is $48,000 per year, based on a 40-hour workweek.

    I am interested in interviewing for the position. Please call me at (609) 712-3111, evenings between 7:00 and 9:30 pm.

    Thank you most sincerely for your consideration.

                           Very truly yours,

                            George C. Scott

Enclosure

FIG. 12.7 Sample cover letter to accompany a résumé.

ets which should be omitted are: (1) the key personnel's addresses and (2) personal data.

The key element which must be addressed is "special competence," which may be the unique qualifications for the assignment under consideration. In the description of the "special competence," the technical expertise must be carefully worded. If it is not, the key personnel's expertise may be exaggerated, or in other instances, the very expertise required in the proposed assignment may be buried rather than highlighted and precisely stated.

The sample format that follows has proven effective because it permits the reviewer to quickly assess the proposed personnel's expertise to find what experience may be applicable to the proposed project. The résumé in Figure 12.8 is the typical format normally used with proposals.

HARRY DOE
MANAGER OF EDP SYSTEMS AND PROGRAMMING

SPECIAL COMPETENCE AND QUALIFICATIONS
Extensive experience in programming and design. Manages complex application development assignments. Provides direction to affect scheduling, simulation, financial planning models, time sharing, graphics, and management information systems.

EDUCATION
Rensselaer Polytechnic Institute B.S.chE, 1960 University of Michigan, M.S.chE, 1963
New York University, Applied Mathematics, 1966

EMPLOYMENT HISTORY
1974-Present   GENERAL ELECTRIC COMPANY Manager of EDP
                  Systems & Programming

1971-1974   INDEPENDENT CONSULTANT Services to the duPont Co.

1968-1971   NORMAN COMPUTER SYSTEMS Vice President

1967-1968   MOBIL OIL CORPORATION Manager of Engineering Methods

1961-1967   DREXEL INTERNATIONAL Manager of Computer
                  Application

SELECTED EXPERIENCE
Actively engaged in the formulation of development and implementation of a large-scale "Utility-Dispatch" which is used for optimizing the selection of other energy storage units.

Developed the "Monte Carlo" statistical methods for use with energy source and storage systems to perform risk analysis studies, based on the anticipation of anticipated fuel costs and capital costs.

Directed the conversion project for transferring 250 programs and systems in the company technical library to the newly acquired Honeywell 66/60. This memory bank has 400,000 statements, mostly on Fortran, which were converted.

Authored several technical papers for technical journals and reports on computer technology and applications.

FIG. 12.8 Sample résumé used with proposals.

SUMMARY CHECKLIST

1. Review plans, samples, and strategies used in business letters, including the 10 guidelines provided for effective letter writing.
2. Prepare your next business letter by using each of the guidelines presented.
3. Review each method of preparing the various types of professional résumés and a letter of application for positions.

SECTION 13
GRAMMAR AND USAGE

Most technical writers are so concerned with the engineering or scientific phases of their work that they have little time to consider the niceties of grammar, spelling, and good usage. But to be effective in writing, you must use correct English. To help you write better English, we'll devote this section to a quick review of troublesome areas in spelling, usage, and grammar. Study this section now; refer to it later when specific questions arise. You'll find it solves many of the problems encountered in normal technical writing tasks.

AUTHORITIES

Webster's New International Dictionary, the University of Chicago's *A Manual of Style,* the Government Printing Office *Style Manual,* and the U.S. Geological Survey *Suggestions to Authors* are excellent guides for technical writers. For technical words not found in *Webster,* use the forms accepted by the industry, trade, profession, or science to which they have specific application. When you are still in doubt as to the correct spelling or usage, refer to a standard textbook or glossary of words and terms used in a given industry or profession.

SPELLING AND COMPOUNDING

Webster is the road to correct spelling and compounding. So consult that dictionary and this section when there is any doubt. *Webster* not only gives correct and preferred spelling but also shows, in the *Collegiate* edition, preferred practice in the compounding of many words. In addition, it includes a section on compounds not specifically listed. To help you overcome some of the common hurdles in technical spelling and compounding, a quick summary of helpful rules is given below. Study these rules now. They will help you become a more confident technical writer.

Final Consonants. To double or not to double? The rule is clear. When words of one syllable end in a single consonant preceded by a single short vowel, double the final consonant before a suffix beginning with a vowel. The final consonant is not doubled if the vowel is long.

Stop, stopping; flit, flitting; rot, rotted

But: Boat, boating; stoop, stooping; read, reading

What about words of more than one syllable? Again the rule is clear. When words of two or more syllables end in a single consonant preceded by a single vowel, the consonant is doubled only when the accent falls on the last syllable.

Befit, befitting; control, controlled; occur, occurred

But: Combat, combatant, combating; benefit, benefiting; travel, traveling; focus, focused

Mute e. Keep the *e* or drop it before a suffix? Rule: Drop the *e* before a vowel; keep it before a consonant.

Make, making; excite, excitable; sale, salable; like, likely

There are two exceptions. The *e* is retained before the vowels *a* and *o* to preserve the soft sound of the preceding *c* or *g*.

Change, changeable (*but* changing); singe, singeing (otherwise it would be "singing"); gage, gaging, gageable

Note that participles are given with the verb in the dictionary. To repeat, consult *Webster* if in doubt. Note that *judgment* and *ageing* are preferred, even though *Webster* also shows *judgement* and *ageing*.

ei *and* ie. Which after *c*? Which after *l*? The key word is *Alice;* that is, *ie* after *l*, *ei* after *c*.

Believe, relief, deceive, receipt

A more inclusive rule, covering all consonants, is to write *i* before *e* when pronounced *ee,* except after *c* or when pronounced like *a*.

Yield, siege, sieve, weight

Adjective to Adverb. When adjectives ending in *l* or *ll* are changed to adverbs ending in *ly,* the double *ll* is always used.

Especial, especially; purposeful, purposefully; dull, dully

Double n. Nouns ending in *ness* made from adjectives ending in *n* retain both *n*'s.

Common, commonness

Nouns Ending in ful. Add *s* to the final syllable to form plurals.

Spoonfuls, *not* spoonsful

Nouns Ending in y. When the *y* is preceded by a consonant, change *y* to *i* and add *es.* But where nouns end in *y* preceded by a vowel, the usual practice is to add *s*.

Chimney, chimneys; money, moneys; monarchy, monarchies; butterfly, butterflies

Plurals of Compound Nouns. Add *s* to the principal word.

Engineer-in-charge, engineers-in-charge

Latin Plurals. Do not use where the English equivalents have become good practice.

Abscissa, abscissas (*not* abscissae); antenna, antennas (*not* antennae); formula, formulas (*not* formulae); stadium, stadiums (*not* stadia)

Examples Of Technical Spelling And Compounding

The list below gives the preferred spelling and compounding of words frequently used in technical publications and having more than one spelling or form.

[In this list the following abbreviations are used: (a) indicates adjective; (adv) adverb; (v) verb. All other words are nouns. Spelling and form shown is that approved by the industry using the word.]

A

abscissas (*not* abscissae)
acidproof (a)
adviser (*not* advisor)
aftercooler
aftereffect
aileron
airbound (a)
air chamber
air compressor
air-cool (v)
air-cooled (a)
aircraft, airplane, airship
air-dry (a, v)
air duct
airfield, airport
air-hardening (a)
air lift
air line, air liner
air port (opening)
airport (an airfield)
airproof (a, v)
air pump
air shaft
airtight (a)

align (v)
alkalis (*not* alkalies)
all-round (a)
angle iron
angle valve
antennas (*not* antennae)
antifreeze
antifriction
arc weld (n)
arc-weld (v)
autogiro
autotransformer

B

backfill
background
backlash
back pressure
baffle plate
baffle wall
baghouse
balance wheel
ball-and-socket joint
ball bearing

ball mill
ball race
band saw
band wheel
bar iron
baseplate
bedplate
bedrock
bell crank
beltman
belt wheel
benchboard
beside (next to) (adv)
besides (in addition to) (adv)
bilateral (a)
bimetallic (a)
bird's-eye
bivalent
blast furnace
blowdown (n)
blow down (v)
blowhole
blowoff
blowout
blowpipe
blueprint
boilermaker
boilerplate
boiler room
bolthead
bone black
bookkeeper
boring mill
borrowpit
bottleneck
boxcar
box girder
brasswork
breakdown
brickwork
bridgework
briquet (n, v)
briquetting (a, v)
brush holder
bucketful
bulkhead

bull's-eye
bull wheel
bus bar
buses (*not* busses)
butt joint
butt weld
butt-weld (v)
bylaw
byline
bypass
by-product

C

camshaft
camwheel
candlepower
cannot
cape chisel
caprock
cap screw
carburetor
carload
caseharden
cast iron
catalog
catch basin
centerline
check valve
checkweigher
circuit breaker
clamshell
cleanup
close-up (n, a)
clutch shaft
coal bed (dust, field, gas, tar, vein)
cockpit
coefficient
cofferdam
cold chisel (n)
cold-chisel (v)
cold-drawn (a)
cold-process (v)
cold-processed (a)
connecting rod
control stick

Co-op, co-op (in referring to organizations)
cooperate (v)
coordinate (n, v)
coproduct
core box
cornerstone
counterbalance (n, v)
counterbore (n, v)
countercurrent (n, v)
countershaft
coworker
crankcase (crankpin, crankshaft)
crank wheel
cribwork
criticize
crossarm
cross axle
crossbar
cross-connect (v)
cross-connection (n)
cross conveyor (n)
crosscurrent
crosscut
crossflow
cross girder
crosshead
crossover
crosspiece
crossrail
cross-section (n)
cross-section (v)
cross slide
crosstie
crowbar
crown sheet
customhouse
cut-and-fill
cutoff (n, v)
cut out (v)
cutout (n, a)

D

damsite
dashpot
database
daytime
dead center
dead-front (a)
dead load
de-energize (v)
deepwell (as applied to pumps)
detin (v)
dew point
dezinc
die block
die-cast (a, v)
die cast (n)
die head
die holder
diemaker
direct current
disk (*not* disc)
distributor
divalent (a)
downflow
downgrade
downhand
downstream (a)
draft (*not* draught)
drafting room
dragline
drawback
drawbar
drawbridge
drawing board
dried (v)
drier (machine or substance)
drill hole
drill press
drillstock
drop forge
drop-forge (v)
drought (*not* drouth)
dry dock
dry-dock (v)
duraluminum
dust catcher
dustproof (a)
dyehouse
dyestuff

E

earthfill
earthwork
electrochemical (a)
electrolytic (a)
electromagnet (magnetic)
electromotive
electroplate
electropneumatic
electrotype (n, v)
embed (v), embedment
employee
enclose (*not* inclose) (v)
end thrust
engine room
enjoin (v), *but* injunction
enmesh (v)
everyday (a)
everything
everywhere (adv)
explosion-proof
eyebar (*not* I-bar)
eyebolt
eyepiece

F

faceplate
farsighted (a)
feed pump
feed screw
feed valve
feedwater
ferroalloy
fiber (*not* fibre)
field-paint (-rivet, -weld) (v)
filter press
firebox
firebrick
fire clay
firedamp
fire door
fireman
fireproof (a, v)
first-class (a, adv)
firsthand (a, adv)
fishplate
flameproof
flammable (*not* inflammable)
flare-up
flashboard
flashlight
flashover
flash point
floor beam
floorplate
flowchart
flowmeter
flowsheet
flue sheet
flyash
flywheel
focuses, focused
follow-up
foodstuff
foolproof (a)
footbridge
footcandle
foot-pound
foot valve
footwall
footway
force pump
formulas (*not* formulae) (do not use when equation is meant)
formwork
four-cycle
framework
frameworker
freeboard
freehand (n, a)
free hand (right of action)
friction head
fulfill (v)
further (*not* distance) (a, v, adv)
fuselage

G

gage (*not* gauge) (n, v)
gage block

gage board
gage glass
gangway
gas engine
gasholder
gashouse
gas-weld (v)
gas welder
gasworks
gatehouse
gate valve
gearbox
gear cutter
gear motor
gearshift
gear wheel
generator room
globe valve
goodwill
gram-equivalent (a)
gram molecule
gram-molecular
groundwater
groundwork
guesswork
guncotton
gunpowder
gunmaker
gunshot
gutta-percha

H

hacksaw
half hour (30 min)
half-hour (a)
half-tone (n, a)
halfway
handbook
hand drill
hand-finish (v)
hand-fire (v)
handhole
handsaw
handwheel
hardpan

hardwood
haulageway
headframe
head gate
headgear
headlamp
headlight
headroom
headstock
heat-treat (v)
heavy-duty (a)
heavier-than-air (a)
heavyweight
homemade (a)
hookup
horsepower
hot-drawn (a)
hot well
hydrocarbon
hydroelectric
hydrometer
hygrometer
hygroscopic (*not* hydroscopic)
hypotenuse (*not* hypothenuse)

I

I-beam (*not* eye-beam)
inflow
infrared
inflammable (*see* flammable)
inhibitor
in-line engine
input
intercooler
interrelation
ironclad (n, a)

J

jackhammer
jackscrew
jackshaft
jig saw

journal bearing
journal box

K

keynote
keyseat
keyway
keyword
kilowatthour
knee brace
knife switch

L

lac
lacquer
ladderway
lag screw
landmark
layoff
layout
layover
lead screw
left-hand (a)
lengthwise (*not* lengthways)
lifetime
lighter-than-air (a)
lightweight (n, a)
lineshaft
linework
liter
live load
livestock
lockout
lock nut
lock washer
loophole
loose-leaf (a)
louver, louvered (n, a)

M

machine screw
machine shop
machine tool

mainshaft
mainspring
mainstay
maintenance-of-way (a)
makeup
manhole
manyfold (a)
may be (v)
maybe (adv)
melting point
metal-clad (a)
metalize (v)
metalizing (a)
meter
microchemical (a)
micrograph
micrometer
microorganism
microphotograph
midpoint
midsummer
milliammeter
milliampere
millman
moistureproof
moisture-resisting (a)
mold (*not* mould)
monkey wrench
monovalent (a)
motorboat
motorbus
motorcar
motor coach
motor generator
motor shaft
motor truck
mud drum
multicylinder (a)
multiform (a)
multijet (a)
multipolar (a)
multirange (a)
multispeed (a)
multispindle (a)
multistage (a)
multivoltage (a)

N

naphtha
narrow gage
nearby (a), *but* "He stood near by"
nevertheless (adv)
nitroglycerin
no-load (a)
nonacid (a)
noncombustible (a)
noncondensing (a)
noncooperative (a)
noncorrosive (a)
nonessential (a)
nonnegotiable (a)
nonnitrogenous (a)
nonparallel (a)
nonproductive (a)
nonunion (a)
northeast
northwest
notebook
nowadays (n, adv)

O

offhand (a, adv)
offset (n, v)
oil burner
oilcup
oilhole
oilless
oil-temper (v)
oiltight (a)
oil well
open cut
opencut (opencast) (a, adv)
open shop
open-shop (a)
orange-peel (a)
ore bed
ore body
ore zone
outboard (a, adv)
outlay
outlet
outline
out-of-stock (a)
outpour
output
outright (adv)
overall (n, a)
overburden (n, v)
overflow (n, v)
overhand (n, v)
overhand (a, adv)
overlap (n, v)
overload (n, v)
overlook (n, v)
overproduction
overrule (n, v)
overshot (n, v)
oversize (n, v)
overspeed (a)
overstate (v)
overturn (v)
oxyacetylene
oxyhydrogen

P

packing ring
panel board
parcel post
parcel-post (a)
patternmaker
paymaster
payroll
payroller
peacetime
peephole
penstock
percent
percentage
petcock
photochemical
photochemistry
photomicrograph
pickup
piece rate
piecework
pig iron

pile driver
pinch bar
pinchcock
pinhead
pinhole
pin-type (a)
pipe cutter
pipe fitting
pipeline
pipe tongs
pipework
pipe wrench
piston rod
plaster of paris
plate girder
plug switch
polyphase (a)
postwar (a)
pothead
pour point
power factor
powerhouse
preeminent (a)
preheat (v)
preignite (v)
preset (v)
press fit
presswork
prewar (a)
program
pump rod (shaft)
pump house (room)
push button
push rod

Q

quasi-legal (a)
quicklime

R

radioactive (a)
radio frequency
radiogram
radiograph
radiophone
radiotherapy
railroad (railway)
rainfall
rawhide
reach rod
reenter (v)
reestablish (v)
refuse (n, v)
re-fuse (electrical) (v)
reinforce (v)
relief valve
right-of-way
riprap
riveted (v, a)
roadbed
road roller
rock burst
rock drill
rock dust
rocker arm (shaft)
rod mill
rough-turned (v, a, adv)
roundhouse
runoff
runway

S

safety valve
salable (a)
sandblast
sandpaper
sawmill
screwdriver
screw eye
screw machine
seaboard
sea level
seaward
seawater
searchlight
second-class (a, adv)
second-feet
secondhand (a, adv)

self-contained (a)
self-evident (a)
selfsame (a)
semiannual (a)
semiautomatic (a)
semichrome
semicircle
semisquare
semisteel
series-multiple (a)
series-parallel (a)
series-wound (a)
setback
set back (v)
setscrew
setup
set up (v)
sheet pile
shipyard
short circuit
short-circuit (v)
short-circuited (a)
shortcut
short-cut (v, a)
shunt-wound (a)
shut down (v)
shut-down (a)
shutdown
shutoff
shut-off (a)
shut off (v)
sidelight
sideline
sidetrack (n, v)
sight-feed (n, a)
single-phase (n, a)
single-stage (a)
skillful
slip fit
slip ring
slip-ring (a)
smokebox
smokeflue
smokestack
snap switch
so-called (a)

soda ash
someone
soybean (*not* soyabean)
spillway
splice bar
spray nozzle
squirrel-cage (a)
stamp mill
standard-gage (a)
standby (n, a)
standpipe
stay bar
stay bolt
stay-bolt (v, a)
steamboat
steam chest
steam engine
steamtight (a)
steelwork
stiffleg
stiff-legged (a, adv)
stockholder
stockpile
stockpile (v)
stockroom
stopcock
straightedge
straightforward (a, adv)
streambed
streamline
stub shaft
stud bolt
stuffing box
subassembly
subbase
subbasement
subcommittee
subcontractor
subdivide (subdivision)
subgrade
sublevel
subsoil
substandard
substation
sulfur (*not* sulphur)
superheat

superhighway
switchboard
switch house
switchkeeper
syrup (*not* sirup)

T

T-bar
T-rail
T-slot
T square
tail block
tailrace
tailstock
takedown
takeoff
taxpayer
templet (*not* template)
test tube
textbook
theater (*not* theatre)
thermocouple
thermodynamics
threefold (a)
three-phase (a)
three-way (a)
third-class (a)
thumbnail
thumbscrew
tidewater
tie bolt
tie-in
tie rod
tie-up (n, a)
timekeeper
time sheet
time study
time-study (a)
timetable
time-temperature (a)
tin plate
tin-plate (a, v)
today
toggle joint
tomorrow

tonight
toolholder
toolmaker
toolroom
tool steel
topsoil
toward (*not* towards)
towboat
townsite
tracklaying
trackwork
trademark
trainload
tranship (transhipment)
traveler
traveling (a, v)
try cock
tube mill
tube sheet
tugboat
turbine-generator
turbogenerator
turnbuckle
turnout
turnover
turntable
turret lathe
tuyere
twist drill
two-phase (a)
twofold (a, adv)

U

U-bend
U-bolt
U-tube
ultramarine (n, a)
ultraviolet (n, a)
underconsumption
undercut (n, v, a)
underfeed (v, a)
underground
understatement
underway (adv)
unheard-of (a)

unidirectional (a)
unilateral (a)
unionize (v)
unipolar
upkeep
upset (n, v)
upstairs
upstream (a)
uptake
up-to-date

V

V-belt
V-connected (a)
V-groove
valve gear
valve seat
valve stem
vaporproof (a)
viewpoint
volt-ampere
voltmeter

W

wallboard
wartime
washout
water column
water-cool (v)
watercourse
waterfall
waterfront
water gage
water hammer
water jacket
water leg
water mill
water power
waterproof (n, v, a)
water-resistant (a)
water seal
watershed
water supply

water tank
watertight (a)
water tube
waterwheel
waterworks
watt-hour
watt-hour meter
wattless (a)
wattmeter
wavelength
weatherproof (a)
weekday (n, a)
weekend
well-being
well-known (a)
wheelbase
wheelwright
widespread (a)
wilful (a)
wing nut
wing wall
wiredraw (v)
wire edge
woodwork
worldwide (a)
worm gear
worm wheel
worthwhile
wrist pin
wrist plate
wrought iron
wrought-iron (a)

Y

Y-connect (v)
yardstick
yearbook
year-round (a)

Z

zigzag (n, v, a, adv)
zinc
zinciferous (a)

***Words Beginning with* in *and* en.**   Consult *Webster*. Preferred practice is to write certain commonly used words as follows:

> enclose, endorse, enforce (*but* reinforcement), enjoin (*but* injunction), enroll, entitle
>
> *Also:* Embed, embedment, embody, embodied, etc., where *im* and *em* are involved.

Compounding

Avoid indiscriminate use of the hyphen. Use the dictionary as a guide. In brief, the hyphen is used as follows:

1. In spelled-out compound numbers of less than 100, as *ninety-two*.
2. In spelled-out fractions, as *two-thirds, four-fifths*.
3. Between two nouns denoting a combination of equal things, as *secretary-treasurer*.
4. Between the words of a compound unit made by combining single units, as *car-miles*.
5. In prepositional phrases forming nouns, as *task-force-commander*.
6. To avoid ambiguity, as *new-business department, fifty one-dollar bills*.
7. Between two or more words used as a single adjective before a noun, as *constant-speed motor, adjustable-voltage system, compound-wound motor, 110-volt line, 91-ft depth*. However, do not use the hyphen when the first word is an adverb ending in *ly*. Write, for example, "carefully prepared plan." Also do not use a hyphen when the two words form a predicate adjective. Write, for example, "The plant has first-class equipment"; but, "The equipment is first class."

For more detailed information on when and when not to use the hyphen, consult the section on "Punctuation, Compounds, Capitals, Etc." in *Webster's New Collegiate Dictionary*.

Word Division

Check the division of words that break at the end of the type line. Do not depend upon the compositor for correct division. Good sense and readability are the best guides to proper word division. There are a few absolute rules aside from the examples given in *Webster*. One is that a word of one syllable may never be divided (*boat, freight, tongue, wrought,* etc.). Other rules express general principles only and must be used with judgment. Accepted practice, according to Woolley, is summarized as follows:

1. Divide at syllables. To determine the syllables of a word, consult the dictionary.
2. Pronounce the word slowly. This is a fairly accurate method of determining division, although error in pronunciation can lead to error in division. For that reason, the dictionary is the final authority. The rule is not to break combinations of letters sounded together. For example, *na-tion*, not *nat-ion*; *intro-*

duction, not *introd-uction*; *illus-trate*, not *illust-rate*; *finish-ing*, not *fini-shing*; *han-dling*, not *hand-ling*.

3. Do not separate a one-letter syllable, such as *y* or *a*, from the remainder of the word; that is, do not divide *many, against, along, stony, atypical.*
4. Avoid awkward division, such as would result in attempting to divide *every, even, only, eighteen.*
5. Divide between a prefix and the following letter. *Be-tween*, not *bet-ween*; *pre-fix*, not *pref-ix*, but *pref-ace*, not *pre-face*; *rec-ommend*, not *re-commend*.
6. Divide between the suffix and the preceding letter. *Invit-ed*, not *invi-ted*; *strong-er*, not *stron-ger*; *act-ing*, not *ac-ting*.
7. When a consonant is doubled, divide between the consonants in spite of rule 6. *Equip-ping*, not *equipp-ing*; *revet-ting*, not *revett-ing*; *rub-ber*, not *rubb-er*; *expres-sion*, not *express-ion*.
8. Do not divide digraphs or trigraphs, such as *th, ch, tch, ea, oa, ou, ow, eau*. The letters *ck* are divided only when followed by a final *le* (see rule 9). *Batch-ing*, not *bat-ching*; *meth-od*, not *met-hod*; *sign-ing*, not *sig-ning*; but *sig-natories, distin-guish, sin-gly*.
9. In words like *possible, bridle, trifle, buckle, twinkle, staple,* and *entitle,* do not separate *le*. Combine it with the preceding consonant. *Possi-ble*, not *possib-le; enti-tle,* not *entit-le; buc-kle,* not *buck-le.*

Turn Lines. Do not allow the printer to turn an abbreviation or a word or syllable of two, three, or four letters at the end of a paragraph. The following are examples of bad practice:

At the Brilliant plant of the XYZ Mfg. Co., Saco,
Me.
of rough topography where aerial tramways find
their greatest usefulness in the mining industry.
giving curves and tables to clarify the discussion.

CAPITALS

Capital letters, like italics, are employed to emphasize certain words. Fundamentally, proper nouns (names) are capitalized whether used as independent nouns or as qualifiers (adjectives). Common nouns are not.

This principle would be simple to apply were it not that many nouns and adjectives that are common in most connections are proper in others. "Board," "committee," and "commission," for example, are plainly common nouns. Yet they become proper in "National Labor Relations Board," "Committee of Ten for the Coal and Heating Industries," and "Interstate Commerce Commission." They also become proper when used alone as a shortened form of a longer name, as "the Lakes" for "the Great Lakes."

It is impossible to set down rules covering every case, but a closer approach to uniformity can be attained by keeping in mind the purpose of capital letters and the fundamental principles covering their use. Some suggestions follow:

1. Capitalize proper names and their derivatives if used with a proper meaning, as "Rome," "Roman" (engineering); "Italy," "Italian" (aircraft); "Germany," "German" (submarines). But derivatives with an acquired and widely familiar meaning are usually lowercased.

babbitt metal	paris green
bunsen burner	portland cement
german silver	roman type
india ink	stillson wrench
italicize	

2. Capitalize, with the exceptions noted at the end of this section, a common noun used as a short form of a specific common name that is capitalized. Therefore, as previously pointed out, write "the Labor Board" when referring to "the National Labor Relations Board." Other examples are: "the Canal" (Panama Canal), "the District" (District of Columbia). Be sure, however, that your readers know which lakes, which canal, or which district is meant. If there is any doubt, it is best to use the longer form.

Exceptions: Do not capitalize "board," "commission," "association," and the like when used alone, since this leads to excessive use of caps. When first used, write the name in full, "The Interstate Commerce Commission," for example, and thereafter write the "commission" without capitalization.

3. Capitalize the plural form of a common noun that is capitalized. For example, "river," a common noun, becomes the "Hudson River" when used as part of a proper name, and "Hudson and East Rivers" when used in the plural. Other examples are:

> Agriculture and Interior Departments
> Atlantic and Gulf Coasts
> Departments of the Army, Navy, and Air Force

4. Capitalize "the" when it is an essential part of a proper name or title, as "The United Nations." But write "the United Nations Court." Strictly speaking, "the" should be capitalized when it is part of a company, newspaper, train, or other name. Commonly, however, it is lowercased unless, for example, a company or organization makes a point of capitalization, as "The Hudson Coal Co." Lowercase examples are:

the Bristol Co.	the Fourth
the Broadway Limited	the *Atlantic Monthly*

5. Capitalize the full names of incorporated or organized bodies and the distinguishing shortened forms of such names.

> Bureau of Census, the Census Bureau
> National Labor Relations Board, the Labor Board (*but not* the Board)

6. Capitalize names of members of a party or organization to distinguish them from the same words used in a merely descriptive sense. Nixon is a "Republican," but France has a "republican" form of government. Gladstone was a great "Liberal," but Franklin D. Roosevelt was only a "liberal." Harry Bridges was adjudged not a "Communist," but he might be a "communist" if he advocated communistic methods. To make the point absurdly simple, a man may be an "Elk" but never an "elk."

7. Capitalize descriptive terms when used to denote definite regions or localities.

the Gulf States	the Orient
the Eastern Shore	the East
the East Side	the West
the Torrid Zone	the Midwest
the Continental Divide	the Middle West
the Western Hemisphere	the Far East
the Continent (Europe)	Down East

Do not, however, capitalize a descriptive term when it denotes direction or position only.

western Massachusetts	north, east
southern California	south, west
equatorial	oriental, occidental

8. Capitalize formal names of historic events, epochs, holidays, feast days, and fast days.

Fourth of July	World War II
the Fourth	Christian Era
Renaissance	Feast of the Passover

9. Capitalize trade names, varieties, market grades, and brands, as:

The Climax machine is produced by the Jones Mfg. Co., Pittsburgh.

Note that the noun following the name is not capitalized.

Caution is advisable in using trade names. Many manufacturers designate products by proprietary (trademarked) names, as Bakelite, Kodak, Jackhamer, and require that these names be capitalized. Best practice in writing or editing is to employ the type or class name of the article, using the proprietary name of a particular make only when the purpose is to mention that one make. Bakelite is a phenolic resin. A Kodak is a camera, but all cameras are not Kodaks. Much rock drilling is done with jackhammers, part of it with Jackhamers.

Many forms of transparent synthetic sheeting are incorrectly called "cellophane" (without capitalization). Cellophane (capitalized) is the product of one manufacturer. Other makes of transparent sheeting are known by other names. Do not call a piece of sheeting "Cellophane" unless you are sure it is Cellophane. Do not call a tractor a "Cat," "cat," or "Caterpillar" unless you are sure that it is a Caterpillar and mean to tell the reader so; otherwise, use crawler, crawling, or track-laying not only when referring to tractors but also to other equipment traveling on similar gear.

A list of names and proprietary terms is given at the end of this section. The list follows current good practice in the use of caps or lowercase and is to be taken as a guide rather than a rule. Some trade names have come to be generally written lowercase even though still subject to proprietary right or claim of such right. In using such names, the writer should carefully consider their status and, when in doubt, capitalize. This will avoid dispute and any possible legal complications.

10. Capitalize fanciful appellations used with or instead of proper name:

the New Deal
the Windy City

the Hub
the Big Four

 11. Capitalize vivid personifications:

> The Court sustained the objection.
> A Committee was appointed by the Chair. (*But* by the chairman.)

 12. Capitalize a personal or official title or designation preceding a person's name. Abbreviate only if the person's given name or initials are used.

Senator Bridges
General Sherman

Pres. O. L. Alexander
Prof. W. B. Plank

 A title immediately following the name of a person, or used alone as a substitute for such name, is not, however, capitalized (except in bylines).

> O. L. Alexander, president, Pocahontas Fuel Co., Inc.
> W. B. Plank, professor of mining engineering
> The president of Princeton University

Exceptions include titles of preeminence or distinction. Dwight Eisenhower was "President" of the United States in 1960 (there is only one President), whereas there are thousands of presidents of universities, manufacturing companies, associations, and so on.

 13. Capitalize the first word of a sentence, an independent clause, a direct quotation, a line of poetry or a formally introduced series of statements following a colon or equivalent.

> The question is, Shall the company declare a dividend?
> Five points made by Mr. Hook in a recent address are: (1) The mutual interest of the engineer and manager should be recognized; (2) The function of engineering....
> Lincoln began to speak: "Four score and seven years ago...."

But the first word of a fragmentary quotation is not capitalized.

> Did not Lincoln say this nation "was conceived in liberty and dedicated to the proposition that all men are created equal"?

Do not capitalize the first word after a colon when what follows is merely supplemental in nature.

> Whether to close down or seek new capital: that is the question.

 14. Where caps and lower case (c&lc) are used, as in titles, subheads, and bylines, capitalize the first word and all following words except articles (a, an, the), prepositions, and conjunctions.

> Food Plants of the Middle West
> Planning for Increased Production
> Looking into the Earth with Diamond Drill
> Henry C. Woods, Chairman of the Board, Sahara Coal Co.

 15. Caps and small caps (c&sc) used in place of caps and lowercase are treated the same as c&lc, the small caps being handled as though they were lowercase.

 16. Do not capitalize the seasons of the year.

spring   summer   autumn   winter

17. Capitalize common nouns used with dates, numbers, or letters.

Exhibit A   Act of 1960   Appendix B   Table III   Chart 2

18. Capitalize the first word of an abbreviation when the full word takes a capital.

Montana, Mont.   September, Sept.

19. Capitalize such abbreviations as "Chap. 2," "Art. 3," "Fig. 1."

20. Beware of using capitals for entire text words. There are a few legitimate places for all caps, chiefly: (a) in table titles and similar isolated semidisplay lines. Usually, however, caps and lowercase (or caps and small caps) are better. (b) In text where an imperative warning is described and the use of caps effectively simulates the sign or sound.

Someone cried FIRE, and the entire crowd....
The warning EXPLOSIVES was painted on the truck.
Some 50 ft ahead loomed the STOP sign.

And (c) in starting an article or a text division, caps may be used for the first one to three words. Caps and small caps may also be used.

Caps or Lowercase?

[For name words and proprietary terms, see the list following this one.]

Abstract D
Act (as in Wagner Act, Revenue Act of 1960)
Act of 1960
Administration, Veterans', Rural Electrification, etc.
administration, Eisenhower
Admiralty, British
Age or Ages, Victorian, Dark, Medieval, etc.
Airport, La Guardia
all-American
Ambassador, British
anarchist
antipodean
Appendix A
Army, U.S., British, etc.
army, Bradley's
Art. 2
Assistant (if preceding a capitalized title)
Associate Justice
Authority, Tennessee Valley, Port of New York, etc.
authority, the leading
autumn

babbitt metal
Badlands (S.D. and Neb.)
badlands (a kind of topography)

Bank (if part of a proper name)
Battle of the Bulge, the Rhine, etc.
battle, the, at the Rhine
Belt, Black, Wheat
bessemer steel, furnace
Bible (sacred work)
bible, the, of the industry, etc.
biblical
Board (if part of a proper name)
board of directors
bordeaux mixture
Bridge, George Washington, Brooklyn, etc.
bridge, New York Central
Bureau, of the Census, Engraving, of the Budget

Cabinet, the President's, British, etc.
Canal Zone
Capes, the Charles and Henry
Capital, National (Washington)
Census, Sixteenth
central Europe
central time (central standard time)
Central European time
Chap. 5
Chart 2
china clay
chinese blue
City (if customarily used as part of the name, as New York City, Mexico City)
city of Boston
Civil Service Commission
Civil War
Colonies, the Thirteen
Comintern, Cominform
Commission, Interstate Commerce, National Waterways, etc.
Committee, of the Whole, Ways and Means, Republican National
Commonwealth, of Australia, of Kentucky
communism
Communist (party member)
Confederacy (of the South, etc.)
Confederation, Swiss
Conference, First Hague, Limitation of Armament, Summit
Congress, U.S., Tenth Geodetic, etc.
Congressional Library
Congressional (action, authority, etc.)
Congressman (only when used as title with proper name)
Continent (continental Europe)
continent, the American, African, etc.
corliss engine
Corn Belt
Cotton Belt
Court (capitalize official titles of all courts)

Dam, Boulder, etc.
Day, Labor, Navy, etc.
Delta, Mississippi, etc.
delta (electrical connection)
Democrat (party member)
democrat, democracy, democratic
Department, of Commerce, Labor, State, etc.
department, executive, legislative, judicial, etc.
Deputy (if preceding a capitalized title)
District, Chicago Sanitary, of Columbia
District, First Congressional
Division, First, etc. (of the Army)
division, New York, of the Reading R.R.
Drawing A

earth (except when used with names of planets, as Saturn, Mars, Earth, etc.)
East (section of the U.S., etc.)
Eastern Shore (Chesapeake Bay)
eastern New York
eastern time
easterner
election day
embassy, British
Equator (of the earth)
equator (middle circumference, as of a man)
Equatorial Region (of the earth)
equatorial
escalator (moving stairway)
Executive, Chief (President of the United States)
Executive Mansion
executive departments
Exhibit A
ex-President (of the United States)
ex-president (of a company, association, etc.)

fall (season)
Falls, Niagara, etc.
Far East, far-eastern
Far West
Federal Reserve Board
Federal government, law, etc.
Fifth Ave.
Fig. 1
flag, British, U.S., etc.
Fleet, Grand, High Seas, U.S., etc.
Foundation, Engineering, Rockefeller
Fourth of July, the Fourth
Frisco (write without apostrophe)
fuller's earth

General Order 14 (*not* General Order No. 14)

general order, a
german silver
Geological Survey
George VI, the Sixth
glauber salt
Gold Coast (Africa, Chicago)
Gothic architecture
gothic type
government, British, French, Federal
Graph 3
great circle
Greater New York
Gulf of Mexico, etc.
Gulf Stream

harveyized steel
Hemisphere, Eastern, Western
hessian fly
High School (if part of proper name)
Highway 1
His Excellency
His Majesty
House of Representatives

independence, in the year of our
india ink
india rubber
Inquisition, Spanish
international law
interstate
Isthmian Canal
italic type
Ivory Coast

jacquard
jamaica ginger
Jersey cattle
jersey (fabric)
Journal, of the ARS, House, Senate

kraft paper

Lakes, the (Great Lakes)
Law, Wagner, Fair-Trade Practices, etc.
Left (political belief)
legation, Chinese
Legion, American, Foreign, etc.
Legislature, Michigan, Ohio, etc.
Levant, the
levant (leather)
Librarian of Congress
Loop, the (Chicago)

lower California
lower house (of Congress)

macadam, macadamized road
Magna Charta
mandate
manila rope
Maritime Provinces (Canada)
mason jar
mercerized cotton
merino sheep
Midcontinent
midcontinent region
Middle Ages
morocco leather
Mountain States
mountain time

Nation, Osage
nation (in general)
nationwide
National Guard
national customs
Naval Academy, U.S.
naval appropriation
Navy, U.S., British, etc.
Navy Yard, Brooklyn
navy blue
Near East
New Year's Day
new year, the
Nordic
North, the
North Atlantic States
North Pole, North Star
northern New York State
northerner

Occident
occidental
Office, Government Printing
Old South
Orient
oriental

Pacific coast, seaboard, slope, time
Pacific Coast States
Pact, Kellogg
Pan-American
Panhandle, Texas, West Virginia
Parliament, House of
papal

paris green
Part 4
party, Democratic, Republican
pasteurized milk
peacetime
pitot tube
plaster of paris
Plate 2
Pole Star (Polaris)
Pope
portland cement
power plant
Powers, Central, Allied, etc.
Premier (as part of a title)
President (of the United States)
president, Long Island R.R.
Prime Minister, British
Province of Ontario
Proving Ground, Sandy Hook
proving ground (Buick)
prussian blue
Puritan (member of sect)
puritan (in general)

Rebellion, Mexican, Russian
Red (Communist)
Representative (only when used with proper name)
Republic, The French
Right (political leaning)
River Plate
roentgen rays
roman type
Rule 6
Ruler of the Universe

Schedule 2
scotch plaid
scriptural
Scriptures
Secretary of State
secretary, Interstate Commerce Commission
Section 1
Senate (U.S.)
Senator (only if proper name follows)
senatorial
Service Air, Customs, Air Transport, etc.
servicecenter, service station
Socialist (party member)
socialist, socialism, socialistic
South Pole
South, the

southerner
Soviet Russia
soviet, a
spring (season)
standard time
Stars and Stripes
State of New York
statehood, statehouse, state's rights, statewide, etc.
stillson wrench
summer
sun (see earth)
Supreme Bench (U.S. Court)
System, Federal Reserve
system, the Pennsylvania

Thanksgiving Day
Tidal Basin (D.C.)
time, standard, eastern, central, mountain, etc.
transatlantic
trans-Canada
Tropic of Cancer
turkey red
turkish bath
turkish toweling
Twentieth Century Limited
Twin Cities (Minneapolis and St. Paul)

ulster (coat)
Under-Secretary of State
Union (synonym for the United States)
Union of Soviet Socialist Republics (Russia et al.)
union shop
upper house (of Congress)
utopia

venetian blinds, color, fabric
vice consul
Vice President, United States
vice president, McGraw-Hill Publishing Co.
Volume 2

War, Civil, Great, World, etc.
Ward 1
wartime
watt
West End (London)
West, the
Western Hemisphere
western part
westerner
White House (in Washington)

World Court

X ray

Your Excellency, Your Honor, Your Majesty, etc.

Zone, Canal, Frigid, Temperate
zone, a low temperature

Typical Product Names and Proprietary Terms

Aerocrete (lightweight concrete)
Alclad (protected aluminum)
Ambursen (dam)
Ameripol (artificial rubber)
Amplidyne (electrical control)
Armco (metal)

babbitt (metal)
Bakelite (plastic)
bascule (bridge)
battledeck (floor construction)
bessemer (steel, furnace, etc.)
brinnell (hardness index)
Buna (artificial rubber)

carborundum (abrasive)
Carryall (scraper)
Caterpillar (crawler tread)
Celanese (fabric)
Cellophane (transparent cellulose sheet)
Celotex (fiber board)
Cement Gun (mortar sprayer)
Copperweld (protected steel)
cordeau (fuse)
Cor-ten (steel)
cravenette (waterproof fabric)
crucible (steel)
Cupaloy (metal)

De-ion (electrical control and lightning arrester)
diesel (engine, oil)
Douglas fir (timber)
Downmetal (metal)
Dresser (pipe couplings)
Duco (enamel)
duralumin (metal)
Durez (plastic)
Duriron (metal)

Enduro (steel)
Escalator (moving stairway)
ethyl (treated gasoline)
Exide (battery)

Fiberglas (glass fiber and fabric)
fresno (scraper)
Frigidaire (refrigerator)

Gelamite (explosive)
gelatin (explosive)
Gelex (explosive)
gelignite (explosive)
gunite (sprayed mortar)

Howe (truss)
Hypernik (metal)

ignitron (rectifier and certain parts)
Imhoff (tank)
Inconel (metal)

Jackhamer (a make of rock drill)
jackhammer (a type of rock drill)

Keene (cement)
Keglined (containers)
Kemtone (wall finish)
Kerite (insulation)
Koroseal (plastic)

Lang lay (rope)
Lastex (fabric)
Leadite (calking compound)
Lincoln (type of milling machine)
Lockbar (pipe)
Lock Joint (pipe)
Lucite (transparent plastic)

Masonite (wall board)
Mayari (steel)
mazda (lamp)
Meehanite (metal)
Megger (resistance meter)
Micarta (plastic)
moly (molybdenum)
monel (metal)

neoprene (artificial rubber)
Nichrome (metal)
nicol (prism)

nylon (fiber and fabric)

open hearth (steel furnace)
osnaburg (fabric)
otto (engine cycle)

Palm Beach (fabric)
Petit (truss)
Pima (cotton)
Pliofilm (rubber derivative)
Plyform (wood)
Polaroid (glass)
portland (cement)
Pratt (truss)
preformed (wire rope)
Presdwood (wall board)
Pretest (piles)
Primacord (fuse)
Pumpcrete (concrete pump)
Pyrex (heat-resistant glass)

rayon (fiber and fabric)
Rockwell (hardness scale)
Rosendale (cement)
Rototrol (electrical control)

Sanforize (fabric treatment)
Sanitize (fabric treatment)
scleroscope (hardness tester)
selsyn (motor)
Shelby (tubing)
shotcrete (sprayed mortar)
silicone (chemical)
sisal (rope fiber)
Smooth-on (lute)
southern pine (timber)
Sprabond (metallizing process)
stainless (steel)
Stellite (hard-surfacing compound; also Haystellite)
Sterilamp (sterilizer)
S-twist (left-hand lay)
Swiss (fabric)

tainter (radial gate)
Tenite (plastic)
Thiokol (artificial rubber)
Toncan (metal)
Tournapull (scraper power unit)
Transite (asbestos-cement board, shape or pipe)

Unaflow (engine)
uniflow (type of engine)

venturi (meter)
Versatrol (electrical control)
Vibrolithic (pavement)
Vickers (hardness tester)
vinylite (plastic)

Wallamp (lamp)
Weightometer (conveyor scale)
Whirley (derrick)
wilton (fabric)

Z-twist (right-hand lay)

ITALICS

Italic type is used to promote reader understanding by emphasizing certain material. Italic, however, is harder to read than roman. Long passages wear the reader down and defeat the main purpose, which is emphasis. Therefore, italic should be used sparingly. Examples of correct (1) and incorrect (2) usage are as follows:

1. The budget is unbiased and impersonal, but furnishes a definite interpretation of the productiveness of each department in uniform, one-dimensional terms: *dollars of cost.*

2. Mr. X gave a good talk on sales management. *Yet he appeared unconscious of the connection between his own awakening and the presence in his city of some important new management personalities who have given the old town's executives a powerful hypodermic.*

Only three words were italicized in the first example. These dealt with costs—something at which any reader picks up his or her ears. Italic type emphasized this key phrase and therefore was a logical choice.

Inspection of the second example shows little need for emphasizing the sentence set in italic. It is long and much harder to read. Consequently, the italic loses its force.

Italicize letter symbols in text references to illustrations, drawings, figures, and so on.

Eight bars were stacked on two pins at *A*, Fig. 3. The lever, *B*, was placed near the top and between the two pins. This lever was connected to a mercury switch, *C*, beneath the table.

FOREIGN WORDS

Avoid using foreign words and phrases. Choose the English equivalent wherever possible. For example, why write *vide supra* when "see above" is plain to everybody?

When foreign words and phrases are used, they should be italicized. But, if possible, avoid such mystifiers as *pari passu, quod vide,* and so on. A few Latin words and some abbreviations have become anglicized and should be printed in roman. Typical ones are:

etc. (*et cetera,* and so on)

i.e. (*id est,* that is)

e.g. (*exempli gratia,* for example)

viz. (*videlicet,* namely)

vs. or v. (*versus*)

et al. (*et alibi,* and elsewhere, or *et alii,* and others)

Certain foreign words have been incorporated into the English language and therefore should be set in roman. An example is "cafe" without the accent. The distinction is easy to make. If a word or expression is so commonly used that it is understood by most readers, there is no justification for setting it in italic.

ABBREVIATIONS

Technical and business publications use abbreviations extensively because they make for better understanding through more compact statement. Nevertheless, abbreviations should not be used unless they are clearly within the scope of reader knowledge and understanding. Puzzling abbreviations should be avoided.

The list at the end of this section contains abbreviations (except common-language abbreviations) recommended for use in technical publications generally. Authorities, with some exceptions, are *Webster* and the *American Standard Abbreviations for Scientific and Engineering Terms,* approved by the American Standards Association. Additional abbreviations appropriate to the special field of a publication should be used where necessary.

Punctuation of Abbreviations. Use the period in abbreviations of proper names, personal or official titles, and other terms (except chemical and certain other symbols) that normally are capitalized.

Mont., U.S., Mr., Dr., Gen., Fig., No., etc.

Do not use periods in scientific or engineering abbreviations. Do not punctuate trigonometric, chemical, and certain mathematical and other symbols, as "sin," "cos," "Na," "Fe," "log," "*M*," and the like.

Do not punctuate abbreviations consisting of the capital letters of association names and names of certain government offices and agencies.

ASME, *not* A.S.M.E. (American Society of Mechanical Engineers)

NLRB, *not* N.L.R.B. (National Labor Relations Board)

Plurals. Abbreviations of plural terms (with specific exceptions) are the same as those of singular terms: "in," *not* "ins"; "kv," *not* "kvs." Exceptions are "Figs.," "Nos.," "Sts.," "Aves.," "Gens.," "Drs.," "Profs.," "Messrs.," etc.

Quantity Designations. Units following a numeral are always abbreviated, as "5 hp," a "2-in bar." *But not* "five hp," a "two-in bar."

Dates and Series. Names of months and series units accompanying numerals are abbreviated, as "Feb. 22," "Art. V," "Fig. 7," "14th St." *But not* "the first of Feb.," "Art. Five," "Fifth Art.," "Seventh Fig.," "Fourteenth St." (See also the section on "Numbers and Numerals" for rule on numbered streets.) Street numbers carry an ordinal suffix; dates do not. Write "March 15," "42nd St.," "25th Ave."

Personal and Official Titles. When a name follows a title of rank or position, abbreviate the title, as "Dr. Brown," "Dr. J. C. Brown," "Dr. James Brown," *not* "Doctor Brown," "Major-General Robinson," "Mister Smith." However, where abbreviation would be awkward, spell the title out. Write, therefore, "Chairman Herzog," "Ambassador Smith," etc.

Countries and States. Abbreviate "United States" in the name of a government department or bureau, as "U.S. Bureau of Mines," "U.S. Department of Commerce." Do not, however, precede a department, bureau, or agency abbreviation with the abbreviation "U.S." Therefore, do not write "U.S. SEC"; write "U.S. Security and Exchange Commission" instead. The same rule applies where state agencies are involved, as "W.Va. Department of Mines."

In addresses, abbreviate "U.S.," "Can.," "Mex.," and their states and provinces following the name of a city, as "Toronto, Ont., Can." Use "D.C." and "D.F." as the equivalent of state names. "U.S.S.R." may be used for the Union of Soviet Socialist Republics.

Name Prefixes. Abbreviate "Mount" and "Fort," as "Ft. Ticonderoga," "Mt. Taylor." Do not abbreviate "Port" and "Point" since the meaning of the abbreviation may not be clear. Write "Port Jervis," "Point Pleasant," not "Pt. Jervis," "Pt. Pleasant." Abbreviate and lowercase "Mexican" when qualifying a stated sum of silver money as "$6 mex."

Sentence Openings. Do not, with certain specific exceptions, begin a sentence with an abbreviation. Spell out the word. Exceptions are abbreviations of personal and official titles and Fig., No., or other term of enumeration when followed by a numeral, as, "Dr. Brown opened the discussion." "Fig. 1 shows the direction of attack." "No. 2 engine failed on take-off."

Directional Prefixes. Abbreviate the prefixes "East," "South," etc., in street and other names, as "E. 42nd St.," "S. Platte," and so on.

Capitalization. Capitalize the first letter of an abbreviation when the word would be capitalized if spelled out.

Spacing. Do not letter-space compound abbreviations. Write "Btu," "ASME," *not* "B t u" or "A S M E."

Connection and Quantity Symbols. Use the lowercase x, without spacing, between figures denoting dimensions, as "2 × 12 in," "17 × 40 ft," etc. Use the ampersand (&) in company and partnership names, as "Williams & Jones," "Morgan & Co.," "Bell & Zeller Coal & Mining Co." Use the hyphen or the symbol @ in place of the word "to" in reporting market quotations, as "Saturday's price was 5-5¼c," or "5@5¼c."

Except in tables and similar condensed statements, do not use symbols for degrees, minutes, and seconds. Use "deg," "min," "sec." Where temperatures are involved, deg may be omitted, as 92F, 135C. Use the symbol for percent, as 83%, *not* 83 percent, in all text, body, and other material. Symbols may be used for certain terms or as a part of compound abbreviations, as "φ" for altitude and "μa" for microampere (see ASA Standards for other examples). Abbreviations of frequently recurring terms may be employed after first use in fully spelled out form in the same piece, as "ac" for "alternating current," "50 mgd" for "50 million gallons per day." Where writing a large number of ciphers is not desirable, money sums, such as $2,000,000,000 may be written $2 billion.

Special Practices. Tables, construction reports, and the like often require the use of abbreviations not permissible in ordinary text. For such cases, the rules given here do not apply.

ABBREVIATIONS OF FREQUENTLY USED WORDS

acre	Spell out
acre-foot	acre-ft
afternoon	P.M.
Alabama	Ala.
Alaska	Spell out
Alberta	Alta.
alternating current	ac
ampere	A
ampere-hour	Ah
amplitude modulation	AM
and	& (company and firm names only)
and so forth	etc.
Angstrom	Å
ante meridian	A.M.
antilogarithm	antilog
April	Spell out
Arizona	Ariz.
Arkansas	Ark.
Article	Art.
Assistant	Asst.
Associate	Assoc.
Association	Assn.
audio-modulation	AM

August	Aug.
Avenue	Ave.
avoirdupois	avdp
barrel	bbl
Baume	Be
Birmingham wire gage	Bwg
board feet (feet board measure)	fbm
boiler horsepower	boiler hp
brake horsepower	bhp
brake horsepower-hour	bhp-h
British Columbia	B.C.
British thermal unit	Btu
Brothers	Bros.
Brown & Sharpe (gage)	B&S
Building	Bldg.
Bulletin	Bull.
bushel	bu
California	Calif.
calorie	cal
Canada	Can.
Canal Zone	C.Z.
candlepower	cp
center to center	c to c
Centigrade	C
centimeter	cm
cents	c
Chapter	Chap.
chemically pure	C.P.
circular mil (mils)	cir mil (mils)
Colorado	Colo.
Committee	Comm.
Company	Co.
Connecticut	Conn.
Corporation	Corp.
cosecant	csc
cosine	cos
cost, insurance and freight	cif
cotangent	cot
County	Spell out

cubic centimeter	cm³
cubic feet (foot)	ft³
cubic feet per minute	ft³/min
cubic inch	in³
cubic meter	m³
cubic yard (yards)	yd³
December	Dec.
degree	deg (°in tables)
degree Centigrade (Fahrenheit)	C (F)
direct current	dc
District of Columbia	D.C.
dollars	$
dozen	doz
elecctromotive force	emf
elevation	el
equation	eq
Fahrenheit	F
February	Feb.
feet per minute	ft/min
Figure (illustration title)	Fig. (Figs.)
Florida	Fla.
for example	e.g.
foot (feet)	ft
footcandle	fc
foot pound	ft.lb
forenoon	A.M.
franc (francs)	fr.
free aboard ship	Spell out
free alongside ship	Spell out
free on board	f.o.b.
frequency modulation	FM
gallon (gallons)	gal
gallons per minute	gal/min
General	Gen.
Georgia	Ga.
grain	gr
gram	g

Hawaii	Spell out
horsepower	hp
horsepower-hour	hp-h
hour	h
hundredweight	cwt
Idaho	Spell out
Illinois	Ill.
inch	in
Incorporated	Inc.
Indiana	Ind.
indicated horsepower	ihp
inside diameter	ID
Iowa	Spell out
January	Jan.
July	Spell out
June	Spell out
Kansas	Kan.
Kentucky	Ky.
kilocycle (kilocycles)	kc
kilocycles per second	kc/s
kilogram	kg
kilogram-meter	kgm
kilograms per second	kg/s
kilometer	km
kilovolt	kV
kilovolt-ampere	kVA
kilowatt	kW
kilowatthour	kWh
latitude	lat
Limited	Ltd.
lineal (linear) foot or feet	lin ft
Philippine Islands	P.I.
post meridian	P.M.
pound	lb
pounds per cubic foot	lb/ft^3
pounds per square inch	lb/in^2

pounds per square inch (absolute, gage)	psia, psig
pounds per square foot	lb/ft^2
pounds sterling	£
power factor	pf
Professor	Prof.
Puerto Rico	P.R.
quart	qt
Quebec	Que.
Railroad (Rail Road)	R.R.
Railway	Ry.
revolutions per minute	r/min
Rhode Island	R.I.
Saint	St.
Sainte	Ste.
Saskatchewan	Sask.
secant	sec
second	s
second (angular measure)	"
Section	Sec.
September	Sept.
shilling	s
sine	sin
South Carolina	S.C.
South Dakota	S.D.
specific gravity	sp gr
specific heat	sp ht
square	sq
square foot (inch, yard, meter)	sq ft (sq in., sq yd, sq m)
Street	St.
Superintendent	Supt.
tangent	tan
Tennessee	Tenn.
Texas	Tex.
that is	i.e.
thousand (lumber, brick, gas, etc.)	M
ton	Spell out

ton-mile	Spell out
Utah	Spell out
Vermont	Vt.
versed sine	vers
versus	vs.
volt, volts	V
volt-ampere	VA
volume	vol.
Washington (State of)	Wash.
watt	W
watthour	Wh
West Virginia	W. Va.
Wisconsin	Wis.
Wyoming	Wyo.
yard	yd
year	yr

NUMBERS AND NUMERALS

Spell out numbers from one to nine, except where exact dimensions or other quantity designations are involved. Use numerals for numbers above nine. Do not begin a sentence with a numeral. Care should be exercised in the use of numbers and numerals in titles of written pieces.

Special Rules

In a series of numbers partly over and partly under nine, use figures throughout, as "The vehicle count was 86 domestic, 8 foreign, and 4 commercial cars." Use figures wherever a number precedes or follows an abbreviation, as "25 ft," "Fig. 3," "42nd St." Spell out numbered avenues up to and including "Ninth Ave."; use numbers beyond, as "10th Ave.," "22nd Ave.," etc.

Use figures in referring to a numbered section, item, or illustration. Thus, when referring to a numbered passage of text, such as "(2) a long row of lathes, (3) a group of milling machines," etc., the reference should read "item 3," *not* "item three." Other like references are written "question 4" (or Q.4), "Fig. 2," and so on, *not* "question four," "Figure Two." Use numerals for mixed numbers, as "2½ years," "3½ times"; but write "A year and a half ago," etc. Use numerals for dates, as "Feb. 22." Tables, semitabular matter, market reports, construction reports, and other matter requiring maximum space reduction for printing are exceptions to the rule calling for the spelling out of numbers. A spell-out fraction takes a hyphen, as "one-half," "three-fourths."

Large Numbers

Numbers of four or more places may be pointed off in groups of three by commas, as "5,000 mi," "$3,000,000." The use of space is also acceptable in pointing off large numbers of five or more digits, for example, "3 000 000." Do not point off house numbers, style or list numbers, telephone numbers, and so on. Write "Style K5280," "2468 Fifth Ave." Do not point off places to the right of a decimal point.

Use only three to six significant figures in giving populations, capital totals, and other large sums, as, "The 1988 population was 4,234,000." "Production in 1990 was 609,000,000 tons." However, where precision is required by the text statement, give all the figures, as, "The surplus increased from $2,168,520 to $2,172,490, or a little less than $4,000."

Money

In stating money in integral dollars, omit the decimal point and the ciphers for cents. Thus, write $3, *not* $3.00. In tables where none of the entries in the same column or comparable columns include cents, omit decimal point and ciphers for cents throughout. Where some entries include cents, use decimal points and figures for ciphers for cents throughout.

Physical and Chemical Data

In figures giving chemical and physical data, 2.3, for example, is not the same as 2.30 and must not be so written where the data call for the latter since the degree of precision would be changed. For a similar reason, a table of analytical constituents does not require uniformity in number of decimal places. Rather, the degree of precision of each constituent is indicated by the decimal places used, as:

	Percent
Iron	74
Nickel	18
Chromium	8
Carbon	0.5
Sulfur	0.02

Other Uses

In a decimal without an integer, place a zero before the decimal point, as 0.268, *not* .268. When expressing sheet, lumber, or other sizes made up to two dimensions, use the lowercase x in place of *by* between the two numbers, as 2 × 4, 60 × 84.

Tables and Captions

Use arabic figures followed by a dash in numbering tables, as "Table 3—Comparative Efficiencies of Wet and Dry Processes." References in the text of the piece are written "Table 3," etc.

Captions carrying figure numbers can be written in the following style, with a dash following the number or numeral, as "Fig. 8—Schematic wiring diagrams showing load centers and circuits in the XYZ plant." Where a series of illustrations shows an operating or manufacturing sequence, large initial-type numbers or numerals may be used. Preferable practice is to permit them to stand alone without a period or a dash following.

PUNCTUATION

Good punctuation clarifies meaning and makes reading easier. However, a dozen commas and semicolons will not cure faulty sentence structure. The answer is a good working knowledge of grammar. Every writer's library should include a good book on grammar. Some hints designed to develop greater proficiency in punctuation are given below. The writers seeking additional information should refer to his or her grammar or to *Webster's New Collegiate Dictionary*.

Period

A period is not used after a heading or title except when the heading or title is on the same line as the subject matter.

Ellipsis. To show that words have been omitted, use three periods separated by em spaces, four periods when the omitted words are at the end of a sentence.

> These publications... will continue to grow in circulation.
> Their circulation has increased consistently....

Abbreviations. Do not use periods after abbreviations, except abbreviations of certain proper names, titles, places, or words which are capitalized in their regular form. See section on "Abbreviations" for rules and examples.

Comma

An original and still valid function of the comma is indicating an elocutionary pause. A second function—and the principal one at the present time—is setting off sentence elements for clarity and easier reading. Major examples of such usage follows.

Compound Sentences. If the coordinate clauses are not directly related—if the subject changes—use a comma.

> A chain is only as strong as its weakest link, and an elevating grader is only as efficient as its elevator.

However, if the clauses are directly related—in other words, deal with the same subject—*do not* use a comma.

> This publication has 30,000 subscribers and covers more than 18,000 plants.

Appositives. Set off by commas, as:

Mr. Edwards, *my partner,* is on the Coast.

If, however, the appositive is so closely connected in thought as to form one idea, *do not* use the comma.

My partner Edwards is on the Coast.

In enumerations, watch out for such appositives as:

Roger Evans, Vice President Crowley, John Clooney, the sales engineer from Ohio, and Chairman Doe spoke on the same program.

How many spoke? Four or five? One way of avoiding the possibility of the reader's not knowing whether John Clooney and the sales engineer from Ohio are two people or one is to italicize *the sales engineer from Ohio.* Otherwise, use semicolons, as:

Roger Evans; Vice President Crowley; John Clooney, the sales engineer from Ohio; and Chairman Doe spoke on the same program.

Parenthetical Material. Set off by commas, noting that in this instance, as in certain others, the commas come in pairs, one before and one after the insertion.

Joe Doakes, who operates one of our milling machines, has been with us 25 years.

Absolute Phrases and Clauses. Such phrases and clauses generally are written with *having* and *being* and usually open the sentence. They should be set off by commas.

The engineering staff having asked for more money, the chairman pointed out that granting the request would require raising the tax rate.
Time being short, the final coat of paint was omitted.

Adverbs and Adverbial Phrases and Clauses. Set off by a comma if they precede the main phrase, clause, or sentence.

If you want results, stick to the job.
Immediately after the missile was fired, he returned to examine the launching pad.
Upon arrival, we will inspect the plant.

If the phrase or clause follows the main clause, the comma usually is omitted.

Stick to the job *if you want results.*
We will inspect the plant *upon arrival.*

If the phrase or clause following the main clause is plainly nonrestrictive, that is, adds a reason or concession introduced by *because, since, as,* or *though,* use a comma.

He is trying to repair the motor, *though I doubt if he knows how.*
The attempt was futile, *since the reserves were inadequate.*

Transitional Phrases and Adverbs Used Transitionally. Set off by commas. The conjunctive adverbs used traditionally include *therefore, moreover,* and *however.*

On the contrary, proper methods are a guarantee of results.

The motor, *however,* was chosen for its ability to operate at high temperatures.

Defining or Restricting Phrases or Clauses. Use commas to set off dependent nonrestrictive adverbial clauses preceding their principal clause, most adverbial clauses containing a verb, nonrestrictive relative clauses, and participial phrases.

The author spent some time at the plant, *which is in South America.*

The engineers favored the expenditure, *as had been forecast.*

That package, *which won first prize,* was designed by Smith.

The engineers, *convinced of success,* went on to their next job.

If a phrase or clause defines or restricts, however, it *must not* be set off by commas.

The package *that won first prize* was designed by Smith.

The man *who operates the milling machine* has been with us 25 years.

Enumerations and Series. Separate elements by commas, using a comma before *and* or *or* if practicing close punctuation.

Steel, concrete, and wood were employed in constructing the plant.

In open punctuation, the comma before the final *and* or *or* is omitted.

Steel, concrete and wood were employed in constructing the plant.

Where confusion or more difficult reading might result, use a comma before the final *and* or *or.*

Nuts and bolts, screws and nails, and wire and rope were employed as needed.

Do not use a comma where the second adjective relates more closely to the noun or where the first adjective modifies the second adjective *and* the noun which are thought of as a unit.

A *modern* high-speed lathe.

A *shallow* meandering stream.

An *able* estimating engineer.

Coordinate Words, Phrases, or Clauses. Separate by commas when used in a series without connectives.

The plant operating executive must make *prompt, rapid* decisions.

If, however, the adjectives are not coordinate, and if each modifies the whole group that follows, do not use commas.

They installed a *modern high-speed* lathe.

A *funny little old* piece of equipment is still being used.

Independent Clauses. When joined by a coordinating conjunction, separate by a comma when closely connected in thought and not broken up by commas.

He seemed not to care, *but missed no detail.*

If both clauses are brief, and especially when the subject is the same, the comma may be omitted.

He will try *but he will fail.*

Other Comma Uses. Before *of* in material dealing with residence, position, or title:

John Smith, *of Arkansas...*
Lieutenant Jones, *of the 12th Regiment...*

Words placed out of their natural position:

Independence is worth the effort, *I agree.*
Accuracy, *not speed,* is the goal.

Short informal quotations:

"Watch out," he yelled.

After *namely, viz., that is, i.e., as, e.g.,* etc., in introducing examples, but not before enumerations:

Three types of steel are preferred, *namely, low-carbon, copper-bearing and high-strength.*

In the second category, *i.e., metal designed to resist corrosion,* we find a number of types.

Adjacent sets of figures:

In *1947, 6,200,000,000* tons of coal were mined.

Semicolon

A semicolon denotes a longer natural pause than a comma. Used properly, it leads to greater precision and exactness in expression.

Compound Sentences. Use the semicolon to separate coordinate members not closely connected in sense.

"Wisdom is the principal thing; therefore get wisdom; and with all thy getting, get understanding."

"The world will little note, nor long remember, what we say here; it can never forget what they did here."

Coordinate Clauses. Use the semicolon between coordinate clauses when there is no coordinating conjunction.

He went fishing; I stayed here to work.

Series. Use semicolons to separate a series of clauses or phrases that have a common dependence on some other phrase or word.

"...here highly resolve that these dead shall not have died in vain; that this nation, under God, shall have a new birth of freedom; and that..."

Enumeration. Use semicolons between successive main divisions of an enumeration.

The ore contained: Gold, $2 per ton; silver, 6 oz per ton; lead, 6%; zinc, 4%.

Semicolon for Comma. Use semicolons in place of commas in series where commas might invite misunderstanding.

The party was composed of George Gray; H. M. Smith, his partner; J. L. Brown, his secretary; and J. Doakes.

Colon

Quotations. Use the colon before a long or formal quotation, and before a quotation that begins a new paragraph.

He began as follows:
"My friends, once more I come to report on the condition of the industry."

Enumeration. Use the colon to introduce an enumeration or a series of items.

The unit has three advantages: (1) adequate power for the job, (2) ability to operate under adverse conditions, and (3) long life.

Try this menu: rice, milk, eggs, fruit.

Elements of Time. Use the colon.

The engineer finished in 0:5:10.6.
He started the job at 10:42.

Ratios. Use the colon.

A 3:1 slope.
The concrete mix was 1:2:4.

Apostrophe

Possessives. Use the apostrophe to form the possessive; also in such expressions as *a year's work, three weeks' vacation, the day's total, a moment's thought.*

Elision. Use the apostrophe to indicate elision of letters, as in *don't, weren't it's.*

Plurals. Use the apostrophe with an *s* to form the plurals of letters, figures, and symbols, as "A's," "9's." Do not, however, write "three's," "five's," when referring to groups of three or five.

Years. Write "the class of '95" if you wish to abbreviate, but write "in the 1990s" or "in the nineties."

Possessive Pronouns. Never use the apostrophe with *hers, its, yours,* etc.

Dash

Avoid excessive use of the dash. Employ it only when it is plainly the most appropriate mark. Usually the comma or semicolon will serve equally well. However, the dash may be used in the following.

Parenthetical Expressions. Senator Reed used satire at times—mocking, Mephistophelean satire—but he also was a master of unassailable logic."

Repetition. "Never is virtue left without sympathy—sympathy dearer and tenderer..."

Unexpected Turn in Sentiment. "He was a good designer, very generous—with other people's time."

Statement or Summary of Particulars. "Reputation, money, friends—all were sacrificed." "A solid has three dimensions—length, breadth, and thickness."

Hesitation, Suspense, or Delay. "The pulse fluttered—went on—fluttered feebly—then stopped."

Before an Author's Name at the End of a Quotation. "...and that government of the people, by the people, for the people, shall not perish from the earth."—Abraham Lincoln.

Hyphen

No universal rule has been accepted for distinguishing compound expressions that should have a hyphen from those that should not. Knowledge gained from long experience, careful observation, and frequent use of the dictionary are the only guide. As a rule, hyphenated expressions tend to coalesce and lose the hyphen as they grow older and more common.

Words united to form compound adjectives usually should be written with a hyphen. Always consider clearness. Be sure the series is attributive. Write, "This is an easy-to-read document," but, "This document is easy to read."

Do not use hyphens in *today, tomorrow, tonight, already;* when two short nouns are joined to make another noun, as, *courthouse, toolroom, baseball;* between an adverb ending in *ly* and an adjective, as, *freshly painted walls.*

For further discussion of the use of the hyphen, see the section on "Spelling and Compounding."

Parentheses and Brackets

Use parentheses to enclose parenthetical material by the author. If a punctuation mark ordinarily would be required after that part of the sentence preceding parentheses, put it after the second curve.

John Jones (a civil engineer), of Boston, was the next speaker.

If an entire sentence is enclosed in parentheses, the period should come before the curve.

The engine burned 8 gal of fuel per hour. (See Table 1 for average fuel consumption of other engines.)

If only the last words of a sentence are enclosed, the period should go after the last curve.

He uses many words incorrectly (for example, *practical* and *practicable*).

Use brackets to enclose matter written in by somebody other than the author.

This problem [fluidation] will be more easily solved.

Quotation Marks

Use quotation marks or italic type to call attention to a new or unfamiliar word or phrase.

"Relativity" is a word known to all of us, but understood by few.

But if the word appears a second, third, and fourth time, don't continue to put it in quotes. Once is enough.

Quotation marks also may be used, but cautiously, to direct attention to technical or other words of unusual application, or to preposterous or absurd assumptions.

References to "galley" slaves have nothing to do with printing.

He visited the plant and bought a "diamond" drill for a dollar.

In direct quotation, use double quotes to indicate a primary quotation, single quotes to indicate a quotation within a quotation.

Two simple rules cover the use of other punctuation marks in conjunction with quotation marks: (1) The period and the comma must always be placed inside the quotation marks. (2) The colon, semicolon, exclamation point, and question mark should be placed outside the quotation marks unless they are a part of the quoted matter.

He writes under the heading of "Notes and Comments": "Nearly everyone can testify to the truth of the old adage, 'It is better to be safe than sorry.'"

The subject of her first lecture was "The Thisness of the That"!

The question is, "Shall we adjourn sine die?"

Question Mark

There should be no doubt about the use of the question mark. Sometimes, however, an author forgets whether a group of words is a question or not and consequently uses a question mark where he or she shouldn't or fails to use one where he or she should. Remember that the question mark is for direct questions only. Avoid the coy use of the question mark in parentheses to express doubt. Do not write "the benefits (?) of the National Engineering License Act."

Exclamation Mark

Avoid excessive use of the exclamation mark. The danger lies in converting a mere statement or question into an exclamation simply because the person who made the statement or asked the question was excited. Usually it is not *how* the speaker says something but *what* he or she says that makes an exclamation.

TEXT DIVISIONS

Divisions or subdivisions of a statement, either in the same sentence or paragraph or in consecutive paragraphs or text sections, often are emphasized by numbers, 1, 2, 3, or by letters, *a, b, c,* A, B, C, etc. Numbers and cap letters are set in roman; lowercase letters may be set in italic to distinguish them from the text.

When the divisions or subdivisions are separate paragraphs, designating figures and cap letters (roman) are followed by a period and a space (no dash and no parentheses); designating lowercase letters (italic) are enclosed in parentheses followed by a space (no period and no dash). Examples are:

   Lines of action suggested are:
   1. Independent action by the several engineering consultants
   2. Concerted action by commercial firms

   Lines of action suggested are:
   A. Independent action by the several engineering consultants
   B. Concerted action by commercial firms

   Lines of action suggested are:
   (*a*) Independent action by the several engineering consultants
   (*b*) Concerted action by commercial firms

When the divisions or subdivisions occur within a sentence or paragraph, both letters and numbers are enclosed in parentheses followed by a space.

   Lines of action suggested are: (1) independent action by the several engineering consultants and (2) concerted action by commercial firms.

GOOD AND BAD WRITING

Here are a number of common errors that should not appear in any kind of technical writing. Individual authors will think of many more. However, the following are offered as helpful suggestions.

A *and* the. Don't begin sentences and paragraphs with *a* or *the* any oftener than necessary. Monotony is the result. On the other hand, beware of bluepenciling *the* to the extent of making your copy telegraphic.

Admit of. Use *admit of* only in the sense of "present an opening" or "leave room for."

This drafter's ability *admits of* no question.

Adverbs. Be careful where you put them. Write "so it can be readily removed," *not* "so it readily can be removed."

Affect *and* Effect. *Affect* and *effect* are not synonyms. Neither can take the place of the other. (See *Webster.*)

All of the. "All of the motors," "all of the operators." Eliminate *of.* Write "all the motors," "all the operators," or in some instances, "all motors," "all operators," eliminating *of* and *the.*

Alternative Subjects, or Or-groups. When the two parts of an or-group differ in number, the verb usually agrees with the nearer.

The question is whether the foreman or the men *are* right.

And/or. As in "dirt and/or dust." Write instead, "dirt or dust, or both."

And which, but which. Use these only when there is a preceding "which" clause. In sentences like "This suggested a bituminous concrete mixed in a central plant, *and which* could be used without heating," omit the *and.*

Balance. Don't use *balance* when you mean "rest" or "remainder."

Because. When a sentence begins with "The reason is" or "The reason why...is," the clause containing the reason must not begin with *because,* but with *that.*

The reason why the men struck was *that...(not* because)

Between. "Between you and I" is illiterate. "Between each post" is obviously impossible.

Both...as well as. Either omit "both" or change "as well as" to "and."

But. Forget the old rule that you must not start a sentence with *but* or *and.* But don't overdo it.

But, However. But is stronger than *however.* Use *but* to start a sentence or a clause. *However* is placed, if possible, after the word that is contrasted to some word in the preceding sentence.

He said he would not use....He would, however, try the...

But...however. Watch for this common redundancy. "But one must take into consideration, however,..." Either *but* or *however,* by itself, is enough.

Can't hardly. Can't hardly slips by once in a while. It shouldn't. Correct to *can hardly.*

Case. Beware of *case,* "*cases,*" "*instances.*" The phrase, "In many cases the motors were..." is much more simply written, "Many of the motors were..."

In the case of. Guard against its use. "In the case of machine tools, operators are paid on a piecework basis," is slipshod writing. What you meant to say was, "Machine-tool operators are paid..." or, "In machine-tool plants, operators..."

Common. The comparative and superlative forms are *commoner* and *commonest.* "Most common feedwater problems" and "commonest feedwater problems" do not mean the same thing.

Compound Subjects. Compound subjects (two or more subjects connected by *and*) take *are*, not *is*, even when the noun nearest the verb is singular.

Their jobs, their families, and their security *are* in danger.

Consensus of opinion. *Consensus* means "unanimity of opinion." Write "consensus," or "consensus of opinion."

Contact. *Contact* is much overworked as a verb in modern technical English, as, "The salesperson contacted the space-buyer." Check with *Webster* before using *contact* as anything but a noun.

Copulative Verbs. Copulative verbs (*be, is, are, was, were, seen, become, look, sound, appear, taste, feel, smell*) take adjectives, not adverbs, when the qualifier refers to the subject.

He felt bad (*not* badly).

Dangling Preposition. Some people used to say it was wrong to end a sentence with a preposition. It isn't. For example, "This is a good plant to work in." Don't change to "in which to work."

Data. *Data* is the plural form of *datum*. You must, therefore, use a plural form of a verb.

Data are (*not* is).

Device. Use *device* sparingly.

Differ (meaning "be different"). *Differ* is followed by *from*, never by *with*. But you can say, "I differ with you," meaning "I have a difference of opinion."

Different than. Common in careless writing. Correct to "different from."

Due to. Beware of treating *due to* like a compound preposition (which is what *owing to* has become). *Due to*, like any participle or adjective, must be attached to a noun expressing cause. Don't write, "The motor would not run, due to sand in the gearbox"; say "because of sand."

Each. Guard against the common error of treating *each* as though it were plural, as, "Each industry will be taxed *their* pro rata share." *Its* is the word to use.

Each other, one another. Use *each other* when there are two, *one another* when there are three or more.

Efficient. Use *efficient* sparingly. "They made efficient use..." You probably mean "effective" use. Or simply "good" use.

Elegant words. Avoid elegant and ponderous expressions. "The operator came to his place of employment in an intoxicated condition." He "came to work drunk," didn't he?

Et cetera, etc., and so forth, and so on. You need not wind up every series with *et cetera* or one of its forms. When you introduce a series by *including, such as, some examples are,* or similar words, you state at once that you are giving a partial list. So *et cetera* is redundant.

Except. Except, not *excepting*, is a good preposition. *Excepting* is a verb form.

Farther and further. Use *farther* when distance is involved. "He walked 10 miles farther." Elsewhere, use *further*.

Firstly, secondly. Drop the *ly* and your copy will sound less like a sermon or legal document.

French and Latin words. Do not use foreign words your reader may not know or fully understand. Better still, do not use foreign words or phrases at all.

Hanging participle. The hanging participle is a common fault. "Going down the street, the sidewalk was seen to be damaged." There must be a noun in the sentence for the participle to modify. "Going down the street, he saw that the sidewalk had been damaged." Better still, "As he went down the street, he saw that the sidewalk had been damaged."

I, for me. How often do we hear or read something like, "The big boss told the foreman and I that...."? The simplest way to determine whether to use *I* or *me* in a sentence like this is to try the sentence without "the foreman and." Since you would never say, "The big boss told I that...," you know *me* is the correct word.

Imply and infer. *Imply* and *infer* are not interchangeable. The one who makes the statement "implies"; whoever is listening may "infer."

In so far as, inasmuch as. Steer clear of *in so far as* and *inasmuch as*. Write *so far as* instead of *in so far as*. *Since* for *inasmuch as* saves three syllables.

Irregardless. Not a standard English word.

It's, its. *It's* is a good colloquial form for "it is," but the possessive of *it* is "its."

Just. Avoid too frequent use of *just*. *Just exactly* is a tautologism. *Just how many* and *just what* are Americanisms. Look on *just* with suspicion.

Lay and lie. *Lay* is a transitive verb (lay, laid, laid). *Lie* is intransitive (lie, lay, lain).

They laid the pipes 6 feet below the ground surface.
The pipes had lain in that position for 10 years.

Less and fewer. Plurals do not take "less." Write "less money," but "fewer dollars."

Norway has fewer (*not* less) ships than Great Britain.
The old machine turned out fewer parts per hour.

Like. *Like* as a conjunction is wrong. Don't write, "They don't work like we used to." The right word is *as*.

Locate. *Locate* is a much abused word. *To locate* means "to select a place for" or "to place in a particular spot." Instead of saying, "The boiler room is located in the basement," why not simply, "The boiler room is in the basement"?

Majority. *Majority* is always singular when it means a "superiority in number," as "the majority was small." It takes either a singular or plural verb after words of multitude, depending on whether the statement relates to the body or to its members. "The majority of the committee was (*or* were) against the bill." But it takes a plural verb when it means "the greater part numerically." "A majority of the employees have chosen...."

Misquotations. If you must paint the lily, paint it, don't gild it. Never screw your courage to the "sticking-point"; it's "sticking-place." If you feel a quotation coming on, look it up and be sure it's right.

More perfect, most perfect. *Perfect* itself is superlative. A thing can't be "more" or "most" perfect.

None. *None* is singular.

None of the employees is (*not* are) on piecework.

Nor. *Nor* goes with *neither;* it does not go with *either* or *not.* Correct: "The company neither makes nor buys its machines." Incorrect: "The company does not make nor buy its machines."

Not only...but also. Watch where you place *not only.* Follow with *but also,* not with *but* alone.

Novel (adjective). Use sparingly.

Number (see also Majority). Nouns of multitude may be treated as singular or plural at discretion, depending on whether they are considered as single entities or as the individuals who compose them. Such words include *committee, company, party, crowd, number, majority.* For example, in, "The committee are agreed that the work is too fatiguing," the plural is used because the writer is thinking of the individuals; it takes two or more to agree. But, "The committee votes..." because here the committee acts as a single entity.

Of. Do not needlessly repeat *of* and other prepositions. "They thought of buying a new machine and of installing it in the shipping room" is better written without the second *of.*

One. Avoid expressions like, "He is one of those men who does things." The verb should be plural to agree with the plural "men." *Does* should be *do.*

Only. *Only* is often misplaced. Compare the meanings of these sentences: (1) "Only I urged him to try this"; (2) "I only urged him to try this"; (3) "I urged only him to try this"; (4) "I urged him only to try this"; (5) "I urged him to try only this."

Possessives. It was formerly customary, when a word ended in *s,* to write its possessive with an apostrophe but no additional *s.* This is still correct practice in verse and in poetic or referential context. But elsewhere add the *s,* as in "Jones's" plant, "Mr. Williams's" office.

Pronouns. Pronouns must have antecedents. Also, (1) the antecedent should not be far off; (2) there should not be two possible antecedents, giving rise to doubt about which one the pronoun represents; and (3) one pronoun should not represent two antecedents on one occasion.

Propensity. *Propensity* is followed by "to do" or "for doing," not "of doing."

Proportion. In phrases like "the greater proportion of buyers," substitute "most buyers."

Proposition. *Proposition* is a poor substitute for *task, job, problem, objective,* or *proposal.*

Puns. A pun almost never has a place in technical writing.

Said. In the sense of "aforesaid," *said* is good only in legal documents. Do not write "the said employees," "said manual," "said product," and so on.

Same. Be careful to avoid use of *same* in the sense of "the aforesaid thing(s) or person(s)." Example: "The company bought new machines and hired additional employees to run same." Write *them* in place of "same."

Scarcely. Should not be used in a phrase such as "without scarcely knowing."

Should have liked, would have been possible, would have been the first. After past conditionals, the present infinitive is almost invariably the right form. Say, "I should have liked to go," *not* "I should have liked to have gone."

Subjunctive mood. Always use the subjunctive mood in clauses expressing a hypothesis that is not a fact; for example, "if he were here now" (but he isn't).

Superfluous words. Watch for *obviously, obtain, viewpoint* or *standpoint, available,* and other superfluous words. *Obviously* almost never adds anything to the sense and frequently insults the reader's intelligence. Why write, "Excellent silica removal has been obtained," when you could write simply, "Silica removal has been excellent"? Now look at, "From the creep-strength viewpoint..." It's enough to say, "For creep-strength..." As for *available,* it's always creeping into copy. For example, "These treatments are available to prevent" means "These treatments prevent."

Tautology. Tautology is needless repetition of the same idea. "The cinders fall down into the trap." *Down* is not needed. "Business was again made the subject of another attack." Omit either *again* or *another.*

That *and* which *clauses.* Many people think *that* is colloquial and *which* is elegant writing. Not so. Each has its place. And the rule is easy: *That* is a defining relative pronoun; *which* is nondefining. A defining relative clause is one that identifies the person or thing meant by limiting the denotation of the antecedent. For example, "The plant that Smith built" or, "The four conveyors that move the product from assembly line to shipping room." Not any old plant, but a specific plant; not all the conveyors in the plant, but certain ones. On the other hand, in, "Their new plant, which was finished last February, is of steel and concrete construction," the clause does not limit "their new plant"; it simply adds a fact. Decide, therefore, whether a relative clause is defining (or limiting) or not. If it is, use *that* and use no commas. If it is not, use *which* and set the clause off by commas.

These, this. "These kind of people" is illiterate. "This kind of a job" is nearly as bad. Leave out the *a.*

Times. "Three times six is eighteen." A singular verb is generally used, regardless of the quantity.

Unique. Unique can be applied only to a thing that is in some respect the sole existing specimen. Don't use *unique* as a substitute for *rare, remarkable, exceptional,* and the like. A thing either is or isn't unique. You may write *nearly, quite,* or *almost* unique. But don't make the error of writing *very, more,* or *somewhat* unique.

Usage. Usage is a good word when used correctly, as in "according to modern English usage." But don't use it in place of *use* merely because you think it sounds better. For example, "The supervisor made good use (*not* usage) of the information." "The conveyor was in good condition after two years' use" (*or* "service," but *not* "usage").

Use (verb). A good word because it's short. But don't use it superfluously, as in, "Three indicators are used to show water level." You mean, "Three indicators show water level." Also, don't overwork *use.* If it appears several times in the same paragraph, choose a substitute to avoid monotony.

Very. Use *very* sparingly. In fact, blue-pencil *very* nine times out of ten.

View of, view to. Statements of reason or motive are often introduced by *in view of* (this or that condition or fact); statements of purpose by *with a view to* (such or such result). Both are correct. But avoid *with a view of* and *with the view of,* which are incorrect.

GRAMMATICAL WRITING

Some of us have painful memories of our early struggles with grammar. Perhaps our troubles were caused by childish memorization of rules we didn't understand, and still may not understand. Good writing is grammatical writing. Ungrammatical writing is poor writing, no matter how accurate the data and facts presented. When we write in an ungrammatical way, we offend our readers in much the same way that an ill-spoken person might embarrass listeners at a formal meeting of educators. Readers may be so strongly repelled by ungrammatical writing that they disregard the technical facts presented and think only of how poorly qualified the writer is. Good technical writing is an important door to success in many fields. Ungrammatical technical writing can harm a writer's otherwise excellent reputation. You owe it to your readers and yourself to learn, and to use, the simple rules of grammatical writing.

To some, the word *grammar* means complicated diagrams of long sentences, involved tense rules, and endless warnings about the horror of split infinitives. Actually, grammar is a valuable tool that helps you write better sentences with greater speed and confidence. When you look upon grammar as a useful tool, you find it more interesting and challenging. The rules that once were stuffy and difficult are now logical and natural. By using these rules as you write, they become a part of your writing skill. Soon you are writing better material and also with less effort.

GRAMMAR GLOSSARY

You needn't memorize the meaning of *sentence, subject, verb,* and the like. But read the definitions in the special glossary below whenever you have a chance. After a few readings you will know the meaning of most of these 32 key terms. Then if you begin to misuse one of the parts of speech, a phrase, or clause, your mind will become more alert and you'll correct the error before it occurs. The terms selected for this glossary are those you'll need most often in technical writing. The definitions are short and easy to understand and use. Study this glossary; see how useful grammar can be.

Active Voice. Form of verb used when the subject of the sentence acts. (This form is preferred for direct, personalized expression.)

    The pump *discharges* 30 gal/min.

    The new bridge *collapsed* during a storm.

    Adjust the engine timing once a month.

Adjective. Word that modifies a noun or pronoun; used to describe, tell number.

    The *large noisy* aircraft took off quickly. (Two adjectives, both modifying *aircraft*.)

    Business was *good* during the *last* month of the year.

    Three ships sank off the reef.

Adjective Clause. Subordinate clause used as an adjective; describes or limits noun or pronoun.

    He is the engineer *whom we met in Texas.*

    They use carburetors *that have automatic chokes.*

Adverb. Modifies verb, adjective, or other adverbs; shows manner, degree, time, or place.

The burner ignited *quickly*. (Manner)
There is *only* one chief engineer. (Degree)
Begin your writing *now*. (Time)
The plant will be built *here*. (Place)

Adverbial Clause. Subordinate clause used as an adverb; answers a question of how, why, when, where, how much.

The plant will be built *as the engineers specified*. (How)
A fire started *after the motor failed*. (When)

Antecedent. Word or group of words to which a pronoun refers.

The *motor* that failed was new. (*Motor* is antecedent of the relative pronoun *that*.)

Article. Either definite (*the*) or indefinite (*a, an*); used as adjectives.

Auxiliary. Verb used to form other verbs; *will, shall, can, may, have, be, do, must,* and *ought* are typical auxiliaries.

The report *had been* reviewed by the writer.
She *will* design the radar set.

Clause. Group of words containing subject and verb, used as part of a sentence. May be independent (a sentence) or dependent (as subject and predicate but does not express a complete thought).

The bridge fell but *it didn't injure anyone*. (Two independent clauses; each is a simple sentence.)
When the missile was fired, the crew cheered. (Dependent clause; has subject and verb, but does not express a complete thought.)

Complex Sentence. Contains one independent and one or more dependent clauses.

The airplane crashed *while it was approaching the field*.

Compound Sentence. Contains two or more independent clauses and no dependent clauses.

The missile was fired and *it went into orbit*.

Conjunction. Word used to connect clauses, phrases, or words. *And, but, or, for, nor, so, yet* (coordinating conjunctions) connect clauses, phrases, or words of equal importance.

She went to the plant *but* she returned immediately

After, although, as (not *like*), *because, if, since, that, unless, when, while* (subordinating conjunctions) connect subordinate (dependent) clauses to independent clauses.

The report will be read *if* it is acceptable

Either...or, neither...nor, both...and, etc. (correlative conjunctions) are used in pairs.

Either the engineer *or* his assistant will write the report.

Dependent Clause. Contains subject and verb but does not express a complete thought.

Ellipsis. Omission of words from a sentence or clause without affecting the clarity of the expression.

Use gas turbines when (it is) necessary.

Mercury is heavier than lead (is heavy).

Gerund. Noun formed by adding *ing* to a verb. Used like a noun but can take an object or an adverbial modifier.

Writing a book is good mental discipline.

Good *piloting* requires skill and training.

Independent Clause. Contains subject and verb, and states a complete thought; can be a simple sentence; may be introduced by a coordinating conjunction.

All engineers can write. But not all engineers are good writers.

Infinitive. Verb preceded by *to* or by another verb form; the simple form of the verb. Can be used as a noun, adverb, or part of a verb.

Her greatest desire is to *design* airplanes.

To *plan* the job, he worked late.

This plane can *carry* 70 passengers.

Modifier. Word or group of words used as an adjective or adverb to qualify, limit, or describe another word or group of words.

The new airplane flew *successfully*.

Articles *of a technical nature* should be written by people *well-trained in technical writing*.

Noun. Name of a person, place, or thing.

The *engineer* wrote the *report* at his office.

A *freighter* collided with a *tanker* in the *channel*.

Object. Word or words (noun, pronoun, phrase, or clause used as a noun) which receives the action of a verb or is governed by a preposition.

They placed the *missile* on the launching *pad*. (*Missile* is the object of the verb; *pad* is the object of the preposition *on*.)

Participle. Word derived from a verb but having the characteristics of both verb and adjective. Present participle ends in *ing;* past participle can end in *ed, en, t, n, d.*

The *rising* pressure caused a head loss.

The *recorded* pressure was 20 lb in².

Parts of Speech. Names of the words used in a sentence. There are eight parts of speech—noun, pronoun, verb, adjective, adverb, preposition, conjunction, and interjection. The interjection is probably the least used in technical writing.

Passive Voice. Verb form which indicates the subject is being acted upon.

The report *was written* by the engineer. (The same thought expressed in active voice should be: The engineer *wrote* the report.)

Person. Relation between the verb and subject showing whether a person is speaking (first person), spoken to (second person), or spoken of (third person).

I am a technical writer.

You are a technical writer.

He (or *she*) *is* a technical writer.

Phrase. Group of related words without a subject or predicate; used as a part of speech.

Hurried writing of *engineering reports* causes confusion. (*Of engineering reports* is a prepositional phrase used as an adjective.)

Writing specifications is a task that requires utmost care. (*Writing specifications* is a participial phrase used as a noun.)

Predicate. Word or words in a sentence or clause that make a statement about the subject.

The engineer *surveyed the property.*

Most well-written technical books *follow a logical line of thought from the known to the unknown.*

Preposition. Relation word that connects a noun or pronoun to some other element in the sentence.

The aircraft crashed *into* the hangar.

Make an outline *of* every proposed report.

Pronoun. Word used in place of a noun. The six types of pronouns are: (1) personal (*I, we, you, he, she, it, they*); (2) interrogative (*who, which, what*); (3) relative (*who, which, that*); (4) demonstrative (*this, that, these, those*); (5) indefinite (*one, nobody, someone, anything,* etc.); (6) reflexive (*myself, yourselves, himself,* etc.).

I collected data in the laboratory.

Mr. Jones was the engineer *who* spoke at the meeting.

Sentence. Group of words expressing a complete thought; contains a subject and a predicate.

He smiled. (Subject is *he;* predicate is *smiled.*)

Mr. Jones, the speaker, smiled broadly as he began to address the audience. (Subject in italic; predicate in roman.)

Subject. Word or group of words naming the person or thing about which something is said in a sentence.

Some *engineers* in industry write their own reports. (Simple subject)
Some engineers in industry write their own reports. (Complete subject)
Some *engineers and scientists* write their own reports. (Compound subject)

Tense. Change in verb form to show the time of the action. The English language has six tenses: present, past, future, present perfect, past perfect, future perfect.

Verb. Word or group of words expressing action or a state of being.

Technical writing *is* a new profession.
He *wrote* the paper for a technical society.

Note: A verb should always agree with its subject in number and person. Examples: *I am; You are.*

Verbal. Having the nature of or derived from a verb, but used as a noun or adjective. In the English language, infinitives, gerunds, and participles are verbals.

Voice. Form of a verb that shows whether the subject acts or is acted upon.

The junior engineer *opened* the throttle. (Active voice)
The throttle *was opened* by the junior engineer. (Passive voice)

SECTION 14
THE LANGUAGE OF PRINTERS AND PUBLISHERS

Throughout your writing career, you will associate with publishers, editors, illustrators, and printers. As a group, men and women in the publishing business are dedicated, sincere, hard-working people. Like any other specialists, they have a unique language. The special terms used in this language can be confusing if you do not know their meaning. To help you understand the language of publishers, editors, illustrators, and printers, a glossary of 236 common terms is given in this section.

Definitions are useless unless you refer to them every time a new term is met in the text. Get into the habit of referring to the glossary in this section whenever you meet a new word in the field of publishing or printing. Study the definition carefully. By referring to the glossary regularly, you will soon acquire a better knowledge of the special language of publishing.

If you are just beginning in the world of writing, study the glossary now, term by term. This may seem like a chore, but you will be repaid handsomely. For by studying the glossary now, you will prepare yourself for many of the terms used later in this book. Your progress will be faster, surer.

GLOSSARY

Acknowledgments: An author's statement listing and giving thanks to organizations or persons from whom he or she has received assistance or material for the preparation of his or her book, article, manual, etc. The statement may appear as a separate section of the front matter, in the preface, or in a footnote.

ACs or AAs: Author's corrections to material that has been set in type.

Align: To arrange in a straight line, as several lines of type, edges of cuts, headings with text, etc.

Appendix: Supplementary matter following the final page of text in a book and preceding the index; appears at the end of an article or paper.

Arabic numerals: The figures 1, 2, 3, 4, 5, 6, 7, 8, 9, 0 as distinguished from roman numerals.

Ascender: The part of a letter above the body of the letter. In the letter *b,* the vertical upstroke is the ascender.

Asterisk: The symbol * used as a reference mark in text and as the first of several symbols for footnotes. See *Reference marks.*

Author's corrections or alterations: Author's proof corrections that constitute changes from the manuscript; often abbreviated ACs, AAs.

Backbone: The edge of a book that is visible as it stands on a shelf. Generally the backbone bears the title, the author's name, and the publisher's name. Also called *shelfback, spine*.

Back matter: All printed matter following the end of the text. It may include appendixes, notes, vocabulary or glossary, bibliography, supplementary exercises, problems and the like, and index, folioed in sequence following the text.

Bastard title: See *Half title*.

Battered type: Type characters that are broken, worn, flattened, or otherwise damaged.

Bearer: A protective band of wood or metal put around type areas for platemaking. The broad black lines around foundry-proof pages are the marks of bearers. These do not show in final printing.

Ben Day process: The application of mechanical shading tints to plates or artwork to give a variety of tones to line drawings. The method is named for its inventor, and the term is loosely applied to other similar shading devices.

Bibliography: A list of books on a selected subject; specifically, the list of references at the end of a book or chapter, or end of an article or paper, citing the literature consulted by the author or recommended for study.

Bleed illustrations: Illustrations that extend to the trimmed edge of the page. Copy for such treatment is prepared with an allowance of $\frac{1}{8}$ to $\frac{1}{4}$ in to be cut off when the page is trimmed.

Blind stamping: The impression of a die or tooled design on a binding, without ink, leaf, or foil; also called *blind tooling, blanking out, blocking*.

Blurb: (1) A brief piece of advertising copy describing a book, particularly the copy printed on a book jacket. (2) Editor's introduction for an article.

Body: (1) The metal block on which a type character is cast. (2) The text.

Body type: The typeface used for the main text, as distinguished from type used for headings, footnotes, extracts, and the like.

Boldface: A typeface that is heavy and black, generally used for headings and symbols to be distinguished from the rest of the type matter. It is contrasted with *lightface*, generally used for the body of the text. To direct the printer to set in boldface, a wavy line is drawn under the matter to be set. In proof the notation "bf" is made in the margin opposite the wavy underscore. The entries in this glossary are set in boldface.

Box: An arrangement of rules forming a square or rectangular section to set off a self-contained unit of copy. See *Rule*.

Box head: The heading of a tabular column, often enclosed within rules.

Brace: The type character { which may appear vertically or horizontally to combine items that are equivalent or simultaneous.

Brackets: Square parentheses used in pairs [ ] to enclose interpolated or extraneous matter, for parentheses outside parentheses, and for mathematical quantities included in the same operation.

Broadside page: A page that must be turned sidewise to be read, used chiefly for illustration or tabular material that is too wide to stand vertically on the page.

Built-up fraction: A fraction with the numerator above and the denominator below a horizontal line, as

$$\frac{x}{y}$$

Bulk: The thickness of the total number of pages in a book. Consideration of the desirable bulk for usability and appearance enters into the selection of paper for a given book.

Caps: Abbreviated form of the word *capitals*.

Caption: The title or descriptive text accompanying a table or illustration.

Card plate: A page in the front matter listing other books in the same series or by the same author.

THE LANGUAGE OF PRINTERS AND PUBLISHERS 14.3

Caret: The symbol used as a mark of omission in manuscript or proof, showing where added matter is to be inserted in a line.

Case: (1) A tray divided into compartments in which a font of hand type is kept. See *Lowercase, Uppercase*. (2) A book cover into which the stitched and trimmed pages are bound.

Character: Any letter, symbol, or mark of punctuation. In counting characters in a line of manuscript, the spaces between words are included as characters.

Clean proof: Proof with few corrections.

Colophon: (1) A brief notice at the end of a book, giving information about the design, printing, and production. (2) A publisher's emblem used on title page, spine, jacket, etc.

Combination plate: A plate on which both line and halftone work are combined.

Composition: Typesetting.

Condensed type: A typeface with narrower letters than those of the parent face of a type family.

Contents: A section in the front matter of a book, listing chapters and sometimes parts, sections, and important subheadings, together with their page numbers.

Copy: Manuscript from which type is set or illustrations from which reproductions are made for printing; the material to be copied in the process of printing.

Copyholder: (1) In proofreading, a person who reads aloud from the copy while the proofreader marks the proofs. (2) In printing, a device that holds the copy for the compositor.

Credit line: A line of type acknowledging the source of an illustration, usually set directly below the illustration or with its legend. Also called *courtesy line*.

Crop: To cut off, as the edge of a cut or a picture for reproduction. Cropping may be done to lessen the reduction necessary to fit the page, to eliminate unwanted areas of the illustration, etc.

Cross-reference: A direction to the reader to refer to related matter elsewhere in the book, article, or paper.

Cut: (1) A metal plate from which an illustration is printed, It is prepared by photoengraving from the original illustration copy and bears a relief image of the original reversed as in a mirror. The image is transferred to the page by printing, which restores the normal attitude of the picture. (2) A printed illustration. (3) To delete or shorten text.

Cut dummy: Illustration (or cut) proofs arranged and numbered in correct order to facilitate correct page makeup.

Dagger: The symbol † used as a reference mark in text and footnotes, next in order after the asterisk. See *Reference marks*.

Dead matter: Manuscript, type, or cuts after printing, or prepared for printing but eliminated in proof stage. See *Kill*.

Deck: See *Blurb*.

Dedication: An inscription sometimes printed in the front matter of a book, addressing the work to a person or group of persons and often bearing a statement of affection, respect, or appreciation.

Dele: The proofreading mark which directs the printer to delete indicated matter.

Descender: The part of a letter extending below its body. In the letter *p* the vertical downstroke is the descender.

Die: A metal stamp bearing lettering or design to be impressed on a book cover.

Dirty proof: Proof with many corrections.

Displayed: Set on a separate line; term applied to headings, equations, etc., so treated.

Display type: Typefaces of 16- or 18-point size and up, used for major headings such as titles of articles, chapters, or books.

Drop folio: A page number set at the foot of the page.

Dummy: (1) A detailed layout showing the size and position of all elements on the pages of a magazine, book, or part of either. Proofs of type and illustrations may be pasted into the layouts (*pasteup dummy*). (2) A sample book, usually of blank pages, made to show the size, binding, quality of paper, etc., for a projected book.

Edition: All copies of a book published in substantially the same form, as distinguished from *impression* or *printing,* which covers all copies issued from a single printing. When significant changes have been made in content or format, reprinting is called a *new edition.*

Electrotype or electroplate: A metal plate exactly duplicating a book page, made by processes of molding and electroplating, and used in printing instead of the original type and engravings. Its advantage is that it is generally more durable than the original and can be duplicated as needed for large printings.

Em: A type measure equal to the square of the body of the type. Originally so called from the letter *M,* the width of which determined the measure.

Em dash: A dash 1 em in length; the regular dash in punctuation.

En: The measure of one-half an em.

En dash: A short dash used such as that used in ranges in place of the word *to,* e.g., 1980–1990.

Endpaper: A folded sheet of paper at the beginning or end of a book, half of which is pasted as a lining against the inside face of the binding, and the other half of which forms a flyleaf. Endpapers may be blank or printed with a pattern or illustration.

Engraver's proof: Proof of line cuts or halftones submitted by the engraver to show how cuts will print.

Errata: A list of errors discovered after the paper, magazine, or book has gone to press, inserted or bound in as a separate sheet to show the corrections.

Expanded type: A typeface with wider letters than those of the parent face of a type family.

Extract: An excerpt quoted from another work, usually set in type smaller than the text, indented, or separated from preceding and following material by space.

Face: (1) That surface of a type character which is inked and pressed against the paper in printing. (2) The style of a type alphabet, as Caledonia, Futura, boldface.

Figure: An illustration.

Flush: Even with the margin. The turnover lines in this glossary are flush left.

Flyleaf: A blank sheet preceding or following the printed pages of a book; the unpasted side of the sheet forming the endpaper.

Folio: (1) A page number. When placed at the foot, called *drop folio.* Page 1 and all odd numbers are always on the right-hand pages. (2) To number the pages of manuscript.

Follow copy: A direction to compositor or proofreader to retain the spelling and punctuation of the manuscript in every detail.

Font: The complete assortment of characters of one typeface in one size.

Footnote: A note of explanation, citation of source or reference, or the like, appended at the foot of a page or table, or occasionally at the end of a chapter or in back matter. Footnotes are referred to in the text by reference marks or superior numbers and are usually set in smaller type than the text.

Foreword: A preface; often a statement by someone other than the author or editor.

Form: All the pages (type, cuts, or plates) that are to be printed at one time on one side of a sheet of paper, assembled and locked up in proper order for the press.

Format: The whole physical form of a magazine or book, including its dimensions, page design, type style, etc.

Foul proof: Corrected proof from which a revised set has been made.

Foundry proof: A proof of type that has been approved and locked up. A broad black line shows all around the type where metal bearers have been placed for protection during the casting. This proof is seldom sent to authors.

Four-color process: A method of reproducing illustrations in full color with a wide range of graduated tones and colors. By photographing the original through color filters, the original colors are separated so that they can be printed from four plates with four inks—black, yellow, red, and blue. In successive runs through the press, the four colors, one for each plate, are combined at varying intensities to reproduce a range of tones like the original.

Front matter: All matter preceding the first page of text. Front matter may include a half title, card plate, frontispiece, title page, copyright, dedication, foreword, preface, acknowledgments, table of contents, list of illustrations, introduction, etc. Sequence of front matter varies somewhat with different publishers. The pages are generally numbered with lowercase roman numerals.

Furniture: Pieces of wood or metal used to fill out blank spaces around type and cuts in page makeup. Since the furniture does not stand as high as matter to be printed, it escapes the ink roller and makes no mark on the page. Occasionally in early proof stages some pieces of furniture may be out of position and show on the page as squares or strips of black.

Galley: The long, shallow tray in which lines of type are placed after being set and from which the first proofs, *galley proofs,* are pulled.

Gatefold insert: A sheet of paper wider than the book or magazine page width, bound in so that its fold is at the outer edge and the folded leaf can be opened out like a gate. Such an insert is occasionally used to accommodate large maps, long charts, or illustrations which cannot be reduced to page size. It is an added expense factor.

Gravure printing: A method of printing used more for newspapers and magazines than for books. The printing surface is in intaglio, with the design of letters and illustrations depressed in the plate. The ink goes into the depressed areas, the surface is wiped clean, and only ink from the depressions is carried to the paper. All matter to be printed, including type, is screened.

Guard: (1) Same as *Bearer.* (2) A reinforcing strip for a tip-in.

Gutter: The inner margins of two facing pages; the space between two columns.

Hairline: The finest of the lines used in printing or engraving. Hairline rules are often used in tabular material.

Half title: (1) The title, sometimes in abbreviated form, appearing on the first printed page of a book, preceding the full title page. Also called *bastard title.* (2) A part title.

Halftone: A photoengraved plate or printed illustration in which a range of solid tones is reproduced by a pattern of dots, more or less concentrated according to the degree of darkness of the original tones. The dots are produced by photographing the original copy through a screen and then transferring the screened image to a sensitized metal plate.

Head or heading: A title or caption standing at the head of a chapter, section, column, list, table, etc. It may be centered, flush with one side, or indented to align with other page elements. The value of the heading in the organization of the article, book, or paper is usually established by its type style and position so that a visual outline of subordination of topics is presented.

Headpiece: An illustration or decorative motif placed above the opening of a text unit, as over a chapter title.

Imposition: The placement of pages in a form so that they appear in correct order, spacing, and position when the printed sheet is folded to page size.

Impression: (1) The contact of printing surfaces with the paper. (2) A printing; all the copies made in one press run. An edition may have several impressions run off at different times from the same plates, as stock is exhausted.

Imprint: (1) Publication data, including the publisher's name and location and sometimes date of publication, usually given on the title page. (2) The publisher's name on the backbone of a book. (3) The name of the printer appearing on any printed matter.

Indention: Holding in from the margin. The first word of a paragraph is usually indented; extracts may be indented from both margins. *Hanging indention,* which is used in most dictionaries, is the holding in of all lines following an entry. In outlines, hanging indention may require additional indention for each subordinate step.

Index: A list of all important subjects treated in a book or manual, arranged in alphabetical order and accompanied by the page numbers on which the subjects occur. It is placed at the end of the book or manual.

Inferior letter or number: A character of a size smaller than the text size, aligning lower than the text type line; a subscript. It is specified for the printer by marking a caret above the character.

Initial: The first letter of a chapter or section, set in a size larger than the text to give accent or decoration to the opening. Initials are called *stick-up* when they align at the bottom with the first line of type and extend upward; they are called *inset* or *cut-in* when they align at the top with the first line and the following lines of type run around the depth of the initial.

Insert: (1) Additional matter to be set in type at an indicated position on manuscript or proof. (2) A page or group of pages printed separately from the rest of the book or magazine and added to it in the binding process.

Intaglio: Engraved below the surface. See *Gravure printing.*

Intertype: A typesetting machine that casts a full line of type on a single metal slug; a competitor to the Linotype.

Introduction: An explanatory statement of the subject matter, scope, or portent of a work by its author, editor, or literary sponsor. An author's introduction, if necessary for full understanding of the work, may comprise the opening chapter in a book. An editor's introduction more often forms part of the front matter.

Inverted caret: The symbol used to direct the printer to set a character in superior position.

Italic: A slanting typeface. *This is italic.* To direct the printer to set a word or character in italic, a single underline is used in manuscript; in proof the line is drawn under the correct word or character and a marginal notation, "ital," is made.

Jacket: The separate paper dust wrapper covering the binding of a book, usually bearing the title and the author's and publisher's names. Jackets for trade books sold to the general public are designed to attract attention and to convey a sales message.

Justify: To space out a line or lines of type to a given measure.

Kill: Destroy; a direction to the printer marked on proofs of type or cuts to be eliminated.

Layout: A scale plan drawn to show the size and position of material to be printed.

Leaders: A line of dots, dashes, or other typographical characters used in lists and tables to guide the eye across a space from one item to the next.

Leading: Spacing between lines of type. The space is measured in points and inserted by the use of narrow metal strips or, in machine setting, by casting the letters on a slug that is wider than the point size of type.

Leaf: (1) One sheet of the paper in a magazine or book, consisting of two pages, one on each side of the sheet. (2) A thin sheet of metal, impressed with a die into stamped lettering or designs on covers.

Legend: The title or descriptive matter for an illustration; usually in a different type size from the text and set directly under the cut. Sometimes called a *caption.*

Letterpress printing: Formerly the most widely used printing method for books and periodicals. All matter to be printed is in relief. Only the raised surfaces are inked, and when these are pressed against the paper, the image is transferred. Letterpress is distinguished from planographic, or offset, printing and from gravure, or intaglio, printing.

Letterspacing: The insertion of space between the letters of words. This is most often done with headings and titles in all capitals to extend the length or to lighten the effect of the line. Lowercase letters are seldom letterspaced.

Ligature: Two or more letters designed and cast as a unit, as ff, fl, ffl. Compare *Logotype.*

Lightface: A typeface made up of fine line strokes, giving an overall effect of lightness in weight and tone on the page; the normal text face as distinguished from boldface.

Line cut: A cut made from illustration copy (drawing) that is executed completely in clean-cut lines.

Line drawing: A drawing executed in clean-cut lines or strokes, without intermediate solid tones, so that it is reproducible by line cut.

Lining figures: Arabic numerals that align at the bottom with the baseline of the letters in the text, as distinguished from the *old-style* figures.

Lining papers: Endpapers.

Linotype: A typesetting machine that assembles and casts a full line of type on a single metal slug. The matrices are assembled by the operation of a keyboard and automatically redistributed when the line has been cast. Any correction in such a line of type involves resetting the entire line.

Live matter: Copy, manuscript, type, or plates still to be used.

Logotype: Two or more separate letters that are frequently used in combination, cast on a single type body, as the subscript $_{\text{min}}$. Compare *Ligature,* in which the strokes forming the letters are joined.

Long page: A page carrying a line or two more than others in the book; a column carrying a line or two more than others; sometimes used to avoid bad breaks in makeup.

Lowercase: Small letters of the alphabet, as distinguished from capitals or small caps. Printers keep type in a pair of cases, the upper one for capitals, the lower for small letters. To mark a word or letter for lowercase, "lc" is written in the margin and a slanting line is drawn through the letter.

Make-ready: The process of adjusting type forms for printing so that all printing surfaces are exactly level and transfer ink evenly to the paper.

Makeup: Assembling and arranging the type, headings, illustrations, legends, etc., in page form with correct spacing and position of all elements for printing.

Marginal note: (1) A proof correction marked in the margin. (2) Any type matter that is placed in the page margin outside the general type area. Such an arrangement in letterpress printing requires building out the page edge with furniture to hold the marginal type in position, and thus constitutes an extra expense item.

Margins: The blank border of a page, usually narrowest at head and back (bound edge), wider at the front (trimmed edge), and widest at the foot.

Matrix: A metal mold from which a type character is cast. Often abbreviated "mat."

Measure: The width of type lines across the page, specified in picas.

Modern: A style of typefaces that has straight serifs and marked contrasts in the weight of strokes. Compare *Old style.*

Monotype: A typesetting machine that casts each character on a separate metal block. Composing is in two steps: the operation of a keyboard perforates a strip of paper, which then governs the casting of characters and assembling of justified lines. Correction of single characters can be made in monotype lines without resetting the whole line. Compare *Linotype.*

Nomenclature: The list of names or designations of any specialized subject; term often applied to the labels within diagrams or other illustrations.

Off its feet: A description of a character or line that is not level and fails to print evenly in proof.

Offset: The blurred effect caused by slippage of a blotter or slip sheet laid over a proof before the ink has dried.

Offset printing: A method of printing from a flat surface (without relief or intaglio). Matter to be printed is photographically reproduced on a metal plate so treated that ink adheres only to parts to be printed. There is a double transfer of the ink; first from the plate to a rubber blanket, and then from the rubber to the paper. The method is advantageous for heavily illustrated material. It is also called *photo-offset lithography.*

Old style: A style of typefaces having slanting, rather than straight, serifs, and only slight contrast between thick and thin strokes. It is contrasted with *modern* faces (which see).

Old-style figures: Arabic numerals that have ascenders and descenders, as distinguished from *lining figures.*

Optical center: The position on a page that visually appears to be the center. By measurement, the true center is slightly below this point. Illustrations are sometimes placed at optical center in page makeup.

Overlay: (1) A transparent sheet of paper or film fastened over illustration copy and marked with instructions for cropping, color breaks, engraving treatment, and the like. (2) A similar sheet on which corrections can be marked over type proofs for photographic reproduction.

Page proof: Normally the second set of proofs in the production of a book, magazine, or paper, pulled after page make-up. In it the galley corrections are incorporated, illustrations appear, cross references are made, and from it the index is made.

Part title: The title of one of the main divisions of a book, often set on a separate, right-hand page with a blank back; also called *half title*.

Patch: A piece of metal bearing a correction, cut and soldered into a plate. Only a limited amount of patching is practical because of the strain to which plates are subjected in the press.

Photoengraving: (1) The process of reproducing illustrations photographically on a metal plate and etching the metal to produce a relief printing surface. (2) A reproduction made by this process.

Photo-offset lithography: See *Offset printing*.

Photostat: A photographic copy made directly on sensitized paper without the use of film or a mirror-image negative. The size of the print may be larger, smaller, or same size as the original.

Pi or pie: Type that has been thrown out of order, as by dropping a galley. Resetting and complete rereading of the pied proof may be necessary.

Pica: A printer's unit of measure, approximately $\frac{1}{6}$ in., used for designation of width and length of columns, pages, rules, slugs, etc. It is equal to 12 points.

Planography: Printing from a flat surface, as in offset printing.

Plate: (1) A solid metal sheet, duplicating a page or pages to be printed and used in the press instead of the original type and cuts. See *Electrotype*. (2) A full-page illustration, sometimes an insert, numbered in a sequence apart from the regular text illustrations.

Point: The smallest unit of type measurement, equal to approximately $\frac{1}{72}$ in (actually 0.01384 in), or $\frac{1}{12}$ pica. As applied to type sizes, measurement in points expresses height, not width, of the printed line.

Point system: A standard system of type sizes based on the pica, or 12-point type, and adopted for uniformity by the U.S. Type Founders' Association in 1887.

Preface: That part of the front matter of a book in which the author states the scope and aim of his or her work and sometimes acknowledges his or her indebtedness for source material or assistance. It is in the nature of a "covering letter" to the reader. Compare *Introduction, Foreword*.

Press: (1) The printing machine. (2) A device for proving type or engravings. (3) A bindery device for holding newly bound books in shape until they are dry.

Press proof: A proof taken when the make-ready has been done and the forms are on the press ready for the complete run.

Presswork: Preparation and production of the final impression of type and cuts on the paper.

Progressive proofs: A set of color proofs showing each color separately and then in combination, with each color added in the order of printing.

Proofreading: A careful examination of type proofs to eliminate errors. The standard symbols for correction are shown in Figure 14.1.

Proofs: Sample impressions taken from any material to be printed and submitted for approval. The stages of proof are usually as follows: galley, page, foundry, press, and plate. Normally the author reads only galley and page proofs of a book, sometimes galley proofs of a magazine article or paper. A revised proof may be ordered at any stage to show corrections. Illustrations are shown in engraver's proofs and progressive proofs. See also *Reproduction proofs*.

Quad: A blank piece of type metal used for spacing lines of type. It is available in en, em, or multiple-em widths in all point sizes. Quads have no printing surface and are not seen in final impressions, but occasionally in preliminary proofs appear as black rectangles. See *Workup*.

Query: A question on manuscript or proof addressed to author or editor, often marked "qy" or "au." All queries must be answered and crossed off. Unanswered queries are carried to the next stage of proof.

Recto: A right-hand, odd-numbered page.

Reference marks.: The symbols signaling footnotes in text or tabular matter. They are *, †, ‡, §, ¶, ‖, **, etc., used in that order and set in a superior position at the text point of reference and preceding the footnote.

Register: Exact matching of position; a term applied to type pages exactly backing each other on the sheet; to color impressions in successive runs of process color printing.

Relief: A raised design; a printing surface raised above the background. Letterpress printing makes use of type and illustrations in relief form. Compare *Gravure printing, Intaglio, Planography.*

Reprint: An additional printing of all or part of a published work.

Reproduction proofs: Page proofs carefully made ready and pulled on special white paper or glassine for photographic reproduction in offset or gravure.

Retouching: Hand correction of photographs to eliminate unnecessary details, bring out unclear areas, or cover flaws.

Reverse printing: Negative, or white-on-black, printing. Type or drawings can be photographically reversed so that letters and lines are left white on a printed black or color background for special effect or emphasis.

Revised proof: A new proof incorporating corrections from foul proof.

River: A streak of white space in type matter, resulting from overwide spacing of words. Good typesetting avoids rivers.

Roman: The vertical alphabet as distinguished from the slanting letters of italic; the normal text type. This line is roman. The abbreviation "rom" is marked in the proof margin to order a change to roman letters.

Roman numerals: The numerals making use of i, v, x, I, V, X, etc., as distinguished from arabic numerals. Roman numerals are generally used in lowercase for folios of front matter; in capitals for plate numbers, and sometimes also for chapter and volume numbers.

Routing: Cutting away extra metal and blank areas of a cut so that they will not print.

Rule: A narrow strip of metal bearing a straight-line printing surface. Rules are available in hairline and various point sizes of weight. The long strip is cut to desired length for use in tables, boxes, and decorative elements.

Runaround: The setting of type to fit around an illustration that is less than page or column width. Since the position of illustrations is not usually determined until pages are made up, runarounds involve setting sections of the type twice; once at full measure in galleys and again to the narrower measure for inclusion on pages beside cuts.

Run in: A marginal instruction to continue setting copy in the same paragraph, although the manuscript or proof shows a break. Within the page a curved line is drawn to join the last word before and the first word after the break.

Run-in head: A heading set on the same line as the following text.

Running head: The line of type, usually above the text on a page, giving ordinarily the book title and folio on the left-hand page and the chapter title and folio on the right-hand page. Other titles, such as parts or sections, may sometimes be used instead.

Run on: A manuscript notation to continue setting without a break, often used where the typescript ends short of the page.

Screen: In photoengraving, two sheets of glass ruled with fine vertical and horizontal lines through which illustration copy is photographed for halftone reproduction. The crossings of the screen lines break up the tones of the original into a pattern of dots, more or less concentrated according to the darkness or lightness of the original. The fineness or coarseness of the screen varies with the number of lines per inch. A very fine screen produces an almost indistinguishable dot pattern. Coarse screens, as for newspaper photographs, produce easily distinguishable dots.

Script: Type characters resembling handwriting. Script faces are sometimes used for display lines or chapter titles; certain script characters are needed in composition of technical material.

Section: (1) A subdivision of a chapter sometimes used in technical books, generally numbered in sequence throughout and often abbreviated "Sec." (2) A long paper or article in a journal or magazine; the section may be termed a *special section* when it comprehensively covers a single subject. (3) A signature.

Serif: The small crossbar or tapered stroke at the ends of letters in certain typefaces. This **F** has serifs; this **A** is sans serif (without serifs).

Sheets: Printed matter that has gone through the press but has not yet been bound.

Shelfback: See *Backbone.*

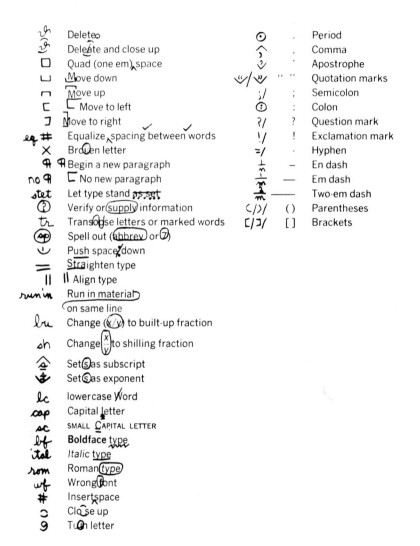

FIG. 14.1 Proofreader's marks. (From McGraw-Hill Book Company, *Working with Proofs: Guidelines for Authors,* New York, 1973.)

THE LANGUAGE OF PRINTERS AND PUBLISHERS 14.11

this/
m/
t/lc

o/

run in

From the end of the Civil War until the start of the great depression, the U.S underwent a period of rapid industrialization which altered virtually all aspects of our lives. During period the mechanization of agriculture transfored us from a rural to an urban country/ improved methods of transporation created National markets and led to the consolidation of industries; and the development of the automobile drastically altered the nature of the city. These material changes, however, were in some ways less important than the social changes which accompanied them/ The wave of industrialization also made basic changes in the nature of work, authority, education, the family, and socialization. Because the next stage of personality development introduces the individual to the world of work, we need to learn more about these changes before we continue with the study of individual development.

Caps //
3/s
;/

3/
3/

FIG. 14.1 *(Continued).*

Shilling fraction: A fraction set all on one line with the numerator and denominator separated by a slanting line or solidus, as x/y.

Short page: A page made up of fewer lines than the normal page, usually to avoid bad breaks in makeup.

Side heading: A heading for a subordinate division of the text, usually set at the left margin of the page, either on a separate line or run in with the opening of a paragraph.

Signature: A sheet of paper printed on both sides and folded to form a section of a periodical or book. The number of pages in a signature may be 4, 8, 16, 36, or 64. Each fold doubles the number of pages.

Sinkage: A measured distance below the top page margin establishing the position of chapter openings and the like.

Slug: (1) A metal strip used for spacing between type lines, usually 6 or 12 points wide, as compared with leading (which see) which is most often 1 to 3 points wide. (2) The solid line of type cast by composing machines such as Linotype or Intertype.

Small caps: Capital letters that are smaller than the regular capitals for a given size of type. They are available in many types used for text composition. A double underline is the mark used in manuscript to direct the use of small caps.

Solid: Without leading between the lines.

Spine: See *Backbone*.

Stamp: To imprint lettering or a design with a die, as on a book cover.

Standing type: Type from which a book is printed or type that is held in storage for further printing.

Stereotype: A duplicate printing plate cast in metal from a mold of paper composition, used mainly by newspapers.

Stet: Literally, "let it stand," a marginal direction to the printer to retain matter that has been crossed off. A row of dots is marked under the words in question.

Straight matter: Text copy uninterrupted by special setups such as tables, formulas, or headings.

Stub: The list of subjects or entries at the left of a table.

Style: A standard of spelling, hyphenation, capitalization, punctuation, abbreviation, use of numerals, typographical setup, etc., to be followed in editing and printing a manuscript. To ensure uniform style throughout the work, a checklist of the approved forms (*style sheet*) is generally compiled. Each publisher has preferred styles for his work.

Subscript: An inferior character.

Subtitle: A secondary title, often an explanation of the main title.

Superior letter or number: A character of a size smaller than the text, aligning higher than the text line; an exponent. It is specified for the printer by marking an inverted caret below the character.

Superscript: A superior character; exponent.

Surprint: To print upon previously printed matter.

Tailpiece: An illustration or ornamental motif at the end of a text unit, often at the bottom of the final page of a chapter.

Text: The body of a book, article, or paper, as differentiated from front and back matter, tabular matter, illustrations, extracts, etc.

Thin space: A piece of type-spacing material, usually ¼ em wide, used between characters in a line.

Three-color process: A color reproduction method making use of separate plates for red, yellow, and blue inks. See *Four-color process*.

Tip-in: To paste a separately printed leaf into a book. See *Insert*.

Title page: The page at the opening of a book or periodical on which the full title and publisher's imprint are carried. In most books it includes the author's name, stands as page iii, unfolioed, in the front matter, and is backed by the copyright notice. In periodicals it usually carries the names of publisher and editors, subscription information, copyright notice, and table of contents.

Tracing cloth or tracing linen: A crisp, transparent linen fabric, usually white or pale blue, used for tracing and finishing drawings for reproduction.

Trimmed size: The dimensions of a page after the folded edges have been cut off; the actual size of the published page.

Turned: Placed broadside on the page, i.e., with the top toward the left margin.

Turnover line: A type line continuing an item not completed in the previous line, e.g., the continuation of a heading or a table stub that is too long for one line.

Type-high: Level with the printing surface of type; said of cuts, ornaments, plates, etc., made to print with type. The actual height is 0.918 in.

Type page: The area of a page in which type is printed; the portion within the margins.

Type stamping: Adding labels within illustrations during the wax engraving process by pressing printing type into the wax that forms a mold for casting.

Uppercase: Capital letters, so called from the upper one of a pair of type cases. To indicate that a letter should be reset as a capital, the term "cap," or sometimes "uc," is written on the proof.

Vandyke: A print from the negative made by an offset printer, sometimes submitted to the publisher as a final check before offset plates are made. It is like a blueprint but is called "vandyke" because it is printed in brown.

Verso: A left-hand, even-numbered page.

Vignette: (1) An illustration gradually shaded off at the edges to merge with the unprinted paper. (2) A small decorative sketch used as decoration at chapter heads or breaks.

Wash drawing: An illustration executed in flowing brushwork of graduated tones. It requires halftone reproduction.

Wax engraving: A method of preparing plates by means of hand-drawing on a wax-coated metal sheet and then using the wax as a mold for casting.

Widow: A single word or a group of words of less than full line measure appearing as the final line of a paragraph and standing at the head of a page or column. Widows are avoided in careful typography.

Workup: An accidental impression of a quad, lead, or piece of furniture that has worked up to the printing surface in proving or printing.

Wrong font: Proofreader's term for a type character of the wrong face or size for the use in which it appears. The abbreviation "wf" is marked in the margin to correct this.

PROOFREADING

Every technical writer does some proofreading. Probably the most common proofreading job you'll meet is the checking of typed manuscripts before they are released to the publisher or printer. You will also read galley and page proofs. Although proofreading is a specialized profession, anyone who is an observant and attentive reader can learn to catch errors. As the author of a written piece you are the most important of the several proofreaders who correct proofs of your manuscript, since the other readers are probably less familiar than you with your subject matter. The proofreaders' marks and the way to use them are shown in Figures 14.1 and 14.2. A little study of these illustrations will enable you to correct proof in the style the printer prefers.

The system of correcting proof is different from the system of correcting manuscript. On manuscript you make changes within the lines at the point of correction. The printer then reads your copy line by line, setting type as he or she reads. But once the copy has been set in galleys, he or she does not read each line to find corrections. He or she runs his or her eye down the margin, where the proofreaders' marks call attention to the lines where changes must be made.

14.14   HANDBOOK OF EFFECTIVE TECHNICAL COMMUNICATIONS

> tr/ The findings consistently showed that learning is increased
> wf/ when audiovisual aids are used. This means that motion pictures and projected still material make the lessons clearer and
> o/cap/ convey new ideas when the students were tested, the results indicated that those taught with audiovisual aids has more
> c/a/ fats and ideas than those whose lesson did not include them
> (/)/9 see Fig. 21. ¶Experiments have been conducted in teaching geography, history, languages, various sciences, and many sub- other jects with and without audiovisual aids.
> run in The increased learning of the groups with whom audiovisual
> #/ aids were used over the learning of the (control) groups without
> les/ such aids justify the conclusion that Audiovisual aid are definitely helpful and should be used when they correlate with stet the teaching under way.

> (/)/ The term $\omega - \omega_0/\omega_0$ is the fractional deviation of frequency from the point of resonance. Furthermore, note that if $|\omega - \omega_0|$ is small
>
> $$\frac{\omega + \omega_0}{\omega} \approx 2$$
>
> Therefore, when this approximation is sufficiently good, extend radical
>
> $$bu/\ sub\ before/\ super/\ ln\ (Z_s/R) = \sqrt{1 + Q_s^2\left(\frac{\omega - \omega_0}{\omega_0}\right)^2}$$
>
> $$\theta = \tan^{-1} 2Q_s \frac{\omega - \omega_0}{\omega_0} \quad w/(qk\ 48)$$
>
> ital/ If Q is used to represent either Q_s or Q_p, the function is plotted as the magnitude curve in Fig. 6-23.

FIG. 14.2 Copy marked with proofreader's marks. (From McGraw-Hill Book Company, *Working with Proofs: Guidelines for Authors*, New York, 1973.)

When correcting proof, use ink or pencil of a different color from the marks already on the proof. Put your marks in the right or left margin, whichever is nearer to the word corrected. If there are several corrections in a single line, place them in order from left to right, separated by a slant line (e.g., tr/cap/r). If the same correction is made several times in the same line with no intervening corrections, make your correction once in the margin, followed by an appropriate number of slant lines. For instance, if you add an *s* to three words in a line, write in the margin "s///." When you wish to insert words, put a caret at the point of insertion and write the additional words in the margin. To delete material without substituting anything, cross out and put a delete sign in the margin. When you delete words and substitute other words, cross out the unwanted material, use a caret within the line, and write the new material in the margin; the delete sign is then unnecessary.

Occasionally you may decide the material you have deleted should be restored. Place a row of dots under the crossed-out material, cross off the delete sign in the margin, and write "stet" (let it stand).

If you find it necessary to insert a substantial amount of new material—several sentences or a paragraph—type it at the bottom of the galley or on a separate sheet attached to the galley and identified by galley number, indicating clearly where it should be inserted. Never cut the galleys apart. Publishers prefer that you make an extensive insert only on galley proof, when it is comparatively simple for the printer to set it and drop it into the galley tray. The publisher's task is more difficult if you make the change in page proof. For instance, if you add six lines to the middle of a page, the publisher must transfer the last six lines of that page to the top of the next page, and so on, to the end of the text. When tables, illustrations, and headings intervene, it may be impossible to balance the pages. And if there is not room for the last six lines at the end of the text, the next article or chapter must be pushed forward. Renumbering of pages means further corrections of cross-references and table of contents, new errors that may be made during resetting, and all in all a horrifying expense of time and money. A long deletion in page proofs may cause just as much trouble.

The quickest and cheapest way to make a change in galley or page proofs is to keep the number of characters in the line the same. Count each letter, digit, punctuation mark, or space between words as a character. Then, if you insert words, try to delete words having an equal number of characters so that you come out even within the line or within a few consecutive lines. Similarly, if you delete a few words, try to replace the characters by adding new words. When you confine the changes to a few lines you avoid the necessity of extensive resetting.

If it seems necessary to add material, and impossible to make the compensation described above, try to add it near the end of a paragraph in galleys, and near the end of an article or chapter in page proof. And if the last page of the text will not accommodate the new material, you are faced with a very unpleasant dilemma. You will have to decide whether the change is sufficiently important to justify the expense of repaging a large portion of the remaining material. Publishers usually set a limit for the expense of resetting authors' corrections; if your corrections are extensive enough to pass this limit, you or your company may be charged with the extra expense.

INDEX

Abbreviations, **13**.30 to **13**.37, **13**.39
 in abstracts, **2**.31
Abbreviations for Use on Drawings, **8**.8
Abstract words in articles, **4**.55, **4**.59
Abstracts, **2**.16
 for papers, **3**.11
 for reports, **2**.30 to **2**.31
Acceptance, report writing for, **2**.19, **2**.23 to **2**.27
Acknowledgments, **2**.34, **3**.12
Action verbs in articles, **4**.56
Actions:
 in directives and procedures, **7**.21
 report writing for, **2**.1, **2**.4 to **2**.5
Active voice for article titles, **4**.48
"Adapted from" in captions, **11**.23
Adjustment instructions, **8**.20
Administrative specifications, **6**.16, **7**.7
Advanced engineering books, **11**.3
Advantages, citing of, in articles, **4**.54
Adverbs, commas for, **13**.40
Advertising:
 and articles, **4**.20, **4**.40
 and catalogs, **10**.1
 for government procurement, **5**.3
 headlines for, **10**.11
 purpose of, **10**.7 to **10**.8
 and sales brochures, **9**.2
 threatening to cut, **4**.45
 writing of, **10**.8 to **10**.11
Advertising agencies:
 editor contacts by, **4**.42
 for sales brochures, **9**.11
Advice (*see* Criticisms and suggestions)
"After" in captions, **11**.23
Agreements (*see* Contracts)
"All rights" to articles, **4**.43
American Association for the Advancement of Science, **3**.9

American Council of Learned Societies, **3**.16
American Petroleum Institute, **6**.14
American Society of Civil Engineers, **3**.9
American Society of Mechanical Engineers, **6**.14
 processing of papers by, **3**.6 to **3**.7
 suggestions by, **3**.9 to **3**.16
American Society for Testing Materials, **6**.14
American Water Works Association, **6**.14
Analyses:
 for instructions, **8**.31
 of magazines, **4**.39
 and outlines, **2**.12
 for proposals, **5**.11
 for RFPs, **5**.32 to **5**.33
 for specifications, **6**.13
Anderson, Chester, writing hints by, **1**.16
Annual reports in proposals, **5**.28
Apostrophes, **13**.43 to **13**.44
Appellations, capitalization of, **13**.17 to **13**.18
Appendixes:
 in directives and procedures, **7**.25
 in instructions, **8**.28
 in multivolume reports, **2**.53
 in papers, **3**.12
 in proposals, **5**.29
 in reports, **2**.33, **2**.35, **2**.44 to **2**.46
Applied Mechanics Abstracts, **2**.30
Appositives, **13**.39 to **13**.40
Approaches:
 for papers and reports, **3**.2
 for proposals, **5**.12 to **5**.13
 for specifications, **6**.11
Architects as specification writers, **6**.2, **6**.3
Arrangement of reports, **2**.24, **2**.27 to **2**.28, **2**.54

Articles, **1**.3
  and books, **11**.8
  checklist for, **4**.60 to **4**.62
  compared to papers, **4**.1 to **4**.2
  editors for, **4**.39 to **4**.47
  examples of, **4**.16 to **4**.17
  ideas for, **4**.17 to **4**.26
  outlines for, **4**.26 to **4**.38
  overview of, **4**.2 to **4**.3
  submission of, **4**.57 to **4**.60
  titles and leads for, **4**.47 to **4**.52
  types of, **4**.3 to **4**.16
  writing of, **4**.52 to **4**.57
Assembling of facts for articles, **4**.27
Assembly instructions, **8**.19
Associates, article ideas from, **4**.19 to **4**.20
Attachments for specifications, **6**.22
Attention-getting techniques for directives and procedures, **7**.13 to **7**.14
Authors:
  of books, **11**.5 to **11**.8
  information on, in papers, **3**.11
  as research resource, **2**.56
Automated formats for specifications, **6**.24
Automotive Engines (Crouse), **11**.2
Ayer Directory of Publications, **4**.40

Back fly sheets for proposals, **5**.29
Back matter:
  in directives and procedures, **7**.25
  for instructions, **8**.28
  in proposals, **5**.29
Base-bid specifications, **6**.10 to **6**.11
Base specifications, **6**.8
Basic Television (Grob), **11**.2
Baumeister, T., *Marks' Standard Handbook for Mechanical Engineers*, **11**.4
Bibliographies:
  in articles, **4**.57
  in papers, **3**.12
  in reports, **2**.34
Bidder's-choice specifications, **6**.9 to **6**.10
Bidding compared to RFPs, **5**.3
Bindings for proposal covers, **5**.18
Blake, Richard, *Water Treatment for HVAC and Potable Water Systems*, **11**.3
Blurbs for articles, **4**.49
Body, **1**.12
  of articles, **4**.24 to **4**.28, **4**.52 to **4**.47
  of directives and procedures, **7**.19 to **7**.20, **7**.24 to **7**.25
  of papers, **3**.12

Body (*Cont.*):
  of proposals, **5**.21 to **5**.28
  of reports, **2**.32 to **2**.33
Booklets, **8**.2
Books:
  authors of, **11**.5 to **11**.8
  cautions for, **11**.5
  contracts for, **11**.12 to **11**.14
  handbooks, **11**.24 to **11**.28
  ideas for, **11**.8 to **11**.9
  kinds of, **11**.1 to **11**.4
  outlines for, **11**.9 to **11**.12
  progress notebooks for, **11**.17 to **11**.18
  requirements for, **11**.4
  for research, **2**.56
  schedules for, **11**.15 to **11**.17
  writing of, **11**.18 to **11**.24
Bottom matter in directives and procedures, **7**.20
Brackets, **13**.44 to **13**.45
Brainstorming:
  for multivolume reports, **2**.53
  for outline preparation, **2**.5
  for proposal development, **5**.35
Brochures (*see* Sales brochures)
Bulletins, **7**.5 to **7**.6
  formats for, **8**.30
  (*See also* Instructions and bulletins)
Business cards in business letters, **12**.2
Business procedures, articles for, **4**.11 to **4**.12

Calculation method articles, **4**.7 to **4**.8
  examples of, **4**.16 to **4**.17
  outline for, **4**.34
Calibration instructions, **8**.20
Capitalization, **13**.15 to **13**.29
  and abbreviations, **13**.31
Captions for illustrations, **3**.13, **11**.22
Capture strategies with sales brochures, **9**.4 to **9**.6, **9**.17
Catalogs:
  importance of, **10**.6 to **10**.7
  planning of, **10**.2 to **10**.3
  purpose of, **10**.1
  writing of, **10**.3 to **10**.6
Change proposals, **6**.30
Changes in article reprints, **4**.46
Chapters in books, **11**.9 to **11**.10, **11**.20 to **11**.21
Charts:
  in advertising, **10**.10

Charts (*Cont.*):
  in formulas, **10.6**, **10.7**
  for troubleshooting, **8.22** to **8.23**
Checklists:
  for articles, **4.60** to **4.62**
  for clear writing, **1.8** to **1.15**
  for directives and procedures, **7.35** to **7.36**
  for instructions, **8.41** to **8.42**
  for outlines, **2.9**, **2.13** to **2.15**
  for papers, **3.17** to **3.18**
  for proposals, **5.39** to **5.40**
  for reports, **2.56** to **2.57**
  for sales brochures, **9.17**, **9.19** to **9.20**
  for specifications, **6.31** to **6.32**
Chemical Abstracts, **2.30**
Chemical data, spelling of, **13.38**
Chemical Engineers' Handbook (Perry), **11.4**
Choppiness in articles, **4.56**
Clapp, John, readability formula by, **1.16**
Classification:
  of facts for articles, **4.27**
  of ideas for articles, **4.21**
  of tasks, **1.8**
Classified materials, clearance for, **3.10**, **3.14**
Clauses:
  commas for, **13.41** to **13.42**
  semicolons for, **13.42**
Clearances (*see* Permissions and releases)
Clients:
  and directives and procedures, **7.6** to **7.7**
  illustrations from, **10.3**
  lists of, in sales brochures, **9.14**
  requirements of, and sales brochures, **9.12** to **9.13**
Closed specifications, **6.10** to **6.11**
Coauthors for books, **11.6** to **11.7**
Codes and specifications, **6.14**, **6.31**
Coherence in articles, **4.56**
College texts, **11.2**
Colons, **13.43**
Commas, **13.39** to **13.42**
Comments (*see* Criticisms and suggestions)
Commerce Business Daily, **6.5**
  RFPs in, **5.4**, **5.8**, **5.9**
Company brochures (*see* Sales brochures)
Company history in proposals, **5.28**
Comparisons in articles, **4.50**, **4.54**
Compensation (*see* Payment)
Competition, sales brochures by, **9.5**
Component specifications, **6.7**
Compound sentences, **13.39**
  semicolons for, **13.42**
Compound words, **13.1** to **13.29**
Computer program description articles, **4.5** to **4.6**
  examples of, **4.16**
  outline for, **4.31** to **4.32**
Computers:
  article ideas stored on, **4.22**, **4.23**
  data retrieval services using, **2.30**
  instruction manuals for, **8.2** to **8.4**
  for outlining, **2.7**
  (*See also* Word processors)
Concentration, outlines to help, **2.13**
Conciseness in articles, **4.52** to **4.53**, **4.55**
Conclusions:
  in articles, **4.24**, **4.26** to **4.28**, **4.57**
  in directives and procedures, **7.20**
  in reports, **2.33** to **2.34**
Concrete words for articles, **4.55**, **4.59**
Consistency of technical and cost proposals, **5.15**
Construction projects, specifications for, **6.6**
Construction Specifications Institute, **6.6**
Consultants for specifications, **6.14**
Consulting Engineer, **10.6**
Consulting engineers, **6.3** to **6.4**
Contacts:
  with editors, **4.41** to **4.42**
  establishment of, **1.10**
Continuity, outlines for, **2.12**
Contract negotiations and directives and procedures, **7.6** to **7.7**
Contractors and specifications, **6.3**, **6.10**
Contracts:
  for articles, **4.44**
  for books, **11.12** to **11.14**
  and instructions, **8.31**
  and reports, **2.53**, **2.54**
  and specifications, **6.4** to **6.5**
  (*See also* Proposals)
Contributed technical papers, **3.5**
Contributors for handbooks, **11.7**, **11.24** to **11.28**
Control documents, **7.4**, **7.5**
Coordinates phrases, commas for, **13.41**
Coordination:
  of multivolume reports, **2.53**
  of proposal development, **5.32**, **5.34** to **5.36**
Copy, advertising, **10.8**

Corporate experience matrices in proposals, **5.28**, **5.30** to **5.31**
Corrective maintenance:
  articles for, **4.10**
  instructions for, **8.26** to **8.27**, **8.39** to **8.40**
Correspondence (*see* Letters; Transmittal letters)
Cost proposals, **5.13** to **5.15**
Costs:
  for books, **11.13**
  estimation of, **8.32** to **8.33**
  of papers, **3.16**
  of sales brochures, **9.7** to **9.8**
Countries, abbreviation of, **13.31**
Cover letters:
  for mail questionnaires, **1.10**
  for papers, **3.14**
  for proposals, **5.17** to **5.18**
  for resumes, **12.10**
Covers and cover sheets:
  for directives and procedures, **7.23** to **7.24**
  for instructions, **8.11** to **8.13**
  for proposals, **5.18**, **5.29**
  for reports, **2.29** to **2.30**
  for specifications, **6.18** to **6.19**
Credits, **2.34**, **3.12**
Criticisms and suggestions:
  with books, **11.5**
  in directives and procedures, **7.20**
  invitations for, **2.32**
  with papers, **4.44**, **4.45**
  in reports, **2.5**, **2.33** to **2.34**
Crouse, W. H., *Automotive Engines*, **11.2**
Customized inserts for sales brochures, **9.5**, **9.14**
Cutaway views:
  in advertising, **10.10**
  in manuals, **8.18**

Dale-Chall readability formula, **1.16**
Dashes, **13.44**
Data:
  assembling of, **1.10**
  collection of, **1.9**
  listing and organization of, **1.12**
  retrieval services for, **2.30**
Dates, abbreviation of, **13.31**
Deadlines, determination of, **1.9**
Decimal numbering systems for specifications, **6.16** to **6.18**

Decision-making responsibilities, **1.7**
Decisions, report writing for, **2.4** to **2.5**
Definitions of terms in reports, **2.33**
Departmental feature articles, **4.12** to **4.13**
  examples of, **4.17**
  outline for, **4.37** to **4.38**
Depth of coverage of papers and reports, **3.3**
Descriptions:
  in article leads, **4.50**
  in instructions, **8.16** to **8.17**, **8.36** to **8.37**
Design analyses, articles for, **4.4**
Design ideas for brochures, **9.8** to **9.9**
Design procedure articles, **4.6**
  examples of, **4.16**
  outline for, **4.32** to **4.33**
  (*See also* Directives and procedures)
Design specifications, **6.6**
Development equipment, instructions for, **8.6** to **8.7**
Dictionaries, **13.1**
Direct costs for cost proposals, **5.13** to **5.14**
Directives and procedures, **7.1** to **7.2**
  in client relations, **7.6** to **7.7**
  compared to instructions, **8.3**
  comparison of, **7.5**
  language and style in, **7.27** to **7.30**
  memos, **7.2**
  preparation of, **7.30** to **7.35**
  structure of, **7.14** to **7.27**
  types of, **7.9** to **7.14**
Directories for research, **2.56**, **4.40**
Directory of Technical Magazines and Directories, **4.40**
Discussion:
  in directives and procedures, **7.20**
  of papers, **3.8**
Distribution:
  of directives and procedures, **7.9**, **7.14**
  of sales brochures, **9.8**
Documentation, computer, **8.3**
Documents for specifications, **6.25** to **6.26**
Double-numbering system for books, **11.20**, **11.22** to **11.23**
Drafts, **1.6** to **1.7**
  for catalogs, **10.5** to **10.6**
  for directives and procedures, **7.33**
  for outlines, **2.6**
  for proposals, **5.36** to **5.38**
  for reports, **2.19**
  for sales brochures, **9.16**
  for specifications, **6.27**

Drawings:
 for articles, **4.44**, **4.45**
 in books, **11.20**
 for catalogs, **10.5**
 change notices for, **6.25**
 in instructions, **8.18**
 for maintenance instructions, **8.27**
 for reports, **2.54**
 for specifications, **6.25**, **6.26**
 for troubleshooting, **8.22**, **8.24** to **8.25**
 (*See also* Illustrations)

Editing (*see* Reviewing, editing, and revising)
Editorial schedules, article ideas from, **4.20**
Editors:
 for articles, **4.39** to **4.47**
 for books, **11.14**
Education:
 article ideas based on, **4.18**, **4.20**
 resume sections for, **12.8**
Edwards, T., *Pump Application Engineering*, **11.3**
Eisenhower, Dr. Milton S., **1.3**
Electrical and Electronic Reference Designations, **8.8**
Electrical and Electronic Symbols, **8.8**
Electronic parts, specifications for, **8.6**
Elisions, apostrophes for, **13.43**
Ellipses, **13.39**
Emergency operation instructions, **8.28**
Employment interviews, resumes for, **12.6**, **12.8** to **12.10**
Encyclopedia of Associations, The, **3.17**
Engineering:
 directives for, **7.2**
 procedures for, **6.14**, **7.3** to **7.4**
 specifications for, **6.4**, **6.6**
Engineering Drawing (French and Vierck), **11.2**
Engineering Society Monographs, **11.3**
Engineers:
 on catalogs, **10.6** to **10.7**
 and specifications, **6.1** to **6.3**, **6.27**
 as technical writers, **1.3** to **1.4**
Enumerations, **13.41**, **13.43**
Equipment:
 maintenance procedure articles for, **4.10**
 operating procedure articles for, **4.8** to **4.10**
 releases for, **4.13** to **4.15**
 specifications for, **6.6** to **6.7**

Errors in examples, **3.14**
Evaluation criteria in RFPs, **5.3**, **5.5**, **5.10**
Examples:
 in articles, **4.54**
 in catalogs, **10.5** to **10.6**
 with instructions, **8.33** to **8.34**
 in papers, **3.14**
Exclamation marks, **13.46**
Exclusives for articles, **4.43**
Executive summaries, **2.18** to **2.22**
 for proposals, **5.20** to **5.22**
Exemplary technical papers, **3.16**
Expanded format for directives and procedures, **7.21** to **7.25**
Expenses (*see* Costs)
Experience section on resumes, **12.8**
Explanations in directives and procedures, **7.29**

Fact sheets for sales brochures, **9.5**, **9.14**
Faults in articles, **4.58** to **4.60**
Fees for cost proposals, **5.14**
Field servicing instructions, **8.5**
Field trips, **1.9**
Files for ideas, **4.19**, **4.22**
Financial statements for cost proposals, **5.14** to **5.15**, **5.28**
Firmness of specifications, **6.11** to **6.12**
"First serial rights" to articles, **4.43**
Flesch, Rudolph, readability formula by, **1.16**
Flexibility in directives, **7.2**
Flowcharts:
 in directives and procedures, **7.25** to **7.27**, **7.32**
 for instructions, **8.31**
 in program description articles, **4.5** to **4.6**
 in proposals, **5.24**, **5.25**
 for specifications, **6.13**
Footnotes for references, **3.12**
Foreign words, **13.29** to **13.30**
Forewords in reports, **2.32**
Formal reports, **2.2**
Formats:
 for directives and procedures, **7.16** to **7.27**
 for instructions, **8.29**
 for outlines, **2.6** to **2.7**, **2.12**
 for proposals, **5.15** to **5.29**
 for reports, **2.35** to **2.49**
 for resumes for RFPs, **12.10** to **12.11**
 for specifications, **6.15** to **6.24**

Forms:
  for article ideas, **4.22** to **4.24**
  for cost proposals, **5.13**
  for directives and procedures, **7.10**, **7.13** to **7.14**, **7.19**
  for magazine analyses, **4.39**
Formulas:
  in catalogs, **10.6**, **10.7**
  readability, **1.15** to **1.16**
Freelancers for sales brochures, **9.12**
French, T. E., *Engineering Drawing*, **11.2**
Friendliness and editors, **4.44**
Friends, article ideas from, **4.21**
"From" in captions, **11.23**
Front matter:
  in directives and procedures, **7.22** to **7.23**
  for proposals, **5.17** to **5.21**
  in sales brochures, **9.13**
Frontispieces for proposals, **5.19**
Functional format for directives and procedures, **7.17** to **7.21**

Gale Research, Inc., **3.17**
Gebbie Press All-in-One Directory, **4.40**
General and administrative costs for cost proposals, **5.14**
General specification formats, **6.8**, **6.16**, **6.18** to **6.24**
Ghostwriting of papers, **1.5**, **3.5** to **3.6**
Glossaries:
  of grammar terms, **13.52** to **13.56**
  of printing and publishing terms, **14.1** to **14.13**
Goodier, J. N., *Theory of Elasticity*, **11.3**
Government Scientific and Technical Periodicals, **4.40**
Governments:
  proposals for, **5.1**, **5.3** to **5.5**
  publications by, for research, **2.56**
  specifications by, **6.5**, **6.8** to **6.9**
Grammar and usage:
  for abbreviations, **13.30** to **13.37**
  authorities for, **13.1**
  checking of, **2.54**
  of foreign words, **13.29** to **13.30**
  glossary for, **13.52** to **13.56**
  and good writing, **13.46** to **13.52**
  of italics, **13.29**
  of numbers and numerals, **13.37** to **13.39**
  in papers, **3.9**
  of punctuation, **13.39** to **13.46**
  spelling and compounds, **13.1** to **13.29**
  text divisions, **13.46**

Graphical solution articles, **4.8**, **4.21**
  examples of, **4.17**
  outline for, **4.35**
Graphics (*see* Illustrations)
Grob, Bernard, *Basic Television*, **11.2**
Guide specifications, **6.7** to **6.8**
Gunning, Robert, readability formula by, **1.16**

Handbooks, **8.2** to **8.3**, **11.3** to **11.4**, **11.24** to **11.28**
  for article ideas, **4.20** to **4.21**
  contributors for, **11.7**
  outlines for, **11.12**
  for specifications, **6.15**
Hardware and engineering specifications, **6.4**, **6.12**
Headings:
  in advertising, **10.11**
  in books, **11.19** to **11.20**
  for conclusions and recommendations, **2.34**
  in directives and procedures, **7.34**
  in instructions, **8.21**, **8.32**
  in resumes, **12.8**
Heads for articles, **4.47** to **4.49**
Helpfulness to clients in advertising, **10.10** to **10.11**
Hicks, T. G., *Pump Application Engineering*, **11.3**
Highlighting in reports, **2.33**
Historical article leads, **4.51**
Home-study texts, **11.1** to **11.2**
Honorariums for papers, **3.16**
Hydraulic Transients (Rich), **11.3**
Hyphens, **13.14**, **13.44**

Ideas:
  for articles, **4.17** to **4.26**
  for books, **11.8** to **11.9**
Illustrations:
  for advertising, **10.10**, **10.11**
  and article ideas, **4.24**
  for articles, **4.54**, **4.56**, **4.59** to **4.60**
  for books, **11.5**, **11.21** to **11.23**
  for calculation method articles, **4.8**
  for catalogs, **10.3**, **10.5**
  collection and evaluation of, **1.13**
  for design procedure articles, **4.6**
  for equipment releases, **4.14**, **4.15**
  for graphical solution articles, **4.8**
  for handbooks, **11.28**
  for instructions, **8.19**, **8.20**, **8.34**

Illustrations (*Cont.*):
    for maintenance instructions, **8.**27
    for management technique articles, **4.**12
    for operating procedure articles, **4.**9 to **4.**10
    and outlines, **4.**27
    for papers, **3.**9 to **3.**10, **3.**13
    for parts lists, **8.**6, **8.**7
    for process description articles, **4.**5
    for product function articles, **4.**7
    for program description articles, **4.**5 to **4.**6
    for proposals, **5.**19, **5.**24
    for reports, **2.**24 to **2.**26
    for sales brochures, **9.**15
    for system and plant description articles, **4.**4
    in table of contents, **2.**32
    (*See also* Charts; Drawings; Photographs; Tables)
Imperative mood:
    in directives and procedures, **7.**29
    in operating instructions, **8.**5
Impersonal style in directives and procedures, **7.**28 to **7.**29
Index cards for directives and procedures, **7.**33
Index tabs for reports, **2.**24
Indexes:
    compared to response matrices, **5.**19
    for instructions, **8.**28
    in reports, **2.**35
    for specifications, **6.**19
    of technical magazines, for research, **2.**56
Indicative mood in directives and procedures, **7.**29
Industry news items, **4.**13 to **4.**14
    outline for, **4.**38
Industry projects, specifications for, **6.**5 to **6.**6
Informal reports, **2.**2
Information manuals, **8.**3
Inserts in sales brochures, **9.**5, **9.**14
Inside track for proposal research, **5.**32 to **5.**33
Installation procedures and instructions, **7.**3, **8.**19
Instructions and bulletins:
    checklist for, **8.**41 to **8.**42
    in directives and procedures, **7.**25, **7.**29
    features of, **8.**1 to **8.**4
    and specifications, **8.**8 to **8.**10

Instructions and bulletins (*Cont.*):
    structure and content of, **8.**10 to **8.**30
    types of, **8.**4 to **8.**7
    writing of, **8.**30 to **8.**41
Interpretive article leads, **4.**51
Interviews, **1.**9
Introductions, **1.**12
    for articles, **4.**24, **4.**25, **4.**27 to **4.**28, **4.**50 to **4.**52, **4.**58 to **4.**59
    for instructions, **8.**16
    for proposals, **5.**21 to **5.**22
    for reports, **2.**32
Italics, **13.**29

Job description of technical writer, **1.**5 to **1.**8
Jobs, article ideas based on, **4.**18
Journals of societies, **3.**3
Justification for technical papers, **3.**2

Key words for abstracts, **2.**30
Knowledge requirements for technical writers, **1.**7
Kreider, J. F., *Solar Heating Design Process*, **11.**3

Labor costs for cost proposals, **5.**13 to **5.**14
LaLonde, W. S., *Professional Engineers' Examination Questions and Answers*, **11.**2
Language:
    for directives and procedures, **7.**27 to **7.**28
    for reports, **2.**23
    (*See also* Grammar and usage)
Large numbers, spelling of, **13.**38
Leads for articles, **4.**27 to **4.**28, **4.**50 to **4.**52, **4.**58 to **4.**59
Learned Societies Directory, **3.**16
Legal matters and specifications, **6.**4 to **6.**5, **6.**14, **6.**31
Length:
    of abstracts, **2.**31
    of articles, **4.**46 to **4.**47
    determination of, **1.**8
    of directives, **7.**6
    of papers, **3.**3, **3.**9 to **3.**10
    of reports, **2.**2 to **2.**3, **3.**3
Letters, **12.**1 to **12.**5
    query, **4.**41 to **4.**42
    (*See also* Transmittal letters)
Libraries, **2.**56
Limitations, listing of, in operating instructions, **8.**5

Lists, action, in directives and procedures, **7.21**
Logic, findings based on, **2.33**
Long-form reports, **2.2**

MacLaren, Anson A., on catalogs, **10.2**
Magazines:
  article ideas from, **4.20**
  choosing of, **4.39** to **4.41**
  (*See also* Articles)
Mail surveys, **1.9** to **1.11**
Maintenance procedures and instructions, **4.10**, **7.3**, **8.5**, **8.7**, **8.26** to **8.28**, **8.38** to **8.40**
  examples of, **4.17**
  outline for, **4.36**
Management:
  directives and procedures for, **7.8** to **7.9**
  proposal review by, **5.38**
  reports for, **2.3**
Management technique articles, **4.11** to **4.12**
  examples of, **4.17**
  outline for, **4.37**
Mandated formats for proposals, **5.16**
Manual and Style Guide, **2.7**
Manual of Style, A, **13.1**
Manuals, formats for, **8.29**
  (*See also* Instructions and bulletins)
Manufacturers, **6.3**
Manufacturing procedures, **7.3**
Marks' Standard Handbook for Mechanical Engineers (Baumeister), **11.4**
Master specifications, **6.8**
Masthead, magazine, for contacts, **4.40**
Material specifications, **6.6** to **6.7**
Meeting arrangements for papers, **3.8**
Membership in societies, **3.5**
Memorandum of Agreement for books, **11.12**
Memorandums, **7.2**, **7.5** to **7.6**
Messages:
  in advertising, **10.9**
  for directives and procedures, **7.5**, **7.30** to **7.31**, **7.33**
  in papers and reports, **3.3** to **3.4**
  in sales brochures, **9.4**
MIL-STD-12, -15, and -16, **8.8**
Military:
  instructions for, **8.3**, **8.4**, **8.7**
  parts lists for, **8.6**, **8.7**
  specifications for, **6.5**, **8.8** to **8.10**
  standards for, **6.5**, **6.6**, **6.16**, **8.8**
Missile industry, **1.2**

Models for specifications, **6.13**
Modern MOS Technology (Ong), **11.3**
Modifications and exceptions in proposals, **5.23**
Modules for instructions, **8.29**
Money, spelling of, **13.38**
Monographs, **11.3**
Motion in illustrations, **2.26**
Multivolume proposals, **5.19**
Multivolume reports, **2.49** to **2.53**

Names:
  capitalization of, **13.16**, **13.17**
  in directives and procedures, **7.28**
    prefixes for, abbreviation of, **13.31**
National Academy of Sciences, **3.9**, **3.16**
National Technical Information Service, **2.30**
Newness in article titles, **4.47** to **4.48**
News article leads, **4.51**
News items and releases:
  examples of, **4.17**
  industry, **4.13** to **4.14**
  outline for, **4.38**
Newspapers, article ideas from, **4.21**
Nomograms, **4.8**, **4.9**
Notices in instructions, **8.13**
Nouns, capitalization of, **13.16**, **13.19**
Numbering systems:
  for books, **11.20**, **11.22** to **11.23**
  for specifications, **6.16** to **6.18**
Numbers and numerals:
  spelling of, **13.37** to **13.39**
  for text divisions, **13.46**

Obituaries, **4.14**
Objective section on resumes, **12.8**
Objectives:
  in article leads, **4.51**
  in directives and procedures, **7.24**
Obsolescence of sales brochures, **9.5**
Ong, D. G., *Modern MOS Technology*, **11.3**
Open specifications, **6.8** to **6.9**
Operating instructions and procedures, **4.8** to **4.10**, **7.3**, **8.5**, **8.21**, **8.37** to **8.38**
  examples of, **4.17**
  outline for, **4.35** to **4.36**
"Or equal" in specifications, **6.9** to **6.10**
Oral discussion of papers, **3.8**
Ordering of reports, **2.24**, **2.27** to **2.28**, **2.54**
Organizations, capitalization of, **13.16**

Origins:
  of papers, **3.6**
  of specifications, **6.5** to **6.6**
Outlines, **1.6**
  for articles, **4.26** to **4.38**, **4.52**
  for books, **11.9** to **11.13**
  for catalogs, **10.3** to **10.5**
  for directives and procedures, **7.21**, **7.31** to **7.33**
  drafts from, **2.19**
  for instructions, **8.32**
  for multivolume reports, **2.53**
  for papers, **3.10** to **3.12**
  preparation of, **1.12** to **1.13**
  for proposals, **5.34**, **5.36**
  for reports, **2.5** to **2.15**
  for sales brochures, **9.15**
  for specifications, **6.26**
  for summaries, **2.17**
Overhaul instructions, **8.6**
Overhead rates for cost proposals, **5.14**

Page layout for sales brochures, **9.9**
Paper:
  for articles, **4.44**
  for business letters, **12.1** to **12.2**
Papers, **1.3**, **3.1** to **3.2**
  checklist for, **3.17** to **3.18**
  compared to articles, **4.1** to **4.2**
  compared to reports, **3.2** to **3.4**
  development of, **3.4** to **3.6**
  exemplary papers as guides for, **3.16**
  processing of, **3.6** to **3.8**
  requirements for, **3.16**
  societies publishing, **3.16** to **3.17**
  writing of, **3.8** to **3.16**
Paragraphs:
  length of, in articles, **4.56**, **4.59**
  in reports, **2.23**
Parentheses, **13.44** to **13.45**
Parenthetical material, **13.40**
  dashes for, **13.44**
Parts catalogs and lists, **8.6**, **8.7**
Patents, indexes of, for research, **2.56**
Payment:
  for articles, **4.45** to **4.46**
  for handbooks, **11.25** to **11.26**
  for papers, **3.16**
Performance requirements of technical writers, **1.5** to **1.8**
Performance schedules in proposals, **5.26** to **5.27**

Performance specifications, **6.8** to **6.9**
Periods as punctuation, **13.39**
Permissions and releases, **2.25**
  for articles, **4.26**
  for books, **11.12** to **11.13**
  for illustrations, **11.23**
  for papers, **3.10**, **3.14**
Perry, John, *Chemical Engineers' Handbook*, **11.4**
Personal information on resumes, **12.9**
Personality profiles, **4.12**
Personalization in articles, **4.57**
Personifications, capitalization of, **13.18**
Personnel:
  articles about, **4.14**
  descriptions of, in brochures and proposals, **5.11** to **5.12**, **5.27** to **5.28**, **9.13**
  directives and procedures to benefit, **7.8**
Pestering of editors, **4.45**
Photographs:
  in advertising, **10.10**
  for articles, **4.44**, **4.45**
  in books, **11.20**
  in papers, **3.13**
  in reports, **2.26** to **2.27**, **2.54**
  for sales brochures, **9.11** to **9.12**, **9.14**
  in system and plant description articles, **4.4** to **4.5**
Phrases, commas with, **13.40** to **13.41**
Physical data, spelling of, **13.38**
Picture stories, **4.21**
Planning:
  of brochures, **9.6**
  of catalogs, **10.2** to **10.3**
  of instructions, **8.10** to **8.11**
  of proposals, **5.11**
  of reports, **2.3** to **2.4**
  of specifications, **6.25** to **6.26**
Plurals:
  abbreviation of, **13.30**
  apostrophes for, **13.43**
Policy statements, **7.5** to **7.6**
Pomposity in articles, **4.55**
Possessives, apostrophes for, **13.43**
Power, T. C., *Practical Shop Mathematics*, **11.2**
Practical Shop Mathematics (Power), **11.2**
Practicality considerations for proposals, **5.11**
Precautions in instructions, **8.13** to **8.14**
Precursor specifications, **6.14**

Prefaces:
  for books, **11.12, 11.13**
  for instructions, **8.14** to **8.16**
  for reports, **2.32**
Preprints, **3.3, 3.8**
Prescribed formats for proposals, **5.16**
Preselling, sales brochures for, **9.2**
Presentation methods for papers and reports, **3.3**
Press clippings in reports, **2.25**
Preventive maintenance and instructions, **4.10, 8.5, 8.26, 8.38** to **8.39**
Primary specifications, **6.8**
Principles of operations in manuals, **8.3, 8.5, 8.38**
Printers for sales brochures, **9.8, 9.12**
Printing, terms for, **14.1** to **14.13**
Problems:
  as article ideas, **4.19** to **4.20**
  in books, **11.23**
  identification of, for proposals, **5.33**
  solutions to, in article leads, **4.50**
Procedural flowcharts, **7.25** to **7.27, 7.32**
  in proposals, **5.24, 5.25**
Procedures (*see* Directives and procedures)
Process description articles, **4.5**
  examples of, **4.16**
  outline for, **4.30** to **4.31**
Procurement (*see* Proposals)
Product brochures, **9.3**
Product function articles, **4.6** to **4.7**
  examples of, **4.16**
  outline for, **4.33** to **4.34**
Production directives, **7.2**
Products and magazine selection, **4.40**
Professional brochures (*see* Sales brochures)
Professional Engineers' Examination Questions and Answers (LaLonde), **11.2**
Professional societies (*see* Societies)
Professional technical writers, **1.4** to **1.5**
Profit margins for cost proposals, **5.14**
Program description articles, **4.5** to **4.6**
  examples of, **4.16**
  outline for, **4.31** to **4.32**
Program discussion in proposals, **5.26** to **5.27**
Progress notebooks for books, **11.17** to **11.18**
Project control procedures, **7.4**
Project experience in sales brochures, **9.14**
Project leaders for proposals, **5.34** to **5.36**

Project phases, specifications in, **6.6** to **6.7**
Project procedures, **7.4, 7.5**
Promotional tools, sales brochures as, **9.1** to **9.2**
Proofreading, **14.10** to **14.11, 14.13** to **14.15**
Proposals:
  checklist for, **5.39** to **5.40**
  cost, **5.13** to **5.15**
  development of, **5.29** to **5.39**
  formats and elements of, **5.15** to **5.29**
  importance of, **5.1** to **5.2**
  technical, **5.10** to **5.13**
  types of, **5.2** to **5.10**
Prototype equipment, instructions for, **8.6** to **8.7**
Public relations representatives:
  article ideas from, **4.20**
  editor contacts by, **4.42**
  for illustrations, **11.23**
Publication of papers, **3.8, 3.14, 3.16**
Publications, listing of, on resumes, **12.9** to **12.10**
Publishing, terms for, **14.1** to **14.13**
Pump Application Engineering (Hicks and Edwards), **11.3**
Punctuation, **13.39** to **13.46**
  of abbreviations, **13.30**
Purchase specifications, **6.6** to **6.7**
Purchasing directives, **7.2**
Purpose:
  in article leads, **4.51**
  in directives and procedures, **7.24**

Qualifications booklets, sales brochures as, **9.2**
Qualifications of staff in technical proposals, **5.11** to **5.12, 5.27** to **5.28**
Quality control in specifications, **6.21**
Quantifying in proposals, **5.26**
Quantity designations, abbreviation of, **13.31**
Querying:
  of magazine editors, **4.41** to **4.42**
  of societies, **3.5**
Question and answer articles, **4.11** to **4.12**
  examples of, **4.17**
  outline for, **4.36** to **4.37**
Question marks, **13.45** to **13.46**
Questionnaires, **1.9** to **1.11**
Questions:
  in article leads, **4.50**
  in books, **11.23** to **11.24**

INDEX

Quotation marks, **13.45**
Quotations:
  colons for, **13.43**
  dashes for, **13.44**

"Randomizing" of article ideas, **4.24**
Ratios, colons for, **13.43**
Readability:
  of directives and procedures, **7.13**
  formulas for, **1.15** to **1.16**
  index tabs for, **2.24**
  and proposal introductions, **5.22**
  in system and plant description articles, **4.4**
Readers and users, consideration of, **1.8** to **1.9**, **1.13**
  for advertising, **10.8**
  for articles, **4.28**, **4.43**, **4.51**, **4.53**
  for books, **11.20** to **11.21**
  for catalogs, **10.2**
  for directives and procedures, **7.9**, **7.11**, **7.31**
  for instructions, **8.4**, **8.33**
  for maintenance procedure articles, **4.10**
  for outlines, **2.12** to **2.13**
  for papers, **3.12**
  for reports, **2.3**, **2.23**, **2.33**
  for resumes, in proposals, **5.11** to **5.12**, **5.27**
  for sales brochures, **9.4** to **9.6**
  for summaries, **2.18**
Reference books, **11.2** to **11.3**
  outlines for, **11.11**
  for specifications, **6.15**
References:
  in articles, **4.57**
  in papers, **3.12** to **3.13**
  in reports, **2.34**
  on resumes, **12.8**
  in specifications, **6.21**
Relationships with editors, **4.43** to **4.46**
Releases (*see* Permissions and releases)
Reliability of company in proposals, **5.11** to **5.12**
Relocation instructions, **8.28**
Repetition:
  dashes for, **13.44**
  in reports, **2.33**
Reports:
  action-getting in, **2.4** to **2.5**
  arrangement of, **2.27** to **2.28**
  checking and editing of, **2.53** to **2.54**

Reports (*Cont.*):
  checklist for, **2.56** to **2.57**
  compared to papers, **3.2** to **3.4**
  delivery of, **2.55**
  format for, **2.28** to **2.49**
  multivolume, **2.49** to **2.53**
  outlines for, **2.5** to **2.15**
  planning of, **2.3** to **2.4**
  reasons for writing, **2.1**
  requirements of, **2.4**
  researching for, **2.55** to **2.56**
  style guides for, **2.35** to **2.49**
  summaries for, **2.16** to **2.19**
  types of, **2.2** to **2.3**
  using word processors for, **2.53**
  writing of, **2.19**, **2.23** to **2.27**
Reprints, changes in, **4.46**
Repro paper, **2.36**, **2.40**
Requests for proposal, **5.3** to **5.5**
  analyses of, **5.32** to **5.33**
  resumes for, **12.10** to **12.11**
  and specifications, **6.5**
  and technical proposals, **5.10**
  and unsolicited proposals, **5.9**
Requirements:
  client, and sales brochures, **9.12** to **9.13**
  and outlines, **2.11**
  for papers, **3.9** to **3.10**, **3.15** to **3.16**
  for reports, **2.53**
  in specifications, **6.2** to **6.3**, **6.12**, **6.20** to **6.22**
Research, **1.5** to **1.6**
  for books, **11.5**
  directives for, **7.2**
  for proposals, **5.32** to **5.33**
  for reports, **2.55** to **2.56**
Response matrices for proposals, **5.19**
Responsibilities of technical writers, **1.7**
Restricted specifications, **6.9** to **6.10**
Resumes:
  preparation of, **12.4** to **12.12**
  with proposals, **5.11** to **5.12**, **5.27**
Review request forms, **6.30**
Reviewing, editing, and revising, **1.6** to **1.7**
  of advertisements, **10.7**
  of articles, **4.58**
  of directives and procedures, **7.34**
  of instructions, **8.34** to **8.35**
  of papers, **3.6** to **3.8**, **3.13** to **3.14**
  of proposals, **5.37** to **5.38**
  of reports, **2.53** to **2.54**
  of sales brochures, **9.16**

Reviewing, editing, and revising (*Cont.*):
  of specifications, **6.27** to **6.31**
  of writing, **1.15**
RFPs (*see* Requests for proposal)
Rich, G. R., *Hydraulic Transients*, **11.3**
Risks and uncertainties, listing of, in proposals, **5.23**
"Rule of 2" for specification formats, **6.17**

Sales brochures:
  checklist for, **9.17**, **9.19** to **9.20**
  development of, **9.17** to **9.18**
  features of, **9.2** to **9.4**
  preparation of, **9.6** to **9.13**
  as promotional tools, **9.1** to **9.2**
  structure and content of, **9.13** to **9.15**
  writing of, **9.4** to **9.5**, **9.15** to **9.17**
Salesmen, editor contacts by, **4.42**
Sample material:
  for books, **11.13** to **11.14**
  for magazines, **4.42** to **4.43**
Sample specifications, **6.8**
Schedules:
  for books, **11.15** to **11.17**
  of charges for cost proposals, **5.14** to **5.15**
  of deliverables in proposals, **5.24**
  for instructions, **8.32**
  for proposal development, **5.34** to **5.35**
Scientific and Technical Societies of the United States and Canada, **3.16**
Scientists as technical writers, **1.3** to **1.4**
Scope:
  of directives and procedures, **7.10**, **7.24**
  of instructions, **8.3** to **8.4**
  in specifications, **6.20**
Search terms for abstracts, **2.30**
Seasons, capitalization of, **13.18** to **13.19**
Sections:
  for instructions, **8.34**
  for multivolume reports, **2.49**, **2.53**
  for specifications, **6.16** to **6.24**
Self-improvement articles, **4.12**
Semicolons, **13.42** to **13.43**
Sentences:
  in directives and procedures, **7.29** to **7.30**
  length of, for articles, **4.56**, **4.59**
  openings of, abbreviations in, **13.31**
  in reports, **2.23**
Series:
  commas for, **13.41**
  semicolons for, **13.42** to **13.43**

Service descriptions in sales brochures, **9.13** to **9.14**
Service instructions, **8.5**
Setup instructions, **8.19**
"Shall" in specifications, **6.26**
Shidle, Norman, recommendations by, **1.16**
Shoptalk in directives and procedures, **7.27**
Short-form reports, **2.2**
Simultaneous submissions to magazines, **4.43**
Single-use directives and procedures, **7.9** to **7.10**
Skeleton specifications, **6.8**
Slides for papers, **3.13**
Societies:
  listings of, **3.16** to **3.17**
  presentations to, **3.3** to **3.16**
  and specifications, **6.14** to **6.15**
Sociological Abstracts, **2.30**
Solar Heating Design Process (Kreider), **11.3**
Solicited proposals, **5.3** to **5.5**
Solicited technical papers, **3.4**
Sources:
  of directives and procedures, **7.8**
  listing of, in reports, **2.34**
  for research, **2.56**
  for RFPs, **5.4** to **5.5**
  for sales brochures, **9.9** to **9.11**
  selection boards for, **5.3**
  for specifications, **6.12** to **6.14**, **6.25**
SOW (statement of work) in proposals, **5.4**, **5.10**
Space industry, **1.2**
Special competence, listing of, on resumes, **12.11**
Special Libraries Association, **2.56**
Special warnings in instructions, **8.13**
Specialists:
  and ghostwriting, **3.6**
  for proposal development, **5.35** to **5.36**
Specification Practices, **6.5**
Specifications:
  checklist for, **6.31** to **6.32**
  and contracts, **6.4** to **6.5**
  determination of, **1.9**
  engineering, **6.4**
  formats for, **6.15** to **6.24**
  instruction writing to, **8.8** to **8.10**
  and outlines, **2.11**
  revising of, **6.27** to **6.31**

Specifications (*Cont.*):
  trees for, **6.13**
  types of, **6.5** to **6.15**
  writers and writing of, **6.1** to **6.4**, **6.24** to **6.29**
Spelling, **13.1** to **13.29**
  of numbers and numerals, **13.37** to **13.39**
  in papers, **3.9**
Stand-alone specifications, **6.7** to **6.8**
Standard forms:
  for cost proposals, **5.13**
  for directives and procedures, **7.10**
  for resumes, **12.4** to **12.5**, **12.7**
  for RFPs, **5.4**, **5.6**, **5.7**
Standard nomenclature lists, **8.7**
Standard practices and procedures, **7.2** to **7.4**
Standard report formats, **2.28** to **2.35**
Standardization Policies, Procedures and Instructions, **6.5**
Standards and specifications, **6.6**, **6.14** to **6.15**, **6.25**
Standing directives and procedures, **7.10**
Statement of work in proposals, **5.4**, **5.10**
States, abbreviation of, **13.31**
Steel, E. W., *Water Supply and Sewerage*, **11.2**
"Stories" in sales brochures, **9.7**
Study managers for reports, **2.54**
Style:
  for articles, **4.2**, **4.56**
  for directives and procedures, **7.28** to **7.30**
  for papers, **3.9**
  for reports, **2.6** to **2.7**, **2.23**, **2.35** to **2.49**
Style Manual, **13.1**
Subdivisions, **13.46**
Subheadings:
  in books, **11.19** to **11.20**
  in instructions, **8.17**, **8.21**
  in papers, **3.12**
  in specifications, **6.17**
Subheads for article titles, **4.49**
Submissions:
  of articles, **4.57** to **4.60**
  of papers, **3.14**
  of reports, **2.55**
Subsections in proposals, **5.26**
Subsystems instruction modules for, **8.29**
Subtitles for papers, **3.11**
Subtopics in outlines, **2.8**
Suggestions (*see* Criticisms and suggestions)

Suggestions to Authors, **13.1**
Summaries:
  in articles, **4.28**, **4.50**
  in books, **11.23**
  in instructions, **8.14** to **8.16**
  in multivolume reports, **2.53**
  in reports, **2.16** to **2.19**, **2.30**
  (*See also* Executive summaries)
Suppliers, **6.3**
Supply catalogs, **8.7**
Surveys, **1.9** to **1.11**
  for article ideas, **4.20**
Symbols:
  abbreviations for, **13.31** to **13.32**
  in advertising, **10.9**
Synopses, **2.16**
  for proposals, **5.38**
  of report findings, **2.34**
System analyses for specifications, **6.13**
System specifications, **6.7**
Systems and plant description articles, **4.4** to **4.5**
  examples of, **4.16**
  outline for, **4.29** to **4.30**

Tables:
  in articles, **4.56**, **4.59**
  in catalogs, **10.6**, **10.7**
  in instructions, **8.34**
  in papers, **3.9**, **3.13**
  in proposals, **5.24**
  in reports, **2.24** to **2.26**
  in tables of contents, **2.32**
  in troubleshooting, **8.22** to **8.23**
Tables of contents:
  for appendixes, **2.45**
  for instructions, **8.14**, **8.15**
  for multivolume reports, **2.49** to **2.53**
  and outlining, **2.7**, **2.10**
  for proposals, **5.19**
  for reports, **2.23** to **2.24**, **2.31** to **2.32**, **2.43**, **2.47**
  for specifications, **6.19**
Tailoring of writing to readers (*see* Readers and users)
Teams:
  for instruction development, **8.30**
  for proposals, **5.34**
  for sales brochures, **9.9**, **9.11**
Technical discussion in proposals, **5.22** to **5.23**

Technical journals and societies for research, **2.56**
Technical proposals, **5.**10 to **5.**13, **5.**15 (*See also* Proposals)
Technical requirements in specifications, **6.**20 to **6.**21
Technical terms in directives and procedures, **7.**28
Technical writers, job description for, **1.**5 to **1.**8
Telephone interviews, **1.**9
Telephoning to editors, **4.**42
Test procedures and instructions, **7.**3, **8.**6
Text divisions, **13.**46
Textbooks compared to handbooks, **11.**24
"The," capitalization of, **13.**16
Theory of Elasticity (Timoshenko and Goodier), **11.**3
Theory of operation manuals, **8.**3, **8.**5, **8.**38
Threatening of editors, **4.**45
Time, colons for, **13.**43
Timeliness and time requirements:
  in article titles, **4.**48
  and closed specifications, **6.**10
  and directives and procedures urgency, **7.**12 to **7.**13
  estimation of, **1.**9
  for reports, **2.**2
Timoshenko, S. P., *Theory of Elasticity*, **11.**3
Titles and title pages:
  abbreviation of, **13.**31
  for articles, **4.**47 to **4.**49
  capitalization of, **13.**18
  for directives and procedures, **7.**25
  for instructions, **8.**13, **8.**21
  for papers, **3.**11
  for proposals, **5.**18 to **5.**19
  for reports, **2.**29 to **2.**30
  for specifications, **6.**18 to **6.**19
  for tables of contents, **2.**44
Tone of directives and procedures, **7.**13
Top matter in directives and procedures, **7.**19
Trade associations for research, **2.**56
Trade names, capitalization of, **13.**17
Traditional format for directives and procedures, **7.**16 to **7.**18
Training texts, **11.**2
Transitional phrases, commas for, **13.**40 to **13.**41
Transitional words in articles, **4.**53 to **4.**54

Transmittal data in directives and procedures, **7.**19, **7.**22
Transmittal letters:
  preparation of, **12.**1 to **12.**5
  for proposals, **5.**17 to **5.**18
  for reports, **2.**29
Troubleshooting instructions, **8.**21 to **8.**26
Typing of articles, **4.**44
Typing mats, **2.**36, **2.**40

Unpacking instructions, **8.**17, **8.**19
Unsolicited proposals, **5.**6 to **5.**10
Urgency and directives and procedures, **7.**11 to **7.**13
Usage (*see* Grammar and usage)
Users (*see* Readers and users)
User's manuals, **8.**3

Variability in procedures, **7.**3 to **7.**4
Verification of facts (*see* Reviewing, editing, and revising)
Vierck, C. J., *Engineering Drawing*, **11.**2
Viewpoints for article ideas, **4.**24

Warnings:
  in directives and procedures, **7.**29
  in instructions, **8.**13, **8.**19
Water Supply and Sewerage (Steel), **11.**2
Water Treatment for HVAC and Potable Water Systems (Blake), **11.**3
Webster's New International Dictionary, **13.**1
"Will" in specifications, **6.**26
Word processors:
  for reports, **2.**53
  for specifications, **6.**24
Words:
  in articles, **4.**55, **4.**59
  in directives and procedures, **7.**27
  division of, **13.**14 to **13.**15
  spelling and capitalization of, **13.**1 to **13.**29
World War II and technical writing, **1.**1 to **1.**2
Writer's Market, **4.**40
Writing:
  of advertising, **10.**8 to **10.**11
  of articles, **4.**50 to **4.**60
  of books, **11.**18 to **11.**24
  of catalogs, **10.**3 to **10.**6

Writing *(Cont.)*;
    of directives and procedures, **7**.27 to **7**.35
    good and bad, **13**.46 to **13**.51
    grammatical, **13**.51 to **13**.52
    guidelines for, **1**.13 to **1**.15
    of instructions, **8**.30 to **8**.41
    to magazine editors, **4**.41

Writing *(Cont.)*:
    of papers, **3**.8 to **3**.16
    of reports, **2**.19, **2**.23 to **2**.27
    of sales brochures, **9**.4 to **9**.5, **9**.15 to **9**.17
    of specifications, **6**.24 to **6**.29

Years, apostrophes for, **13**.43

ABOUT THE AUTHORS

Tyler G. Hicks, P.E., is a consulting engineer with International Engineering Associates. He has worked in both plant design and operation in a variety of industries, taught at several engineering schools, and lectured both in the United States and abroad on engineering topics. Mr. Hicks is editor in chief of the *Standard Handbook of Engineering Calculations* and coauthor of the *Standard Handbook of Professional Consulting Engineering Practice* and *The McGraw-Hill Handbook of Essential Engineering Information and Data* due to be published in September 1988. In addition, he is the author of numerous engineering reference books on equipment and plant design and operation. He is a member of ASME and IEEE and holds a bachelor's degree in mechanical engineering from Cooper Union School of Engineering. Mr. Hicks resides in Rockville Centre, New York.

Carl M. Valorie, Sr., has been a professional engineer for over 25 years. He was a consulting engineer for United Engineers and Constructors, Inc., in Philadelphia, Pennsylvania, and was a project quality control engineer for the General Electric Company in both Philadelphia, Pennsylvania, and Daytona Beach, Florida. During his career he has written a wide variety of articles for technical periodicals. Mr. Valorie is a graduate of the University of Pennsylvania. He resides in Voorhees Township, New Jersey.